Quantitative Diagenesis: Recent Developments and Applications to Reservoir Geology

NATO ASI Series

Advanced Science Institutes Series

A Series presenting the results of activities sponsored by the NATO Science Committee, which aims at the dissemination of advanced scientific and technological knowledge, with a view to strengthening links between scientific communities.

The Series is published by an international board of publishers in conjunction with the NATO Scientific Affairs Division

A Life Sciences **B Physics**	Plenum Publishing Corporation London and New York
C Mathematical and Physical Sciences **D Behavioural and Social Sciences** **E Applied Sciences**	Kluwer Academic Publishers Dordrecht, Boston and London
F Computer and Systems Sciences **G Ecological Sciences** **H Cell Biology** **I Global Environmental Change**	Springer-Verlag Berlin, Heidelberg, New York, London, Paris and Tokyo

PARTNERSHIP SUB-SERIES

1. Disarmament Technologies	Kluwer Academic Publishers
2. Environment	Springer-Verlag
3. High Technology	Kluwer Academic Publishers
4. Science and Technology Policy	Kluwer Academic Publishers
5. Computer Networking	Kluwer Academic Publishers

The Partnership Sub-Series incorporates activities undertaken in collaboration with NATO's Cooperation Partners, the countries of the CIS and Central and Eastern Europe, in Priority Areas of concern to those countries.

NATO-PCO-DATA BASE

The electronic index to the NATO ASI Series provides full bibliographical references (with keywords and/or abstracts) to more than 30000 contributions from international scientists published in all sections of the NATO ASI Series.
Access to the NATO-PCO-DATA BASE is possible in two ways:

– via online FILE 128 (NATO-PCO-DATA BASE) hosted by ESRIN, Via Galileo Galilei, I-00044 Frascati, Italy.

– via CD-ROM "NATO-PCO-DATA BASE" with user-friendly retrieval software in English, French and German (© WTV GmbH and DATAWARE Technologies Inc. 1989).

The CD-ROM can be ordered through any member of the Board of Publishers or through NATO-PCO, Overijse, Belgium.

Series C: Mathematical and Physical Sciences – Vol. 453

Quantitative Diagenesis: Recent Developments and Applications to Reservoir Geology

edited by

Andrew Parker

and

Bruce W. Sellwood

Postgraduate Research Institute
for Sedimentology,
University of Reading, U.K.

Kluwer Academic Publishers

Dordrecht / Boston / London

Published in cooperation with NATO Scientific Affairs Division

Proceedings of the NATO Advanced Study Institute on
Quantitative Diagenesis: Recent Developments and Applications to Reservoir Geology
Reading, U.K.
6–9 September 1992

Library of Congress Cataloging-in-Publication Data

Quantitative diagenesis : recent developments and applications to
reservoir geology / edited by Andrew Parker and Bruce W. Sellwood.
p. cm. -- (NATO ASI series. Series C, Mathematical and
physical sciences ; no. 453)
Papers of a meeting held at Reading, Sept. 6-19, 1992.
Includes bibliographical references and index.
ISBN 0-7923-3261-X (acid-free)
1. Diagenesis--Congresses. 2. Petroleum--Geology--Congresses.
I. Parker, A. (Andrew), 1941- . II. Sellwood, B. W. (Bruce
William), 1946- . III. Series.
QE571.Q38 1995
551'.03--dc20 94-39299

ISBN 0-7923-3261-X

Published by Kluwer Academic Publishers,
P.O. Box 17, 3300 AA Dordrecht, The Netherlands.

Kluwer Academic Publishers incorporates the publishing programmes of
D. Reidel, Martinus Nijhoff, Dr W. Junk and MTP Press.

Sold and distributed in the U.S.A. and Canada
by Kluwer Academic Publishers,
101 Philip Drive, Norwell, MA 02061, U.S.A.

In all other countries, sold and distributed
by Kluwer Academic Publishers Group,
P.O. Box 322, 3300 AH Dordrecht, The Netherlands.

Printed on acid-free paper

Printed in the Netherlands

Contents

PREFACE

Andrew Parker & Bruce W. Sellwood

Postgraduate Research Institute for Sedimentology
The University,
Whiteknights,
Reading RG6 2AB
United Kingdom

Sediment diagenesis is one of the main factors influencing permeability heterogeneity within aquifers and hydrocarbon reservoirs. Diagenetic processes, not only in the qualitative way in which they affect sediments, but also in terms of their controlling mechanisms, rates and geochemical mass-balances, have become topics of prime interest over the past ten years. In July 1981, we convened an ASI at Reading on "Sediment Diagenesis" (Parker and Sellwood, 1983). Its emphasis was essentially descriptive. Major themes involved distingushing between facies-influenced diagenetic processes and whether secondary porosity could be generated, in a volumetrically significant way, at a late stage during deep burial within sedimentary basins. To some extent, the intervening years have seen a continuation of these debates and the timing of feldspar dissolution within certain sandstones, or the relative timing of particular types of carbonate dissolution, still remain controversial issues when individual reservoir systems are being evaluated.

However, the newer lines of controversy have centred upon the quantification of processes and responses, not merely on the micro-scale, but sometimes at the scale of entire basins. One of the most ubiquitous cement phases in sandstone, quartz, provides controversy in its own right. Quartz is often recognised, from petrography, as being the first phase of cement (interpreted therefore as being "early"), whereas much accumulated evidence would place the threshold temperature for its formation at close to 80°C (from fluid inclusion and other analyses).

How can such wide discrepancies be rectified? Quartz cement may form in certain types of soil (silcrete), as an early (pre-compactional) pore-filler, but can silcretes alone explain all occurrences of apparently early quartz cements?

In carbonate systems many examples of regional dolomitization still require adequate explanation. Explanations for such occurrences have often depended upon current fashion, seepage-reflux being thrown over for mixing-zone (Dorag) interpretations, these latter themselves being superseded by marine dolomitization models. Marine dolomitization models are currently in the ascendancy. A further problem with carbonates remains the origin of late, deep cements, their timing and the sources of the carbonate itself.

At the heart of many of these controversial issues is quantification. In order to fill the pore-systems within permeable rocks many thousands of pore-volumes of water are needed, to supply the cementing phases. Water, saturated with either carbonate or silica, is still a very dilute fluid. These notions are not new. Indeed, Prof. Robin Bathurst (University of Liverpool) was preaching this very point more than twenty years ago. What is different now is that the global database is broad enough to enable us to offer interpretations that can be tested quantitatively.

To provide a forum for the discussion of these issues a NATO Advanced Study Institute was held at Reading, from September 6 - 19th, 1992. Its title: *Quantitative Diagenesis: Recent Developments and Applications to Reservoir Geology* reflected our aims of consolidating what was already known about diagenetic processes, in carbonates and siliciclastics, and then developing quantitative themes in these broad areas. Around sixty selected scientists from nine NATO and four other countries participated in this Institute. Most of the keynote addresses, which acted as the stimuli for extensive discussion sessions, are presented here.

Each of the main themes (carbonates and siliciclastics) were prefaced by sessions consolidating, qualitatively, current states of understanding, particularly with regard to process (e.g. contributions by Sellwood (carbonate diagenesis), Wright (aspects of pedogenesis), Clayton (microbial processes and organic carbon), and Chamley (aspects of clay mineralogy). Some aspects of quantification in carbonates are addressed by Mountjoy (dolomites), whereas Bjørlykke, Longstaffe and Giles address terrigenous clastic systems. The contribution by Giles represents a wide-ranging overview and will be published separately by Kluwer Academic Publishers.

PRINCIPLES OF CARBONATE DIAGENESIS

B.W. SELLWOOD
Postgraduate Research Institute for Sedimentology
The University
Whiteknights
Reading, U.K.
RG6 2AB

ABSTRACT. The aim of this chapter is to provide an overview of diagenetic processes as they affect carbonate grains and carbonate sediments, thus giving a foundation for other carbonate chapters. Post-depositional changes may begin upon the depositional substrate (e.g. micro-infestation and dissolution). Subsequently, during burial, compaction, further dissolution, cementation, recrystallization and replacement may occur. Some of these processes lead to obvious changes but others are cryptic (e.g. the recrystallization of foraminiferans and nanno-grains), being difficult to recognize even through the closest of scrutiny employing sophisticated analytical techniques. In this chapter marine phreatic, marine vadose, fresh-water vadose, fresh-water phreatic, mixing zone and deeper burial processes are reviewed. The use of isotopic and fluid-inclusion methods of analysis are illustrated in the context of particular case studies. Diagenetic changes in carbonate sediments are the record, very often imperfectly preserved, of successive fluid-rock interactions that have occurred during burial (and uplift). Thus, any methods which enable us to recognise the sequence of diagenetic effects, and refine more closely the succession of diagenetic processes that have occurred, will enable us to interpret the evolution of fluid-types, and fluid migration pathways, in the context of burial evolution within basins. Closer understanding of these processes will enable predictive diagenetic models to be constructed, such models being applicable to a wide range of carbonate facies in many types of basin.

1.0 INTRODUCTION

Following deposition, all the changes that take place at low temperatures and pressures in a deposited carbonate sediment are termed diagenetic. Diagenesis, as a complex of processes, grades into metamorphism under conditions of elevated temperatures and pressures. Diagenetic processes include infestation, compaction, dissolution, cementation, recrystallization and replacement. Some of these processes lead to dramatic, and easily recognizable results (e.g. dissolution of caverns and their partial filling with stalactitic and stalagmitic cements). Other processes are cryptic, involving only the subtle redistribution of ions. Some large-scale changes (such as the dolomitization of entire reef complexes) may not be immediately obvious under field conditions and may go unrecognized for many years. Most marine limestones exhibit few compaction features and appear to have been cemented at an early stage. Experimental studies have shown that compaction affects grainy carbonates less than muddy ones and that a porosity loss of 30% may occur in lime muds without significant breakage of contained shells occurring (Bhattacharya and Friedman, 1979). Diagenetic changes reflect processes that have acted sometimes on a variety of time frames, to move towards the re-establishment of equilibrium states between the sediment (in all its individual facets) and the post-depositional environment. However, post-depositional environments, and the fluids within them, are seldom uniform through time. Changes may be progressive, abrupt or even cyclic in character. So, sediments may

A. Parker and B. W. Sellwood (eds.),
Quantitative Diagenesis: Recent Developments and Applications to Reservoir Geology, 1–32.

bear the hallmarks of multi-phase diagenesis, and may frequently exhibit the effects of reaction processes that failed to reach complete equilibrium.

It is well known that hydrocarbon reservoirs are porous and permeable units. It is also becoming clear that in many basins carbonates retain their best reservoir properties where the "race against time" between cementation and hydrocarbon emplacement was won by the hydrocarbons. It is also clear that the cementing fluids and the hydrocarbon-bearing fluids were sometimes the same.

Several new approaches lead to the prediction that very soon the real-time "plumbing dynamics" within sedimentary basins will be predictable, and thus allow forward modelling. It should be possible to predict, for particular basins, not only where cemented zones exist and what produced them, but also when and why they developed and whether they were likely to have been connected to regional aquifers or hydrocarbon fairways. To these ends this paper introduces the basic observational framework within which diagenesis in carbonate sediments can be both recognized and, by reference to a variety of investigative techniques, referred to possible diagenetic environments. Subsequent papers will discuss the detailed aspects of some of these approaches, often in the context of specific examples.

2.0 NEAR SURFACE BURIAL DIAGENESIS

Most cementation, and some secondary porosity development, in limestones occurs at relatively shall depths (Fig. 1). There are four major near-surface diagenetic environments: marine phreatic; vadose; meteoric phreatic and mixing zone. Each of these zones may be further sub-divided on the basis of the rate at which water flows through it (often termed "active" or "stagnant"), and the degree of saturation of the water with respect to $CaCO_3$. A major paper dealing with processes and responses in each of these settings was that of Longman (1980).

It has been realized, over the past 15 years or so, that early-cement fabrics in marine carbonates may be environment-specific (providing information on both the depositional and early burial history). But successful interpretation of such fabrics relies initially on observational recognition (in thin-section or acetate peels). To these ends, this review will be conducted by reference to the observational petrographic base, which is still the essential pre-requisite for the more sophisticated laboratory analyses that are then used to validate the observational interpretations.

2.1 MARINE PHREATIC ZONE (Fig. 1)

Marine carbonates start here (upon the sea-floor) as grains, the pores within the sediment being entirely filled with sea-water. Active phreatic conditions prevail where pore-water circulation rates are high (e.g. as a result of wave-pumping), in reefs and shoals of lime-sand, whereas stagnant phreatic conditions exist where little or only sluggish movement of pore waters occurs (e.g. in back-reef and platform-interior lagoons).

In modern seas the chemical characteristics of sea-water are affected by depth (controlling temperature, $CaCO_3$ saturation, Eh, etc.), climate and proximity to major rivers. Sandberg (1983, 1985) and Wilkinson et al., (1985) have inferred, from the composition of ancient marine precipitates, that the chemical composition of sea-water may

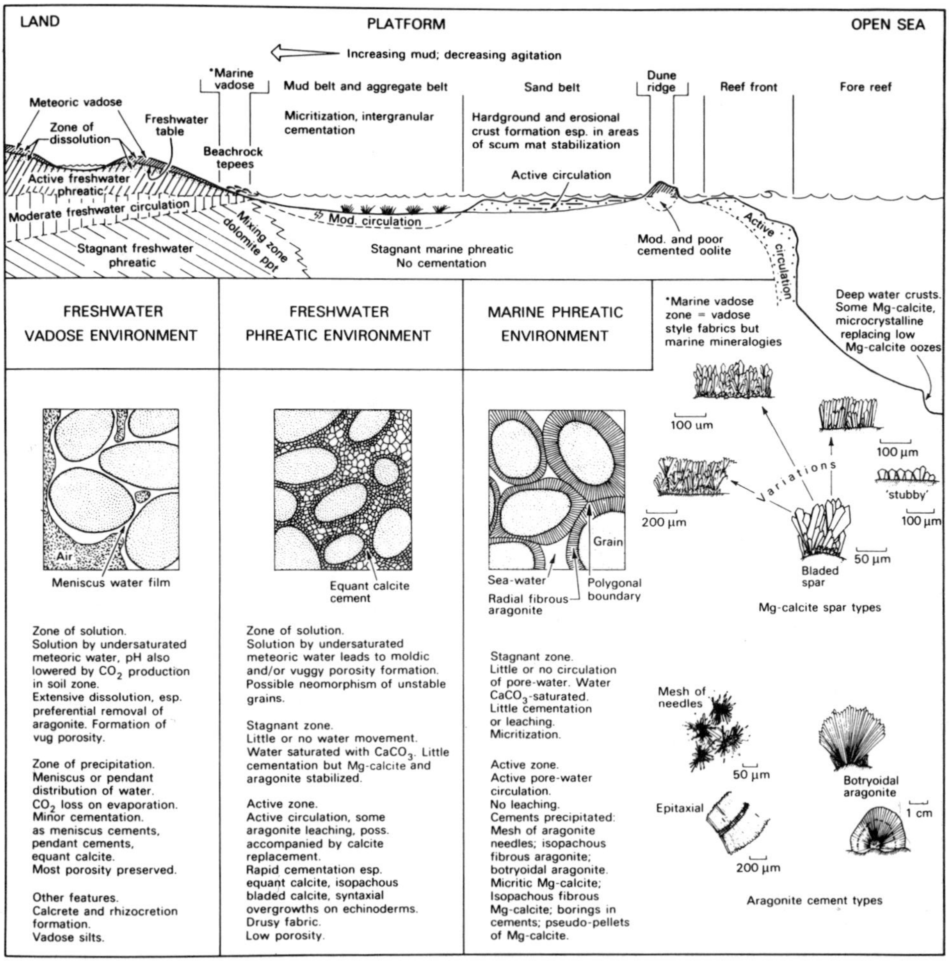

Schematic cross-section of a carbonate shelf to illustrate the relationship between environments, cement types and general diagenetic processes (after many authors, particularly James and Ginsburg, 1979; Longman, 1980; Flügel, 1982; but including: Dunham, 1971; Badiozamani, 1973; Ginsburg and Schroeder, 1973; Hanor, 1978; Schlager and James, 1978; Halley and Harris, 1979; and Marshall and Davies, 1981).

Fig. 1a

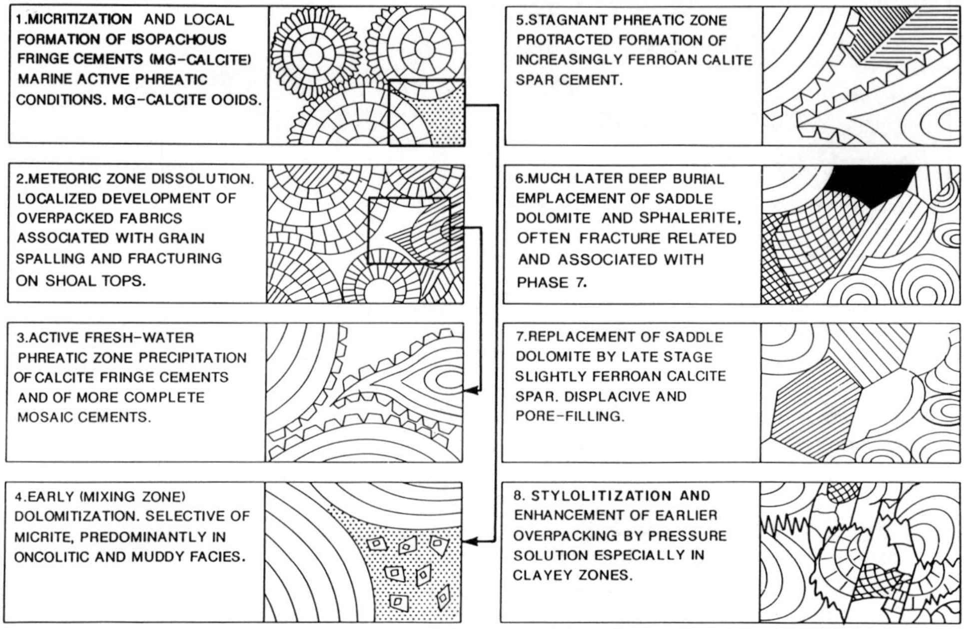

Fig. 1b. Diagenetic sequence generally developed in oolitic grainstone facies of Mid-Jurassic age (Great Oolite Fm.) buried to 1 to 2 km, Weald Basin, Southern England. From Sellwood et al. (1989).

have varied significantly through time. There is also good evidence that the physico-chemical properties of ocean waters (e.g. pH, Mg/Ca ratio, temperature and density profiles), as seen today, may have been different in the geological past, and especially at times when the earth lacked significant ice caps. With these provisos in mind we will now consider marine diagenesis.

2.1.1. *Modern Active Marine Phreatic Environments.* Many carbonate grains begin to undergo diagenetic alteration upon the sea bed, particularly within the photic zone (Fig. 2). This is in sharp contrast to most terrigenous clastic grains. Carbonate grains may be infested by endo-lithic micro-borers of various types (cyanobacteria, sponges, bryozoans), whose effects are to destroy or obscure original textures within grains (May and Perkins, 1979). These processes of micro-infestation occur particularly where grains are at rest for long periods under intense photic conditions, or are only intermittently mobile (as on stabilized sand shoals).

Where porous carbonate sediments accumulate in areas with strong currents and waves, the potential exists for seawater to be pumped through the interstitial pores. Although seawater is not generally supersaturated with $CaCO_3$, precipitation is favoured where agitation, and the warming of cooler oceanic waters, triggers CO_2 degassing. Even so, to fill one pore volume with carbonate cement requires the passage of tens of thousands of pore volumes of watersaturated with respect to carbonate (Bathurst, 1975) and such conditions are best achieved in active phreatic conditions. Cements are of aragonite and high-Mg calcite and, although the processes controlling which mineral is precipitated are still problematic, fluid-flow rates have a major influence, as do the Mg/Ca ratio of the fluid and rate of carbonate ion supply (Given and Wilkinson, 1985; Fig. 2).

In the Arabian Gulf, alternations of lithified and unlithified layers exposed on the sea bed were first recognized as submarine cemented hardgrounds by Shinn (1969) who showed that submarine lithification, manifested by boring and encrustation of exposed hard layers, had taken place since the Flandrian transgression. Cements consisted of successive phases of acicular aragonite separated by micro-crystalline rinds (probably Mg-calcite). The cemented sediments, which contained artifacts, provided dates in a range from 450 to 8000 years BP and provided analogies for hardgrounds recognized in ancient rocks (e.g. those from the Mid-Jurassic of the Paris Basin described by Purser, 1969). Shinn (1969) also recognized how submarine cementation could lead to the development of "submarine anticlines" and "tepee" structures.

In carbonate sand belts (Fig. 1), such as those described by Dravis (1979) scum-mat stabilization allows (possibly triggers) the emplacement of acicular aragonitic cements at the surface of tidal sand ridges during phases of temporary abandonment. Within such ridges former cemented surfaces are found as cemented, but discontinuous layers. Dravis (1979) was unable to resolve whether the role of the microbial filaments seen in the cemented zones were active or passive as far a cement generation was concerned.

Reefs represent modern carbonate environments with highly active phreatic circulations and it is likely that early marine cements play an important part in producing the rigidity and robustness that typifies reef systems (Purser and Schroeder, 1986). An extensive literature exists in which Recent reef cementation is described (e.g reviews by James, 1983; Macintyre et al., 1968; Land and Goreau, 1970; James and Ginsburg, 1979; Marshall and Davies, 1981; Marshall 1986, Schroeder and Purser, 1986; and many others). Although the morphologies of reef cements are variable, the mineralogy is

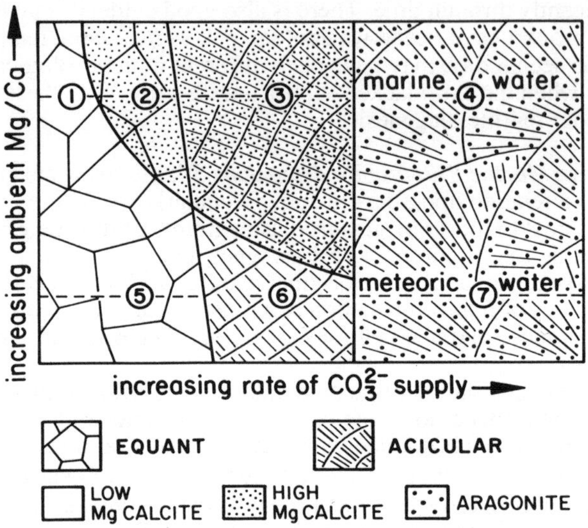

Fig. 2. Schematic representation of the relationship between Mg/Ca ratio, rate of carbonate ion supply, crystal morphology and mineralogy of carbonate precipitates. Mg:Ca ratio of the meteoric water line is 0.3. That of the marine water line is 5.2. 1=low Mg calcite spar in modern cool, deeper water, and shallow water temperatre water carbonates; 2=Mg calcite in reefs (rare); 3=Mg calcite cement in reefs and lime sands; 4=aragonite cements in reefs and lime sands; 5=low Mg calcite in meteoric environments; 6=low Mg calcite in speleothems and travertines; 7=aragonite cement in speleothems (rare). After Given and Wilkinson (1985).

relatively simple as predominantly aragonite, and Mg-calcite. There are recent reports of thin calcian dolomite cements associated with fringing reefs off Jamaica (Mitchell et al., 1987).

Aragonite cements that grow under active marine phreatic conditions are usually acicular crystals forming isopachous fringes around grains, and as needle meshworks. Within reefs there may also be more massive botryoids (James and Ginsburg, 1979; ("mamelons" of Aissaoui 1985)). Mg-calcite cements also occur as isopachous fringes in the form of columnar, bladed and stubby crystals (Ginsburg and Schroeder, 1973; James and Ginsburg, 1979; Aissaoui, 1988). Chafetz (1986) has recently demonstrated that Mg-calcite can form as micro-peloids (20-60 μm diameter) within reefal micro-cavities and surface crusts. The peloids nucleate upon bacterial clumps and the micro-peloids were precipitated in response to the vital activities of the bacteria. Thus, in addition to the physico-chemical controls on cement mineralogy mentioned above, a further control is almost certainly biochemical, due to the presence of particular bacterial communities within the sediment. These factors are addressed by Clayton (this volume).

The large-scale pumping of open seawaters through major carbonate platforms is a phenomenon recently demonstrated by the hazardous cave-diving work of Smart et al., (1988) and Whitaker and Smart (1990) in the Bahamas. In this case, pore-water movements are driven by Kohout convection, reflux and differences in head that are the result of tidal, oceanic and storm currents. As a result, a vigorous interstitial flow is occurring, especially through caves enlarged as a result of Pleistocene low-stand dissolution. The waters passing through the caves are undersaturated with respect to $CaCO_3$, but saturated with respect to dolomite. But, as Machel and Mountjoy (1990) point out, despite this oversaturation neither dolomite nor "protodolomite" have been reported. There is yet more discussion of these problems by Mountjoy (this volume).

2.1.2 *Modern Stagnant Marine Phreatic Environments.* Within platform interiors, currents and waves are more sluggish and so water circulation within the sediment is very restricted. Under these so-called stagnant marine phreatic conditions, intensive cementation of the sea floor does not take place. However, localized cementation may occur and lead to the formation of aggregates ("grapestone" intraclasts) which are often initially bound together by cyanobacterial filaments within which needle meshes of aragonite cements grow. Partial stabilization of the sediment surface by cyanobacterial scum mats lead to the ideal conditions for extensive micro-infestation and micritization of grains.

2.1.3 *Modern Deep Water Marine Phreatic Environments.* Adjacent to platforms, where aragonitic and Mg-calcite peri-platform sediments may accumulate in deep waters (0.7-2.0 km) leaching of aragonitic components occurs. In addition, Mg-calcite grains lose their magnesium and, where sedimentation is arrested, cementation by low-Mg calcite occurs (Mullins et al., 1985; Dix and Mullins, 1988)

In deep waters, where pelagic oozes accumulate, cementation by low-Mg calcite has been recorded on the sea floor down to depths of 3.5 km. Preferred sites for cementation are those with minimal sedimentation. Deep sea waters are cool (close to $0^{o}C$) and the form of the cements (equant crystals of calcite of about 10 μm length) reflect the nature of this habitat.

2.1.4 *Ancient Marine Phreatic Environments.* As we have now seen, the morphologies of marine cements are in many cases very distinctive, and this has allowed their extensive recognition in ancient carbonates. Although morphological comparisons permit first order interpretation, morphological criteria generally require some geochemical re-enforcement. Application of geochemistry and detailed electron-microscopy (including micro-probe analyses) has provided new insights into the mineralogy of ancient non-skeletal grains and cements. These insights have themselves thrown up fundamental questions about the nature and evolution of seawater, the geochemistry of the ancient seawater/atmosphere system and, as a consequence, raised questions about atmosphere/oceanic controls on past and future climate.

Prima Facie evidence that submarine cementation had occurred comes from the presence of hardgrounds with fossilized encrusters. From the detailed petrographic investigation of such beds (e.g classic work by Purser, 1969) have come some of the criteria whereby ancient marine cements may be recognized. Such cements are usually isopachous and with an acicular or columnar fabric. Such cements may be truncated by borings and may be overlain bylater internal sediment. The palisade cements are generally the first cement phase and are often non-luminescent under the CL.

Marshall and Ashton (1980) carefully documented a series of hardgrounds in the Mid-Jurassic of Central England and these may be directly compared with hardground systems from modern environments, in certain aspects (Fig. 3; Sellwood, 1986). In the Mid-Jurassic grainstones, grains close to the hardground surface are seen to be have isopachous fringes of radial-fibrous calcite. Such cements become less abundant (and thinner) downwards, away from the surface. The fringes also exhibit Mg contents significantly higher than those of the grains and later spar cements. This was interpreted as reflecting an original Mg-calcite mineralogy for the fringes, an interpretation supported by the $\delta^{18}O$ values which, within the hardground, remained close to the expected 0 per mil PDB values of seawater (see review by Marshall, 1992 for fuller explanation of the significance of $\delta^{18}O$ values). Subtle evidence that early marine phreatic cementation occurred is provided by some types of syntaxial calcite overgrowths (Maliva, 1989), and especially those that show displacive fabrics and textures where adjacent grains have been moved apart as a result of cement emplacement. Such cements appear to have been precipitated as calcite in the Cretaceous marine sandstones and Devonian limestones studied by Maliva (1989).

Ancient reefs often show excellent marine cements (see reviews in Schroeder and Purser, 1986). Permian and Triassic reef complexes (e.g. Mazzullo and Cys, 1979; Tucker and Hollingworth, 1986; Henrich and Zankl, 1986) exhibit calcitized botryoids with inclusion relics of aragonite, and many other features comparable with those seen in Tertiary and Recent reefs. Later Mesozoic (particularly Cretaceous buildups: Enos,1986) and those from the Carboniferous (Miller, 1986) and Devonian (Hurley and Lohmann, 1989; Kerans et al., 1986) exhibit a range of fabrics consistent with either calcite or Mg calcite precipitation (as non- luminescing isopachous bladed crusts, radiaxial calcite, microcrystalline and micritic calcite). Late Devonian marine cements have characteristic $\delta^{18}O$ signatures of -4.5 per mil PDB and $\delta^{13}C$ signatures of +2.0 per mil PDB, values significantly different from those seen in Mesozoic and later cements. In addition, some Jurassic buildups (Sun and Wright, 1989), and Cretaceous ones (Enos, 1986), exhibit micro-peloidal crusts (some plastering near-vertical surfaces) which have been interpreted as cryptocrystalline calcite marine cements.

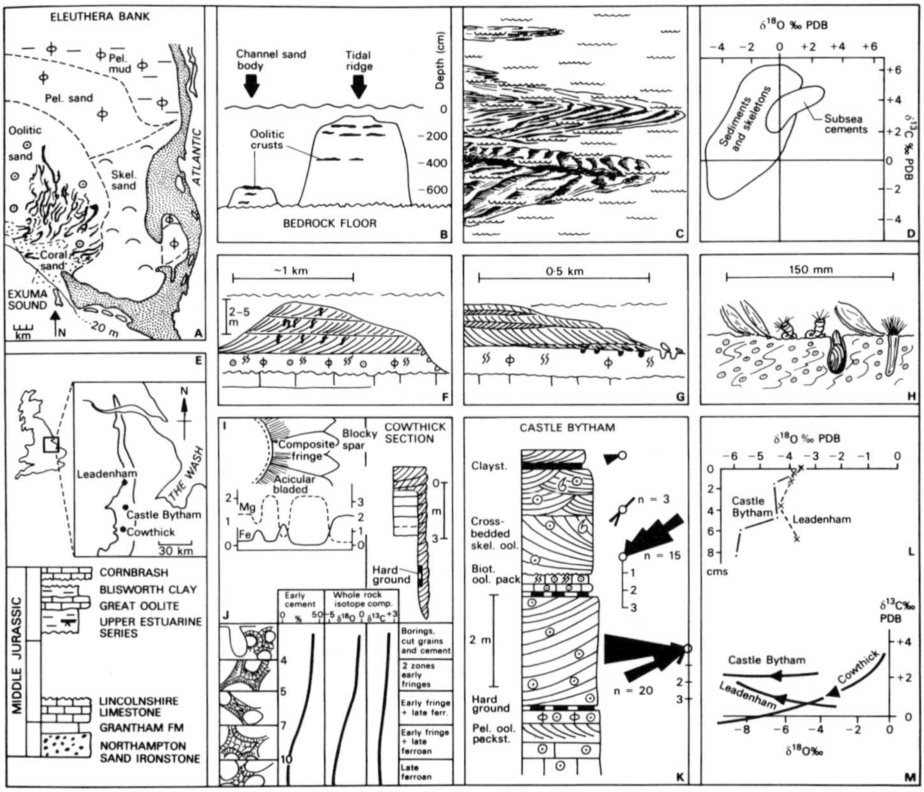

Composite diagram illustrating an application of stable isotope compositions in carbonate sediments and cements to facies interpretation. The lower part of the figure shows a case study of hardground relationships in the Mid-Jurassic Lincolnshire Limestone Fm., England (mostly after Ashton, 1980). The upper part of the figure illustrates the possible facies context of the Jurassic case study by reference to modern environmental analogues. A, Holocene surface sedimentary environments and facies on Eleuthera Bank, Bahamas (after Dravis, 1979). B, Schematic section of an oolitic tidal sand ridge and adjoining channel sand body showing subsurface oolitic crusts. Water depth in cm (after Dravis, 1979). C, Large spillover lobe (subtidal delta) extending bankward into the shelf lagoon. It is covered with abundant flood-oriented sandwaves (based on aerial photograph in Hine, 1977). D, Carbon and oxygen isotope ratios of Holocene sediments, skeletons and cements (after James and Ginsburg, 1979). E, Location of the Lincolnshire Limestone Fm in England and of the sites illustrated. Open sea was to the south, alluvio-deltaic systems to the north and east. Generalized stratigraphic relationships are also shown (after Marshall and Ashton, 1980). F, Hypothetical abandoned oolitic sand shoal, channelled, with a hardground developed on the eroded margin (cf. Cowthick hardground in I). G, Schematic reconstruction of section through Castle Bytham 'spillover lobe' advancing over hardground surface. H, Schematic reconstruction of the most protected hardground at Leadenham (after Sellwood, 1978). I, Schematic representation of $MgCO_3$ wt% (Mg) and $FeCO_3$ wt% (Fe) in grains and cement fabric, Cowthick hardground as determined by microprobe (after Marshall and Ashton, 1980). The sequence at Cowthick is indicated showing relationship between the channel and hardground. Channel cuts into underlying Grantham Fm. J, Schematic representation of fabric and isotopic composition away from the highest energy Cowthick hardground (after Marshall and Ashton, 1980). K, Sequence at Castle Bytham showing relationship between hardground, adjacent facies and cross-bedding dip-directions. L, Oxygen isotopic variation with distance (cm) from hardground surface. M, Carbon and oxygen isotopic compositions (as trends) of Lincolnshire Limestone hardgrounds (after Marshall and Ashton, 1980).

Fig. 3.

Following the work of Sandberg (1975, 1983, 1985) many authors have addressed the question of whether the compositions of non-skeletal precipitates from seawater have remained similar to those of the Recent through Phanerozoic time. Sandberg's pioneering works suggested that there were phases in the geological past when marine precipitates were broadly comparable, both in terms of their morphology and mineralogy, with those of the present (e.g Late Carboniferous to Permian, and post-Cretaceous) and other phases when significant differences existed.

According to Sandberg (1983) the criteria by which aragonite may be recognized in ancient carbonates are, in order of decreasing reliability:

1) grain or cement still aragonite;
2) original feature now a mosaic of generally calcite spar containing oriented aragonite relics;
3) original feature a mosaic of calcite but lacking aragonite relics. High Sr^{2+} contents relative to levels reasonably expected in primary calcite(i.e. thousands of ppm);
4) original feature a calcite mosaic but Sr^{2+} values are either low (hundreds of ppm) or not measured;
5) particular grain types selectively dissolved (moulded) or comprise calcite-filled moulds. This criterion can be strengthened if skeletal grains known to have been originally aragonitic, by analogy with modern forms, have behaved similarly in the same rock (e.g. marine gastropods and cephalopods).

Criteria whereby ancient Mg calcite might be distinguished from calcite (low Mg) are even more ambiguous than those for aragonite. The initial Mg content of skeletal grains may, for example, range from around 4 mole % Mg to more than 30 mole %. Determination of the amount of Mg^{2+} initially present in ancient ooids and cements is currently a matter of dispute. Sandberg (1983) proposed that most ancient radial ooids, such as those commonly seen in the Jurassic, were originally calcite (low-Mg). But this proposal has been disputed by Richter (1983) who notes that such ooids are often ferroan. The iron has been incorporated as Fe^{2+} ions, possibly during early diagenesis, under sub-oxic conditions, and in substitution for originally occurring Mg^{2+} (Richter and Füchtbauer, 1978). Blocky isopachous fringe cements with a ferroan character are commonly encountered within, for example, European Jurassic hardgrounds, especially in carbonate facies that accumulated adjacent to land areas (Sellwood et al., 1985; 1989). Land areas could have supplied the necessary Fe^{2+} ions. A paucity of available iron would, perhaps, have inhibited the potential for such reactions, as in isolated or very extensive platforms.

Micro-dolomites, especially as inclusions within cloudy prismatic calcites, may provide a further criterion for the recognition of former Mg calcite (Lohmann and Meyers, 1977).

During the Early and Mid-Paleozoic (Cambrian to Early Carboniferous) and most of the Mesozoic (Jurassic and Cretaceous) low-Mg calcite appears to have predominated as both cements and non-skeletal grains. Wilkinson (1979) has also argued that the skeletal compositions of major carbonate producers has also changed through time (Fig. 4), and subsequently speculated on the possible implications of a shift in carbonate production from shelves (as aragonite and Mg calcite) to open sea (as pelagic low Mg calcite). These latter changes are a consequence of the explosive radiation of coccoliths (and to a lesser extent globigerinid forams) during the Mesozoic and Cenozoic. As non-skeletal precipitates (ooids), their composition and their distribution both spatially and temporally may also reflect gross changes in seawater composition (Wilkinson et al., 1985). Oolites are well represented in Late Cambrian to Early Ordovician carbonates, and those from the Early Carboniferous, and Late Jurassic to Early Cretaceous. They are particularly poorly represented in the Late Ordovician, Mid-Devonian and Late Cretaceous. Figure 5 gives the temporal distribution of ooids, and their probable compositions. It is well known that modern temperate-water carbonates are predominantly calcite (Sellwood, 1986). It has been noted by James and Bone (1989) that calcite-dominated carbonate sediments, whether more recent temperate-water carbonates or ancient tropical carbonates, will behave differently from aragonitic sediments during diagenesis (particularly meteoric diagenesis) and these problems will be addressed later.

There are no easy explanations for the variations in mineralogy suggested above. Although aragonite-precipitating seas, in the past, broadly correlate with general lowerings of relative global sea-levels (e.g. Sandberg, 1983; Wilkinson et al., 1985; Fig. 6), such a "quick-fix" explanation, and references to "greenhouse" as opposed to "icehouse" earth conditions, appear to be only part of the answer. The geochemical controls are likely to have been both subtle and complex, involving partial pressure of CO_2, the Mg/Ca ratio, less obvious factors such as relative concentrations of organic substances (e.g. citrates, malates and humic acids), SO_4^{2-} and phosphates, and possibly other factors that have yet to be recognized (Beckett, 1991). As Sandberg (1985) recognized, although the Mg/Ca ratio of seawater might have been shifted towards increased Mg as a result of the Late Jurassic coccolith explosion, such an explanation could not be used for earlier times.

Although modern ocean waters are undersaturated with $CaCO_3$ at depths greater than several hundred metres, aragonite does not disappear until depths in excess of 1.5 km (calcite disappears at depths generally in excess of 4 km). Some equivocal evidence exists that during times of calcitic seas (e.g. the Jurassic), aragonite dissolution may have taken place on relatively shallow sea-floors (Palmer et al. 1988), but more work needs to be done to evaluate the significance of these ideas.

2.2 VADOSE ZONE

This is the subaerial zone lying below the land surface and above the water table (Fig. 1). Both air and water may occur within the pores and, depending on the overall setting, the water may be saline brine, seawater, meteoric water or may be any combination of these. Water within the sediment forms a meniscus layer around grains, or sometimes drips downwards through pores, and the morphology of cement fabrics generally reflects this (Longman, 1980: Fig. 1).

2.2.1 *Marine Vadose*. Above the high-water mark, sheet run-off may be triggered by wave-splash. In this case, any precipitates will form laminated sheets, sometimes with complex corrosion/erosion surfaces.

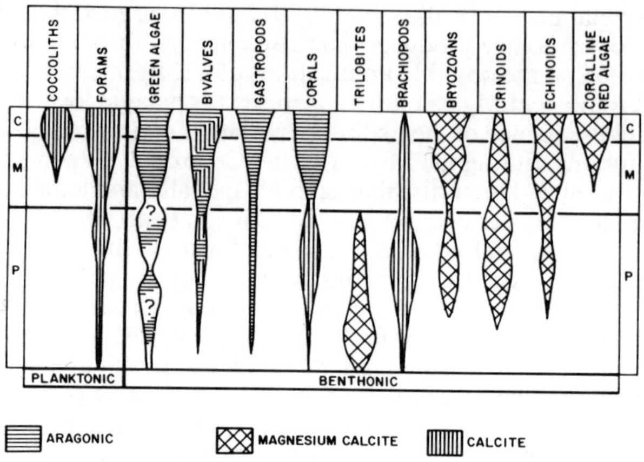

Fig. 4. Approximate diversity, abundance and composition of principal groups of calcareous marine organisms through time (after Wilkinson, 1979).

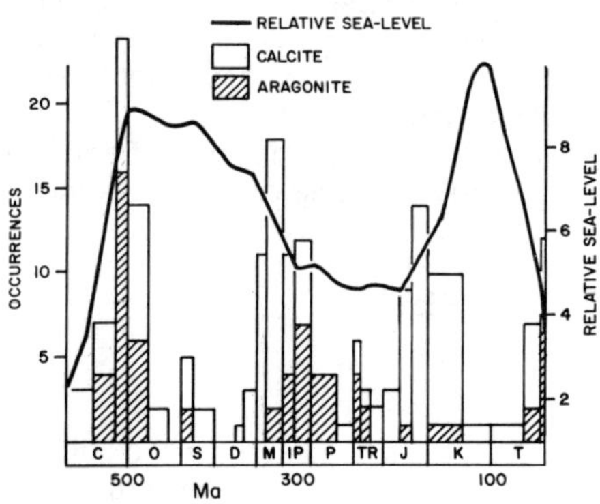

Fig. 5. Occurrence of originally calcitic and aragonitic ooid cortices. There is a general correspondence between abundant aragonite cortices and continental emergence. (After Wilkinson et al. 1985).

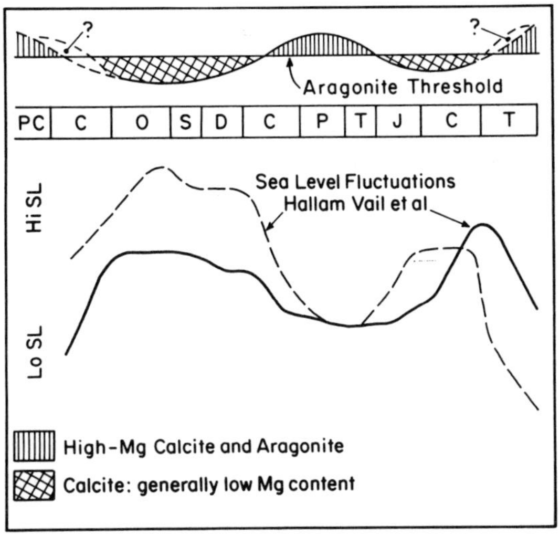

Fig. 6. The pattern of marine precipitates through the Phanerozoic compared with global sea-level (from Sandberg, 1983). Calcite seas correspond to high stands of sea-level and aragonite-high Mg calcite to low stands.

On seashores, marine vadose environments occur. Such areas can be sites of very rapid cementation, the cements precipitating as a result of evaporation and the degassing of CO_2 to produce beachrocks. In addition there may be micro-biological effects. Some beachrocks within shell sands studied on Hawaii by Myers (1987) are cemented by Mg calcite (15.5 mole % $MgCO_3$), forming cryptocrystalline elongate rhombs with well-developed meniscus styles. In addition, hollow rims occur where cements grew around bubbles, and fringing rims with both meniscus and (more rarely) isopachous styles. The fabric appears to be micrite-supported but this micrite is cement, an observation which raises questions over the origins of some ancient micrite-supported carbonates. Indeed, it may now be timely to re-evaluate the origin of material generally interpreted as internal sediment within hardgrounds and beachrocks (c.f. Aissaoui and Purser, 1983). Although commonly reported from beachrocks, aragonite is scarce in the Hawaiian example (except in sheltered pores), probably because the Mg/Ca molar ratio of the pore fluids (6.18) corresponds with an Mg calcite-aragonite threshold of 21 mole % $MgCO_3$ (Given and Wilkinson, 1985). This is believed to inhibit aragonite crystallization and to increase the threshold at which Mg calcite is stable (Myers, 1987).

Cathodoluminescence (CL) studies by Amieux et al. (1989) on some Holocene beachrocks from West Africa have documented the history of such cements (within siliciclastic sands) during shoreline progradation (and thus from marine to marine vadose to meteoric vadose and then to meteoric phreatic environments). Firstly, Mg calcite precipitated (dull orange to dull blue CL colours). Secondly, bright orange-yellow luminescing acicular and equant spar cements grew in a mixed-water phreatic zone. The third stage involved neomorphism of earlier marine cements and resulted in the formation of equant spar with dull blue or medium orange luminescence (fresh-water phreatic in origin).

Although more generally believed to be associated with tropical water carbonate systems, beachrock cementation is also seen in temperate waters too (e.g. Mannin Bay, Western Ireland: Fig. 7).

2.2.2 *Fresh water vadose.* Beachrock cementation is not confined to marine environments. Indeed, as Scoffin (1986) re-iterates, the near-beach environment may offer many sub-environments in which waters may range from hyper-saline to fresh (often the composition depends upon the season). Indeed, from a calcareous lake in Michigan Binkley et al., (1980) have described beachrock fabrics comparable, in their general morphology, with marine cements, but composed of low-Mg calcite.

Thus, marine vadose cements may be distinguished from fresh-water vadose ones by reference to their original mineralogy (Mg calcite and aragonite typifying marine and low-Mg calcite typifying fresh water). The characteristic textures in cements are meniscus, dripstone and whisker, the latter probably resulting from the calcification of fungal filaments. Of course, the same reservations exist when extrapolating these generalizations into the geological record as were made when discussing marine diagenesis.

The meteoric vadose environment (see reviews by Esteban and Klappa, 1983; Tucker and Wright, 1990), which are discussed much more fully by Wright in this volume, receives water from the surface containing dissolved CO_2 from the atmosphere (probably richer in CO_2 during the Early Paleozoic and Mesozoic than in the Cainozoic and Quaternary; (e.g. Berner, 1992), soil-derived CO_2 and organic acids.

In carbonate sediments, dissolution is an important process with Mg calcite and

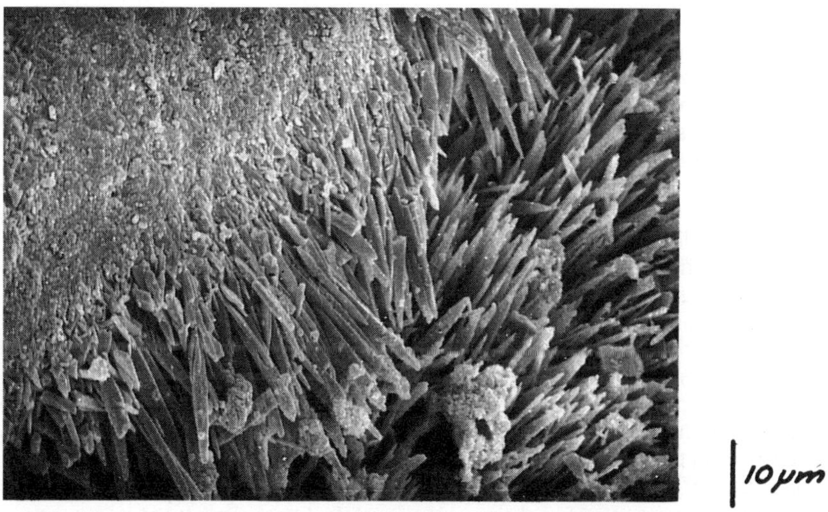

Fig. 7. SEM views to show aragonite cements in recent beachrock, Clew Bay, W. Ireland.

aragonite components being selectively leached. The degree to which leaching occurs is partially dependent upon climate. Calcitization of aragonitic grains may occur, and Mg^{2+} ions are leached from Mg calcite. During Mesozoic times, when meteoric waters were possibly more enriched with CO_2 than those of today, grains exposed on the emergent crests of oolite shoals were micritized and destructured and subsequently acted in a semi-plastic manner during shallow burial (Sellwood and Beckett, 1991; Sellwood et al., 1989).

Contrasting styles of karstification and dissolution, as influenced by climate, have been modelled by Hird and Tucker (1988) and Wright (1988). The details of karstification and pedogenic features (such as rhizocretion formation and calcretes) are best left to Wright (this volume), but it is important to note, at this stage, that processes such as those operating in the vadose zone "condition" carbonate sediments for their later burial evolution. This early conditioning appears to be far more significant in carbonates than in most siliciclastic systems.

The original mineralogy of the carbonate is also important and, as James and Bone (1989) observe, carbonates not dominated by aragonite and Mg calcite are likely to behave quite differently (more robustly) than those that are. These original differences are now known to have had an impact upon the way such carbonates have acted during burial and upon their qualities as both hydrocarbon reservoirs and aquifers. As we have seen already, there is a temporal control on original mineralogy. It is also important to recall that pelagic carbonates of Mesozoic and later age have been dominated by low-Mg calcite. Chalks are important hydrocarbon reservoirs and aquifers, and although they are fine grained their diagenetic behaviour is largely controlled by their original calcite mineralogy (Taylor and Lapré, 1987).

2.3 METEORIC PHREATIC ENVIRONMENTS

The meteoric phreatic zone lies below the water table (Fig. 1). All the pores are filled with meteoric water containing variable amounts of dissolved carbonate. According to the Ghyben-Herzberg model, for every metre that the water table rises above mean sea level the base of the meteoric phreatic zone is depressed 36 m below sea level. But it is a dynamic zone which is once again strongly influenced by climatic and other factors such as tectonic regime, regional permeability etc.

Processes acting within the meteoric phreatic zone are controlled by rates of water flow and the degree of carbonate saturation of the waters. Undersaturated waters will continue to dissolve carbonate whereas saturated waters will precipitate calcite. Longman (1980) re-affirmed the point that waters may be oversaturated with respect to calcite but still undersaturated with respect to aragonite, thus allowing moldic pores to both be generated, and almost simultaneously lined, or filled, with calcite. Calcite linings are generally isopachous, with the most characteristic textures being druses of equant calcite spar. Preferential sites for such cements to grow are echinoderm grains, as inclusion-poor syntaxial calcite. The calcite which grows initially under active, and oxidizing conditions, is generally non-ferroan and non-luminescent. If rates of water flow are sluggish, calcitization of aragonite may occur. Under such conditions, or where stagnant conditions allow Mn^{2+} and Fe^{2+} ions to be liberated, ferroan and manganoan calcite precipitates.

Cement morphologies are by no means definitive in establishing the diagenetic environment, and subtle geochemical tests have been developed to enable petrographically-derived interpretations to be tested. As Allan and Matthews (1982) showed, close to soil zones the δ 13 C values are negative (because soil-generated CO_2 is rich in δ^{12} C. δ^{18} O values are also generally negative, reflecting precipitation from meteoric waters. Meteoric phreatic effects are well known in the literature, but there are few papers documenting the areal distribution of a palaeo-meteoric phreatic lens. One example is that from the Mid-Jurassic of England described by Emery and Dickson (1989).

Overpacking of (generally originally calcitic) ooids may, in ancient times, have continued into the meteoric phreatic zone. In some cases it can be demonstrated that the grains underwent plastic deformation within rigid "jackets" comprising the initial isopachous fringes of calcite (Sellwood et al., 1989). As has been noted by many authors, the emplacement of meteoric fringe cements may not be entirely detrimental to the poro-perm characteristics of the rock during burial. The presence of isopachous calcite fringes may act as a "scaffolding" to the grains which inhibits further overpacking and grain-grain dissolution (e.g. Reekmann and Friedman, 1982). Indeed there are many cases recorded (both in published literature and proprietary oil company archives) of reservoir grainstones in which early-cemented beds have retained higher porosities and permeabilities during deeper burial than those that were not (e.g.several papers in Harris, 1984). These processes are discussed later.

Finally, the depth interval over which meteoric influences may be exerted have also been a function of amounts of sea-level change. Over the past three million years these have amounted to tens of metres of change in tens of thousands of years, but even through those time periods lacking ice caps (e.g. the Jurassic and Cretaceous) tectonically driven sea-level changes of a much smaller order occurred, and probably at slower rates than glacio-eustatic changes. As has been shown by Burchette and Wright (1992), not only will changes of plumbing geometry occur because of highstand or lowstand configurations, but also the plumbing dynamics will be controlled by whether the carbonate system is a ramp or a platform. These latter considerations are particularly important with regard to the nature of the mixing zone at the base of the meteoric phreatic system, and are best left to the next section.

2.4 PHREATIC MIXING ZONE

At the base of the meteoric phreatic zone, and particularly where directional meteoric flow is occurring, a mixing zone will be established between marine phreatic waters below and meteoric waters above (Fig. 1; Badiozamani, 1973; Land, 1980; 1986). If the meteoric waters are saturated with respect to calcite, and the marine waters saturated with respect to both calcite and dolomite, mixtures of these waters will be undersaturated with calcite but supersaturated with respect to dolomite. This apparently paradoxical situation, which will be addressed more fully by Mountjoy (this volume), results from the fact that calcite saturation does not covary linearly as salinity changes (see also excellent summaries in Moore, 1989). Under such conditions, it is argued, dolomitization will be favoured (Badiozamani, 1973). This so called "Dorag" model has been a popular explanation for dolomitization in many ancient successions, and has been used as a base in some recent computer simulations for platform-margin dolomitization (Humphrey and Quinn, 1989). As we have seen in earlier sections, because of climatic and/or eustatic effects the base of the meteoric phreatic zone is potentially dynamic, so extensive zones of massive dolomite might be produced. More recently, some of the Late Quaternary situations cited as key examples of coastal mixing zone dolomitization have been challenged (e.g. Barbados; Burns and Rossinssky, 1989) and the extensive application of the "Dorag" model has been seriously questioned (e.g. Machel and Mountjoy, 1990).

Circumstantial evidence is often used to support a mixing-zone interpretation. For example the following criteria are often cited:

1) dolomites found in an appropriate setting and associated with other fabrics typical of meteoric vadose or phreatic origin (moldic porosity, meteoric phreatic isopachous and equant spar cements);
2) dolomite geochemistry (stable isotope signature, and reduced amounts of Sr and Na).

Machel and Mountjoy (1990) convincingly argue that coastal mixing zones are not likely to form thick successions of dolomite (without secondary porosity), and particularly if the dolomitizing mixtures contain only minor amounts of seawater (5%). The remote possibility of extensive and pervasive dolomitization might exist, however, given appropriate conditions of sea-level, and rate of sea-level change (these arguments are best left for Mountjoy, this volume).

While the jury is still out on the processes discussed above, it is important to reflect that the plumbing dynamics in such a mixing-zone system will also be controlled by the shelf morphology. As Burchette and Wright (1992) have indicated, by virtue of their shallow depositional slopes, ramps will be less likely to develop marine porewater circulatory systems capable of initiating marine dolomitization. However, under appropriate hydrogeological conditions, a hydraulic head may be established and meteoric phreatic conditions may extend considerable distances offshore. Thus ancient ramps may have been preferred sites for mixing-zone diagenesis. Such a situation has been interpreted for several ancient systems (e.g. Moore, 1989).

2.5 DEEPER BURIAL DIAGENESIS

Below the depth at which nearsurface processes have a significant impact, sediments are said to be in the deeper-burial environment. But there is no clear-cut consensus on the depth at which this occurs and there is much ignorance concerning the processes (and their rates) that occur there (Scholle and Halley, 1985).

As we have already seen, given appropriate hydrogeological conditions it would be possible to envisage meteoric phreatic conditions extending very deeply into a contiguous sedimentary basin. The burial environment is influenced by gravity-driven meteoric inflow into aquifers and upward movements of fluids derived from compaction within the basin itself (Fig. 8; after Sellwood et al., 1987).

Although pore fluids deep within basins are generally brines (still of somewhat dubious origin), the sediments that contain them were mostly bathed in marine or meteoric waters at the time of deposition. The evolution of pore waters within a single rock succession may have been very complex. The Chalk of the Weald Basin in England illustrates this point very simply. It originated as a marine pelagic deposit in the Late Cretaceous. During the

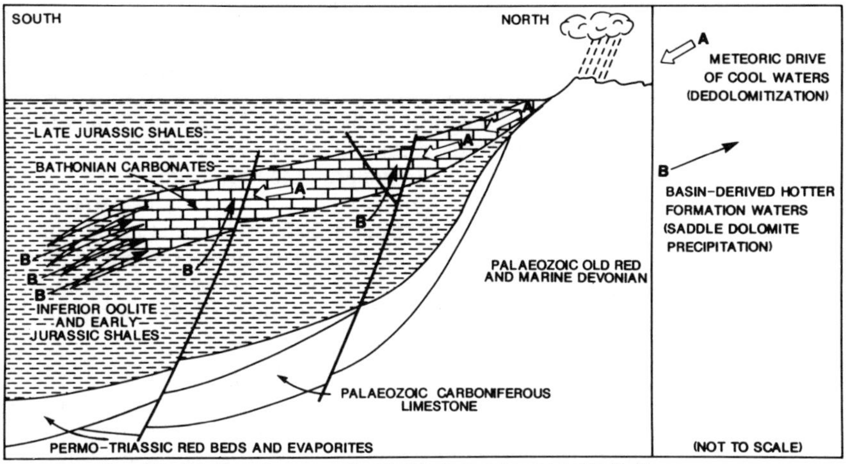

Schematic model for late diagenetic processes of saddle dolomite formation and dedolomitization.

Schematic model for late diagenetic processes of saddle dolomite formation and dedolomitization by mildly ferroan calcite spar. Meteoric drive down an aquifer will not necessarily reach deeper parts of the basin, but will trigger pore-fluid movement at depth.

Fig. 8.

Palaeocene it was uplifted and invaded by meteoric waters. Subsequently, during the Eocene progressive marine onlap led to parts of the Chalk system being re-invaded by marine waters which were then flushed out of the system during the Miocene inversion of the Weald Anticline. Sub-tropical meteoric waters of the Mid-Tertiary were replaced by frigid waters during several phases of Pleistocene glaciation. Today, the Chalk acts as an important aquifer but there are no diagenetic features directly attributable to these intervening drowning and emergence phases. In general, the pore fluids that exist within basins are well known only from either economic aquifers or from hydrocarbon reservoir formations, the latter only being sampled from structural anomalies that are subsurface features with closure. There have been no attempts systematically to sample formation fluids from surface to basement away from subsurface highs. Relative to seawater, subsurface formation waters are generally seen to have salinities of tens or hundreds of parts per thousand. They are often enriched with respect to Ca^{2+} and often depleted with respect to Na^{+}, K^{+}, Mg^{2+} and SO_4^{2-}. As described in later chapters, such brines have complex origins, with contributions from meteoric, marine, evaporative and hydrothermal waters in many cases, as well as from clay mineral transformations within shale successions. There is a general tendency for salinity to increase with depth.

During burial, sediments encounter diagenetic processes influenced by overburden pressures and rising temperatures. Pressure dissolution (Wanless, 1979) generally commences at around 600 to 900 m but may begin at depths as shallow as 300 m (Bathurst, 1983). Pressure dissolution in limestones is usually manifested in the form of stylolites and dissolution seams which are often seen to become less intensely developed as the same rock unit is traced from downflank (water-wet zone) to the crest (oil-saturated zone) in oilfields. This suggests that pressure dissolution is arrested by the arrival of the oil charge (e.g. Johnson and Budd, 1975; and papers in Roehl and Choquette, 1985). Carbonate released by pressure solution can either migrate out of the limestone or may remain for redistribution as cement away from the stylolitic seams. Intensely cemented zones usually occur immediately adjacent to stylolites. The loss in thickness of the original sediment may be of the order of 30% in limestones buried to 2 km or so. Losses in excess of this are reported in Wanless (1979), particularly where tectonic stresses have been involved.

Burial calcites generally form coarsely crystalline mosaics (often poikilotopic) with variable amounts of FeäΩä whose haphazard distribution generally leads to indistinct zonation, as opposed to the crisply-bounded CL zones that are frequently seen in "meteoric" calcites (Emery and Marshall, 1989). Unzoned poikilotopic cements are also well known (e.g Moore, 1985; Sellwood et al., 1989) Clues to their deeper burial origin are provided by their stable isotopic compositions (particularly strongly negative $\delta^{18}O$ values) and from fluid inclusions entrapped within the crystals as they grew. The trapped fluids are generally highly saline brines but may also include hydrocarbons. As we shall see later, the presence of oil inclusions presents *prima facie* evidence for the movement of liquid hydrocarbons along with formation brines (e.g. Sellwood et al., 1989). Fluid-inclusion homogenization temperatures are generally high (40° +, e.g. Figs. 9, 10, 11 & 12). Late cements seriously damage the poro-perm characteristics of deeply buried carbonate reservoirs, but there is a less predictable relationship between porosity and depth in carbonates than in sandstones (Scholle and Halley, 1985; Butler and Pullan, 1990; Figs.

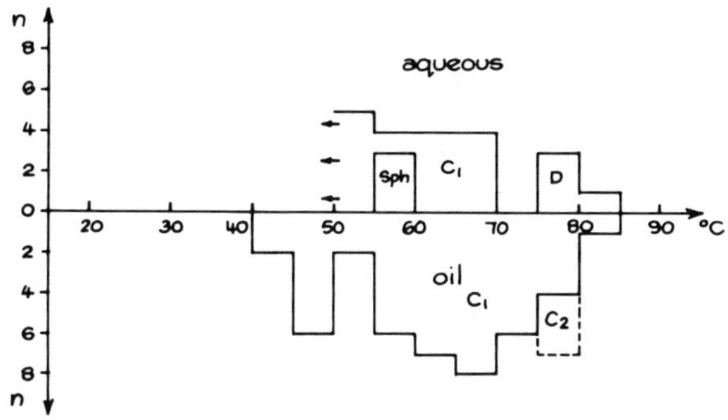

Comparative homogenization temperatures (TH) for type 3 aqueous and type 4 oil inclusions. D = saddle dolomite; Sph = sphalerite; C_1 = late calcite; C_2 = late calcite overgrowth.

Fig. 9.

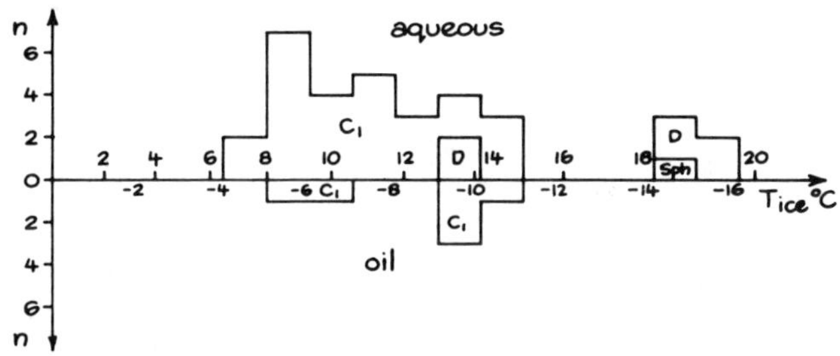

Comparative salinities for type 3 aqueous inclusions and the aqueous phase of type 5 oil–water inclusions in various cements. D = saddle dolomite; Sph = sphalerite; C_1 = late calcite.

Fig. 10.

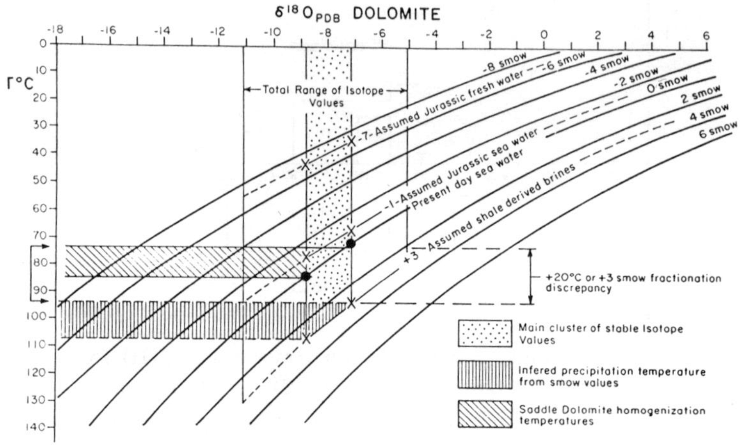

Fig. 11. Homogenization temperatures derived from fluid inclusion analysis plotted on an isotopic fractionation diagram for saddle dolomite. (From Sellwood et al. 1989).

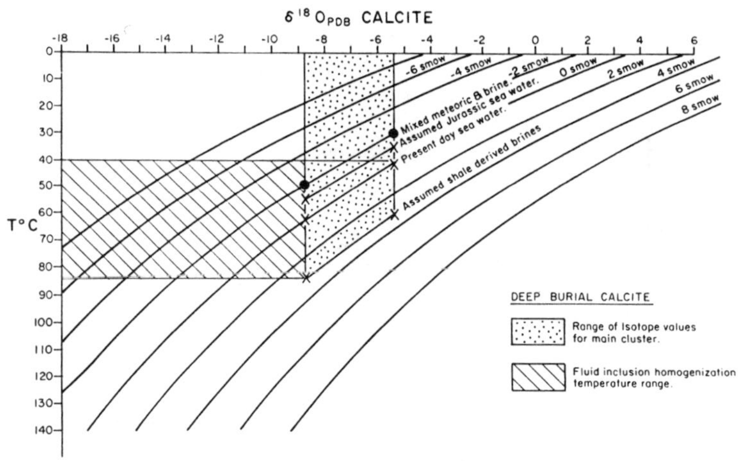

Fig. 12. Homogenization temperatures derived from fluid inclusion analysis plotted on an isotopic fractionation diagram for mildly ferroan calcite. (From Sellwood et al. 1989).

13 and 14). Pore-casting techniques can help in illustrating the pore geometry in highly complex carbonate systems (e.g. Beckett and Sellwood, 1991: Fig. 15).

Dolomite also forms in the burial environment, frequently exhibiting coarsely crystalline, xenotopic textures and homogeneous CL (Mountjoy, this volume; Lee and Friedman, 1987). In addition, "baroque" or "saddle" dolomite is emplaced. Such dolomites also contain fluid inclusions of dense brines and provide high homogenization temperatures and negative $\delta^{18}O$ signatures that are consistent with their precipitation at temperatures in excess of 60°C (quoted temperatures are usually in the 60-150° range: Figs. 9 - 12) (Radke and Mathis, 1980; Woronick and Land, 1985; Machel, 1987; Lee and Friedman, 1987; Sellwood et al., 1989; Kretz, 1992).

Saddle dolomite is a disordered calcian and ferroan dolomite with distinctive curved crystals. It may occur in primary pores, secondary pores, as intragranular displacive crystals, within fractures and as replacement crystals. Associated fractures may contain intimate associations of late calcite, late dolomite and other minerals such as authigenic kaolinite, barytes and other non-carbonates. Petrographic evidence sometimes shows that calcites post- dated dolomites (calcitization of dolomite), but in most reported cases the calcites contain hydrocarbon-bearing inclusions. The interrelationships between these phases can provide significant insights into the nature of fluid evolution within sedimentary basins. The presence of such late-stage cements underlines the necessity to mobilize fluids in order to trigger cementation, and that fluid migration has occurred is confirmed by the presence of hydrocarbon accumulations. In the Weald Basin of Southern England, late-stage dolomite, emplaced locally at temperatures of around 100°C, was succeeded by the regional emplacement of calcite from oil-bearing fluids of both lower temperature and salinities. Thus, the dolomite might mark a significant thermal/tectonic event in the history of the basin, an event that may have had significance in terms of source-rock maturation (Sellwood et al., 1989; Butler and Pullan, 1990).

Fluid inclusions within late cements can also provide information on the timing and longevity of oil migration within sedimentary basins. It has been well established that the colour and intensity of hydrocarbons under UV light is determined by their composition (Figs. 16 and 17; Lumb, 1978). An increase in aromaticity shifts the fluorescence colour towards the red end of the spectrum (longer wavelengths). Oil extracts from source rocks show a shift towards the red end of the spectrum as thermal maturity increases. The opposite trend, a shift towards blue fluorescence with increasing maturity, is observed in reservoir oils (McLimans, 1987). Immature hydrocarbons (those with 19-30ä API) have a fluorescence response in the red-yellow region of the visible spectrum. Oils have a progressively shorter wavelength response (i.e. yellow-green to green-white) with increasing maturity. The separation of a light aliphatic component from a progressively more aromatic residue in the source rock may explain these opposite trends (Guilhaumou et al., 1990). Hydrocarbons may mature within inclusions after entrapment but post-entrapment maturation will create solid-bitumen daughter phases that will be readily visible. Provided no overmaturation has occurred the fluorescence response of included oils within carbonate cements within a basin may be used as a migration indicator. This approach has been used successfully by Burruss et al., (1983; 1985), Videtich et al., (1988) and Sellwood et al., (1993) from a variety of carbonate systems around the world.

From this thin-skin overview we will now move on to the detailed approaches that follow in the rest of this volume.

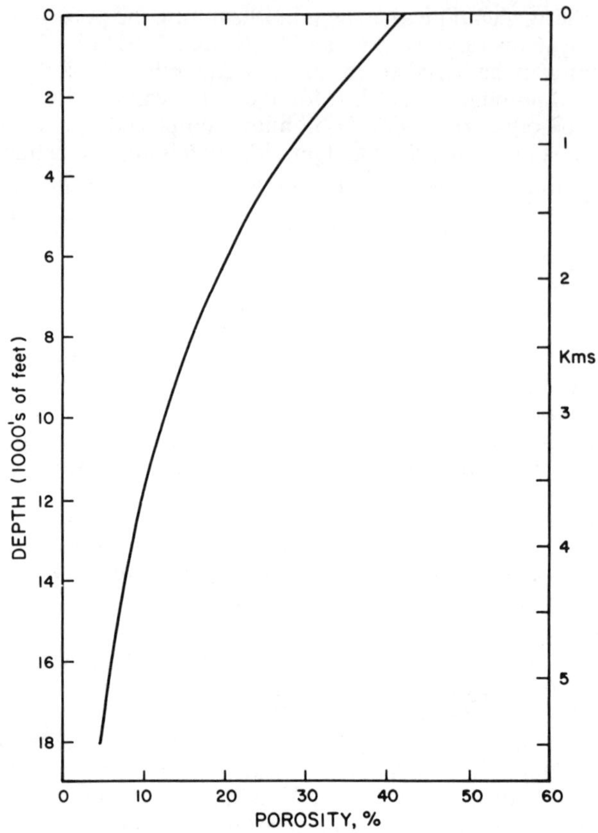

Fig. 13. Porosity-depth values and the least-square exponential fit to all data. This line smooths a wide spread of data. Based on carbonates in Gulf Coast area of the USA. (After Scholle and Halley 1985).

(a) GREAT OOLITE: AVE. POROSITY VS DEPTH

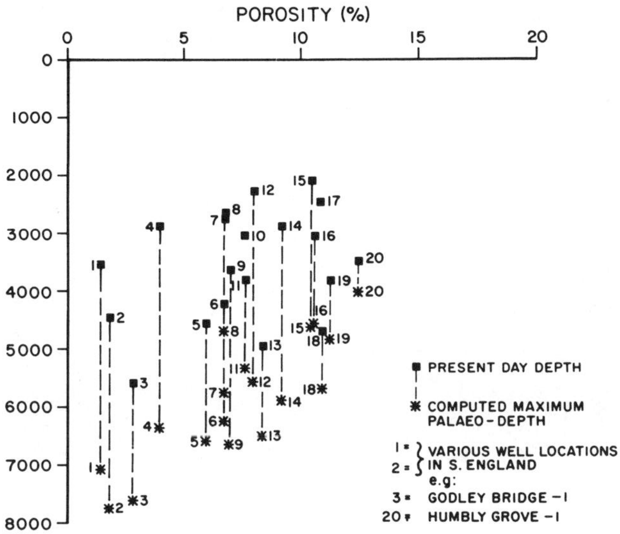

(b) GREAT OOLITE: AVE. POROSITY DISTRIBUTION MAP

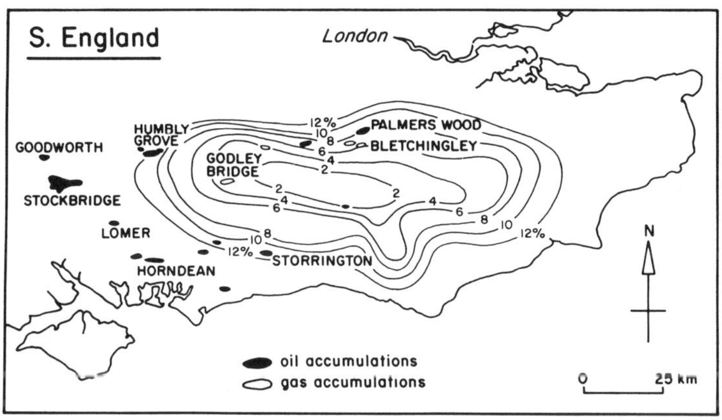

Fig. 14. (a) Plot of Great Oolite average porosity versus present day depth and reconstructed maximum palaeo-depth. (b) Great Oolite average porosity contour map. (After Butler and Pullan 1990).

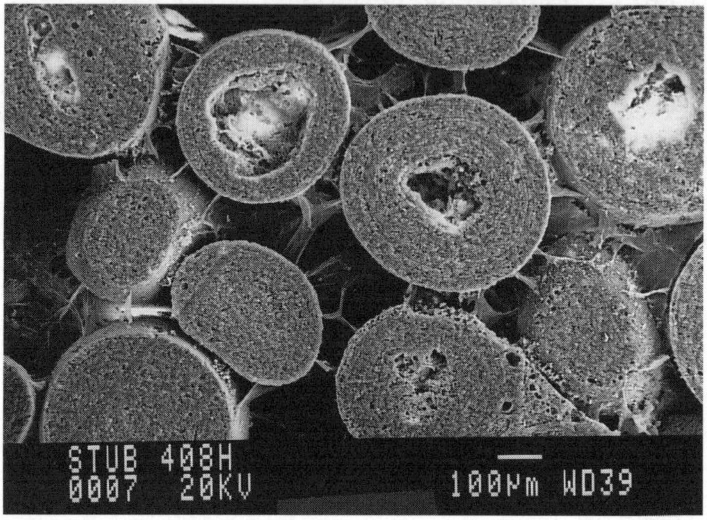

Fig. 15. Porecast of oolitic grainstone (M Jurassic Great Oolite), Horndean Oilfield (1406m depth), porosity 21%, permeability 0.21 md.

General fluorescence response for different gravity oils under ultra violet light

FLUORESCENCE WAVELENGTH(nm)	A.P.I	FLUORESCENCE COLOUR	OIL TYPE
560	19	FULL YELLOW YELLOW/RED	LOW MATURITY OILS
548	25	GREEN	
528	33	GREEN/BLUE	OILS
494	44	BLUE	
450	50+	VIOLET	CONDENSATES

Compiled from data in van Gijzel (1981), Guilhaumou et al. (1990), and Horsfield and McLimans (1984).

Fig. 16. Fluorescence response for different gravity oils under ultraviolet light.

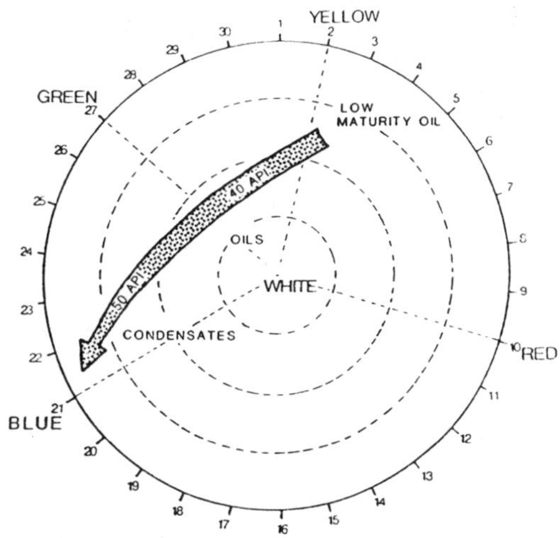

Chromaticity diagram showing fluorescence response of different gravity (API) oils from Table 1 under ultra violet light.

Fig. 17. Chromaticity diagram showing fluorescence response of different gravity (API) oils under UV light

References

Aissaoui, D.M. (1985) Botryoidal aragonite and its diagenesis. Sedimentology, 32, 345-361.

Aissaoui, D.M. (1988) Magnesian calcite cements and their diagenesis: dissolution and dolomitization. Sedimentology, 35, 821-841.

Aissaoui, D.M. and Purser, B.H. (1983) Nature and origins of internal cements of Jurassic limestones of Burgundy (France) and Fnoud Algeria). Sedimentology, 30, 273-283.

Amieux, P., Bernier, P., Dalondeville, R. and Medwecki, V. (1989) Sedimentary Geology, 65, 261-272.

Badiozamani, K. (1973) The Dorag dolomitization model -application to the Middle Ordovician of Wisconsin. Journal of Sedimentary Petrology, 43, 965-984.

Bathurst, R.G.C. (1975) Carbonate Sediments and their Diagenesis. Elsevier, Amsterdam, 658pp.

Bathurst, R.G.C. (1983) Early diagenesis of carbonate sediments. In A. Parker and B.W. Sellwood (Eds.) Sediment Diagenesis. NATO ASI Series D. 349-377. Reidel Dordrecht, 427pp.

Beckett, D. (1991) Aragonite in Jurassic ooids: distribution and significance. Unpubl. Ph.D. Thesis, University of Reading, 307 pp.

Beckett, D. and Sellwood, B.W. (1991) A simple method for producing high-quality porecasts of carbonate rocks. Sedimentary Geology, 71, 1-4.

Berner, R.A. (1992) Palaeo-CO_2 and climate. Nature, 358, 114.

Bhattacharya, A. and Friedman, G.M. (1979) Experimental compaction of ooids and lime mud and its implications for lithification during burial. Journal of Sedimentary Petrology, 29, 1279-1286.

Binkley, K.L., Wilkinson, B.H. and Owen, R.M. (1980) Vadose beachrock cementation along a southeastern Michigan marl lake. Journal of Sedimentary Petrology, 50, 953-962.

Burchette, T.P. and Wright, V.P. (1992) Carbonate ramp depositional systems. Sedimentary Geology, 78, 1-53.

Burns, S.J. and Rossinsky, V. (1989) Discussion - Late Pleistocene mixing zone dolomitization, southeastern Barbados, West Indies. Sedimentology, 36, 1135-1137.

Burruss, R.C, Cercone, K.R. and Harris, P.M. (1983) Fluid inclusion petrography and tectonic-burial history of the Al Ali No. 2 well: evidence for the timing of diagenesis and oil migration, northern Oman foredeep. Geology, 11, 567-570.

Burruss, R.C, Cercone, K.R. and Harris, P.M. (1985) Timing of hydrocarbon migration: evidenced from fluid inclusions in calcite cements, tectonic and burial history. Society of Economic Paleontologists and Mineralogists Special Publication. 36, 277-289.

Butler, M. and Pullan, C.P. (1990) Tertiary structures and hydrocarbon entrapment in the Weald Basin of southern England. In R.F.P. Hardman and J. Brooks (Eds.) Tectonic Events Responsible for Britain's Oil and Gas Reserves. Geological Society Special Publication, 55, 371-391.

Chafetz, H.S. (1986) Marine peloids: a product of bacterially induced precipitation of calcite. Journal of Sedimentary Petrology, 56, 812-817.

Dix, G.R. and Mullins, H.T. (1988) Rapid burial diagenesis of deep water carbonates: Exuma Sound, Bahamas. Geology, 16, 680-683.

Dravis, J. (1979) Rapid and widespread generation of Recent oolitic hardgrounds on a high energy Bahamian Platform, Eleuthera Bank, Bahamas. Journal of Sedimentary Petrology, 49, 195-208.

Dunham, R.J. (1971) Meniscus cement. In O.P. Bricker (Ed.) Carbonate Cements. Johns Hopkins Press, Baltimore, 297-300.

Emery, D. and Marshall, J.D. (1989) Zoned calcite cements: has analysis outpaced interpretation? Sedimentary Geology, 65, 205-210.

Esteban, M. and Klappa, C.F. (1983) Subaerial exposure environment. In P.A. Scholle, D.G. Bebout and C.H. Moore (Eds.) Carbonate Depositional Environments. Memoir of the American Association of Petroleum Geologists, 33, 1-54.

Flügel, E. (1982) Microfacies Analysis of Limestones. Springer-Verlag, Berlin, 633pp.

Ginsburg, R.N. and Schroeder, J. (1973) Growth and submarine fossilization of algal cup reefs, Bermuda. Sedimentology, 20, 576-614.

Given, R.K. and Wilkinson, B.H. (1985) Kinetic control of morphology, composition and mineralogy of abiotic sedimentary carbonates. Journal of Sedimentary Petrology, 55, 109-119.

Guilhaumou, N., Szydlowskii, N. and Pradier, B. (1990) Characterization of hydrocarbon fluid inclusions by infra-red and fluorescence microspectrometry. Mineralogical Magazine, 54, 311-324.

Halley, R.B. and Harris P.M. (1979) Freshwater cementation of a 1000 year old oolite. Journal of Sedimentary Petrology, 49, 969-988.

Hanor, J.S. (1978) Precipitation of beachrock cements: mixing of marine and meteoric waters vs CO_2- degassing. Journal of Sedimentary Petrology, 48, 489-501.

Harris,P.M. (1984) Carbonate sands - a core workshop. SEPM Core Workshop No. 5, San Antonio, 1984.

Henrich, R and Zankl, H. (1986) Diagenesis of Upper Triassic Wetterstein Reefs of the Bavarian Alps. In J.H. Schroeder and B.H. Purser (Eds.) Reef Diagenesis. 245-268. Springer-Verlag, Berlin.

Humphrey, J.D. and Quinn T.M. (1989) Coastal mixing zone dolomite, foreward modelling, and massive dolomitization of platform margin carbonates. Journal of Sedimentary Petrology, 59, 438-454.

James, N.P. (1983) Reefs. In P.A. Scholle, D.G. Bebout and C.H Moore (Eds.) Carbonate Depositional Environments. Memoir of the American Association of Petroleum Geologists, 33, 345-462.

James N.P. and Bone Y. (1989) Petrogenesis of Cenozoic, temperate water calcarenites, South Australia: a model for meteoric/shallow burial diagenesis of shallow water calcite sediments. Journal of Sedimentary Petrology, 59, 191-203.

James, N.P. and Ginsburg, R.N. (1979) The seaward margin of Belize barrier and atoll reefs. Special Publication of the International Association of Sedimentologists, 3, 191pp.

Johnson, J.A.D. and Budd S.R. (1975) The geology of Zone B and Zone C, Lower Cretaceous limestone reservoirs of Asab Field, Abu Dhabi. 9th Arab Petrol. Congress Proceedings, 1-24.

Kretz, R. (1992) Carousel model for the crystallization of saddle dolomite. Journal of Sedimentary Petrology, 62, 190-195.

Land, L.S. (1980) The isotopic and trace element geochemistry of dolomite: the state of the art. In D.H. Ethington, J.B. Dunham and R.L. Zenger (Eds.), Concepts and Models of
Dolomitization. Society of Economic Paleontologists and Mineralogists Special Publication, 28, 87-110.

Land, L.S. (1986) Limestone diagenesis - some geochemical considerations. In F.A. Mumpton (Ed.), Studies in Diagenesis. U.S. Geological Survey Bulletin, 126-137.

Land, L.S. and Goreau, T.F. (1970) Submarine lithification of Jamaican reefs. Journal of Sedimentary Petrology, 40, 457-462.

Lee, Y.L and Friedman, G.M. (1987) Deep-burial dolomitization in the Ordovician Ellenberger Group carbonates, West Texas and Southeastern New Mexico. Journal of Sedimentary Petrology, 57, 544-557.

Lohmann, K.C. and Meyers W.J. (1977) Micro-dolomite inclusions in cloudy prismatic calcites - a proposed criterion of former high magnesium calcites. Journal of Sedimentary Petrology, 47, 1078-1088.

Longman, M.W. (1980) Carbonate diagenetic textures from nearshore diagenetic environments. Bulletin of the American Association of Petroleum Geologists, 64, 461-487.

Lumb, D.M. (1978) Organic luminescence. In D.M. Lumb (Ed.) Luminescence Spectroscopy. Academic Press. 93-148.

Machel, H.G. (1987) Saddle dolomite as a by-product of chemical compaction and thermochemical sulphate reduction. Geology, 15, 936-940.

Machel, H.G. and Mountjoy E.W. (1990) Coastal mixing zone dolomite, forward modeling and massive dolomitization of platform-margin carbonates - discussion. Journal of Sedimentary Petrology, 60, 1008-1012.

Macintyre I.G., Mountjoy, E.W. and D'Anglejan B.F. (1968) An occurrence of submarine cementation of carbonate sediments off the west coast of Barbados, WI. Journal of Sedimentary Petrology, 38, 660-664.

Maliva, R.G. (1989) Displacive calcite syntaxial overgrowths Journal of Sedimentary Petrology, 59, 397-403.

Marshall, J.D. (1992) Climatic and oceanographic isotopic signals from the carbonate rock record and their preservation. Geological Magazine, 129, 143-160.

Marshall, J.D. and Ashton M. (1980) Isotopic and trace element evidence for submarine lithification of hardgrounds in the Jurassic of England. Sedimentology, 27, 271-289.

Marshall, J.F. (1986) Regional distribution of submarine cements within an epicontinental reef system: central Great Barrier Reef, Australia. In J.H. Schroeder and B.H. Purser (Eds.) Reef Diagenesis. Springer-Verlag, Berlin, Heidelberg, New York, London, Paris, Tokyo, 8-26.

Marshall, J.F. and Davies P.J. (1981) Submarine lithification on windward reef slopes: Capricorn-Bunker Group, southern Great Barrier Reef. Journal of Sedimentary Petrology, 51, 953-960.

May, J.A. and Perkins, R.D. (1979) Endolithic infestation of carbonate substrates below the sediment-water interface. Journal of Sedimentary Petrology, 49, 357-377.

McLimans, R.K. 1987. (1987) The application of fluid inclusions to migration of oil and diagenesis in petroleum reservoirs. Applied Geochemistry, 2, 585-603.

Moore, C.H. (1985) Upper Jurassic subsurface cements: a case history. In N. Schneidermann and P.M. Harris (Eds.) Carbonate Cements. Society of Economic

Paleontologists and Mineralogists Special Publication, 36, 291-308.
Moore, C.H. (1989) Carbonate Diagenesis and Porosity. Elsevier, Amsterdam, 338 pp.
Mullins, H.T., Wise, S.W., Gardulski, A.F., Hinchey, E.J., Masters, P.M. and Siegel, D.I. (1985) Shallow surface diagenesis of Pleistocene periplatform ooze: northern Bahamas, Sedimentology, 32, 473-494.
Myers, J.H. (1987) Marine vadose beachrock cementation by cryptocrystalline magnesian calcite - Maui, Hawaii. Journal of Sedimentary Petrology, 57, 558-570.
Purser, B.H. Syn-sedimentary marine lithification of Middle Jurassic limestones in the Paris Basin. Sedimentology, 12, 205-230.
Purser, B.H. and Schroeder, J.H. (1986) The diagenesis of reefs: our present understanding. In J.H. Schroeder and B.H. Purser (Eds.) Reef diagenesis. Springer-Verlag, Berlin, Heidelberg, New York, London, Paris, Tokyo, 1-7.
Radke, B.M. and Mathis, R.L. (1980) On the late formation and occurrence of saddle dolomite. Journal of Sedimentary Petrology, 50, 1149-1168.
Reekmann, A. and Friedman, G.M. 1982. Exploration for carbonate petroleum reservoirs. John Wiley áSons, New York, Chichester, Brisbane, Singapore. 213pp.
Richter, D.K. (1983) Calcareous ooids: a synopsis. In T.M. Peryt (Ed.) Coated Grains. Springer-Verlag, Berlin, Heidelberg, New York, London, Paris, Tokyo, 71-99.
Richter, D.K. and Fuchtbauer H. (1978) Ferroan calcite replacement indicates former magnesium calcite skeletons. Sedimentology, 25, 843-860.
Roehl, P.O. and Choquette, P.W. (Eds.) Carbonate Petroleum Reservoirs. Springer-Verlag, New York, Berlin, Heidelberg, Tokyo, 622pp.
Sandberg, P.A. (1975) New interpretation of Great Salt Lake ooids and of ancient non-skeletal carbonate mineralogy. Sedimentology, 22, 497-538.
Sandberg, P.A. (1983) An oscillating trend in Phanerozoic non-skeletal carbonate mineralogy. Nature, 305, 19-22.
Sandberg, P.A. (1985) Aragonite cements and their occurrence in ancient limestones. In N. Schneidermann and P.M. Harris (Eds.) Carbonate Cements. Special Publication of the Society of Economic Paleontologists and Mineralogists, 36, 33-57.
Schlager, W. and James N.P. (1978) Low-magnesian calcite limestones forming at the deep sea-floor, Tongue of the Ocean, Bahamas. Sedimentology, 25, 675-702.
Schroeder J.H. and Purser B.H. (Eds.) (1986) Reef diagenesis. Springer-Verlag, Berlin, Heidelberg, New York, London, Paris, Tokyo, 455pp.
Scholle, P.A. and Halley, R.B. (1985) Burial diagenesis: out of sight, out of mind. In N. Schneidermann and P.M. Harris (Eds.) Carbonate Cements. Society of Economic Paleontologists and Mineralogists Special Publication, 36, 309-334.
Scoffin, T.P. (1987) An introduction to carbonate sediments and rocks. Blackie, London, 274pp.
Sellwood, B.W. (1986) Shallow-marine carbonate environments. In H.G. Reading (Ed.) Sedimentary Environments and Facies. Blackwell Scientific Publications, Oxford, London, Edinburgh, Boston, Palo Alto, Melbourne, 283-342.
Sellwood, B.W., Scott, J., Mikkelsen, P. and Akroyd, P. (1985) Stratigraphy and sedimentology of the Great Oolite Group in the Humbly Grove oilfield, Hampshire, S. England. Marine and Petroleum Geology, 2, 44-55.
Sellwood, B.W., Scott, J., James, B., Evans, R. and Marshall, J. (1987) Regional significance of "dedolomitization" in Great Oolite reservoir facies of Southern England. In J. Brooks and K. Glennie (Eds.) Petroleum Geology of North West Europe. Graham

and Trotman, 129-137.
Sellwood, B.W., Shepherd, T.J., Evans, M.R. and James, B. (1989) Origin of late cements in oolitic reservoir facies: a fluid inclusion and stable isotopic study (Mid-Jurassic, southern England). Sedimentary Geology, 61, 223-237.
Sellwood, B.W. and Beckett, D. (1991) Ooid microfabrics: the origin and distribution of high intra-ooid porosity; Mid Jurassic reservoirs, S. England. Sedimentary Geology, 71, 189-193.
Sellwood, B.W. and Wilkes, M. (1993) Hydrocarbon inclusions in late calcite cements: migration indicators in the Great Oolite Group, Weald Basin, S. England. Sedimentary Geology, 84, 51-55.
Smart, P.L., Dawans, J.M. and Whitaker, F. (1988) Carbonate dissolution in a modern mixing zone. Nature, 335, 811-813.
Sun, S.Q and Wright, V.P. (1989) Peloidal fabrics in Upper Jurassic reefal limestones, Weald Basin, Southern England. Sedimentary Geology, 65, 165-181.
Taylor, S.R. and Lapré, J.F. (1987) North Sea chalk diagenesis: its effect on reservoir location and properties. In J. Brooks and K. Glennie (Eds.) Petroleum Geology of North West Europe. Graham and Trotman, 483-495.
Tucker, M.E. and Hollingworth, N.T.J. (1986) The Upper Permian Reef Complex (EZ1) of North East England: diagenesis in a marine to evaporitic setting. In J.H. Schroeder and B.H. Purser (Eds.) Reef Diagenesis. 270-290. Springer-Verlag, Berlin.
Tucker, M.E. and Bathurst, R.G.C. (Eds.) (1990) Carbonate Diagenesis. Reprint Series, 1, International Association of Sedimentologists, Blackwell Scientific Publications, Oxford, London, Edinburgh, Boston, Paris. 312pp.
Tucker, M.E. and Wright V.P. (1990) Carbonate Sedimentology. Blackwell Scientific Publications, Oxford, London, Edinburgh, Boston, Melbourne, 482pp.
Videtich, P.E., McLimans, R.K., Watson, H.K.S. and Nagy, R.M. (1988) Depositional, diagenetic, thermal, and maturation histories of Cretaceous Mishrif Formation, Fateh Field, Dubai. American Association of Petroleum Geologists Bulletin, 72, 1143-1159.
Wanless, H.R. (1979) Limestone response to stress: pressure solution and dolomitization. Journal of Sedimentary Petrology, 49, 437-462.
Whitaker, F.F. and Smart, P. (1990) Active circulation of saline groundwaters in carbonate platforms: evidence from the Great Bahama Bank. Geology, 18, 200-203.
Wilkinson, B.H. (1979) Biomineralization, paleoceanography and the evolution of calcareous marine organisms. Geology, 7, 524-527.
Wilkinson, B.H., Owen, R.M. and Carroll, A.R. (1985) Submarine hydrothermal weathering, global eustacy, and carbonate polymorphism in Phanerozoic marine oolites. Journal of Sedimentary Petrology, 55, 171-183.
Woronick, R.E. and Land, L.S. (1985) Late burial diagenesis, Lower Cretaceous Pearsall and Lower Glen Rose Formations, South Texas. In N. Schneidermann and P.M. Harris (Eds.) Carbonate Cements. Society of Economic Paleontologists and Mineralogists Special Publication, 36, 265-276.

DOLOMITIZATION AND THE CHARACTER OF HYDROCARBON RESERVOIRS: DEVONIAN OF WESTERN CANADA

ERIC W. MOUNTJOY
Department of Planetary and Earth Sciences
McGill University
3450 University St.
Montreal, Quebec H3A 2A7, Canada

ABSTRACT Dolomitization of the Devonian carbonates of western Canada occurred in two main phases; 1) pervasive replacement dolomites that form 80 to 90% of all dolomites during early to intermediate burial (500 to 1500m), and 2) minor coarse crystalline and saddle dolomites that formed from hydrothermal fluids (>90oC). In some of the deeper parts of the basin (> 3000 m present depths) small amounts of late-stage dolomites were precipitated after oil generation and presumably were related to thermochemical sulphate reduction (TSR) reactions.

The distribution of porosity in the original limestones strongly controlled the distribution of replacement dolomitization. Where the original limestones were porous and permeable they were extensively dolomitized, but only if they were connected to regional conduit systems during shallow burial. Thus extensive and complete replacement dolomitization, appears to be part of, or connected to, regional conduit systems oriented perpendicular to the flanks of the basin (e.g. Rimbey-Meadowbrook reef trend, and Presqu'ile barrier) that conducted dolomitizing fluids (probably modified seawater) either downdip or updip. Later, during burial, these conduit systems allowed other dolomitizing fluids and TSR-generated fluids and hydrocarbons to move updip.

Generally where limestones have been changed to replacement dolostones with intercrystalline porosity the rock textures are more uniform and the permeability has been improved. These textures have been modified, and locally enhanced, by dissolution events associated with hydrothermal fluids and fluids related to TSR reactions. These fluids dissolved some dolomites and any remaining early calcites increasing porosity in intercrystalline pores, vugs and along fractures, thus improving the reservoir quality, and locally resulting in solution collapse. Locally late-stage anhydrite and carbonate cements have considerably reduced porosity and permeability updip from where the TSR reactions took place.

In general, during early burial, dolomitization and burial diagenesis reduced the overall porosity of most buildups by 1 to 10% compared to precursor limestone buildups. Dolostone reservoirs at depths >3000m have higher porosities and permeabilities compared to adjacent limestone buildups, presumably because limestone buildups were subjected to greater pressure solution.

A. Parker and B. W. Sellwood (eds.),
Quantitative Diagenesis: Recent Developments and Applications to Reservoir Geology, 33–94.

The Western Canada Sedimentary Basin is unique in terms of the enormous amounts of dolomites present compared to Devonian basins in western Europe, Australia, and China. A key difference in the character of these other basins is that they lack extensive evaporites. In the Western Canada Sedimentary Basin a series of evaporite deposits formed along the eastern flank of the basin during much of the Middle and Upper Devonian. Mg-rich fluids from these deposits may have seeped downwards into the basin resulting in the early replacement dolomitization.

1. Introduction

The importance of diagenesis to the formation and modification of hydrocarbon carbonate reservoirs is well established. In many regions dolomitization is a major control on the origin and character of these reservoirs. It is essential that geologists and petroleum engineers understand how diagenesis has affected the primary facies and controlled reservoir geometry and quality in order that hydrocarbons can be extracted as efficiently as possible.

Most of the data summarized and discussed below comes from the Middle and Upper Devonian reservoirs of the Western Canada Sedimentary Basin, which together contain more than 14.25 billion barrels of oil and 35 trillion cubic feet of natural gas. Despite the similarities of depositional facies in different Devonian reefs, the reservoir quality varies considerably from one reef or platform to another. These disparities are often due to diagenetic modifications of the depositional fabrics. Hence it is essential to map both the diagenetic and the depositional facies.

A number of publications have addressed carbonate diagenesis in Devonian carbonates of Western Canada, but few have attempted to examine diagenetic facies within a more regional context or to compare the diagenetic histories of Devonian reefs of varying ages. This has been attempted by Mountjoy and Krebs (1983) and Walls and Burrowes (1985, 1990). In addition there are few papers that summarize and discuss quantitative data about porosity and permeability of these reservoirs.

This paper covers the following topics:

1. Some current models of dolomitization and their relationship to porosity development.
2. Facies and diagenesis of limestone reefs and their porosity - Rainbow, Swan Hills, Golden Spike (as examples of early diagenesis prior to dolomitization).
3. Examples of dolomitized reefs -
   Middle Devonian - Winnipegosis, Rainbow, Presqu'ile
   Upper Devonian - Miette, Swan Hills, Leduc.
4. Reviews of the processes controlling porosity in dolomitized reefs.
   - early chemical reactions
   - later solution of calcite components
     1) at time of dolomitization
     2) post dolomitization
5. Development and causes of secondary porosity by:
   1) suggested origins of dissolution during burial.
   - fluids and gases generated during organic maturation
   - clay mineral reactions

- thermochemical sulphate reduction (TSR)
- upward cooling of CO_2-rich warm brines

2) fracturing of sealed reservoirs by thermal cracking of pooled oil.

1.1. MODELS OF DOLOMITIZATION

There are many ways in which dolomites can form. These include: 1) near surface processes of hypersaline sabkha, reflux, normal seawater, anoxic, cannibalization of high Mg-calcite, and freshwater-seawater mixing (Dorag), and 2) burial processes of compaction, hydrothermal, thermal sulfate reduction (TSR), thermal convection, etc. (see Morrow, 1990a,b; Machel and Mountjoy, 1986; Hardie, 1987).

All models require suitable sources of Mg, a hydraulic pump and suitable porosity and permeability for adequate flow of the dolomitizing fluids. One volume of limestone with about 40% porosity requires about 600 volumes of seawater (at 100% efficiency) to form a dolomite with similar porosity. Clearly dolomitization requires the flow of enormous amounts of fluids. Thus dolomitization, among other factors, is related to unconsolidated or weakly consolidated sediments in near surface environments, or porous and permeable carbonates in older rocks buried in a basin. Although seawater is supersaturated with respect to dolomite, dolomite is a complex crystal that needs considerable time to precipitate under surface temperatures, requiring the chemistry of fluids to be relatively uniform over a considerable length of time. Many modern and Pleistocene carbonates are not dolomitized because there was little or no flow of seawater through them, i.e. the essential hydraulic pump was missing. In the subsurface it becomes difficult to supply adequate Mg (Land, 1980) because of the progressive decrease in porosity and permeability with burial and the limited Mg available in the pore waters, in addition to having a suitable pumping mechanism. In short, it is difficult to form dolomite in both modern and subsurface environments.

Chemically there are a number of thermodynamic and kinetic factors that control dolomitization (see reviews for example Morrow 1990a,b; McIlreath and Morrow, 1990; Machel and Mountjoy, 1986, 1987; Hardie, 1987, Tucker and Wright, 1990), and Land (1991). Of the thermodynamic factors, temperature and Mg/Ca activity ratios are most important. Modern dolomites exhibit considerable variation in cation disorder and thus will have different thermochemical properties and will react very differently from well ordered dolomites (Hardie, 1987).

In many ancient rocks it is difficult to determine which dolomitizing process was involved because dolomitization tends to overprint and obliterate primary features and textures (Mazzullo, 1992; Kupecz et al., 1992). This has led to considerable discussion and debate in the literature as to which models are most suitable, which over the years has given rise to various bandwagons (Fig. 1). In many cases it is often impossible to determine how a particular dolomite formed. One can only place broad limits on the processes involved. This is particularly true for those rocks which have been completely dolomitized. Without some remnant limestones present it is difficult to determine the paragenetic sequence and the relative timing of dolomitization.

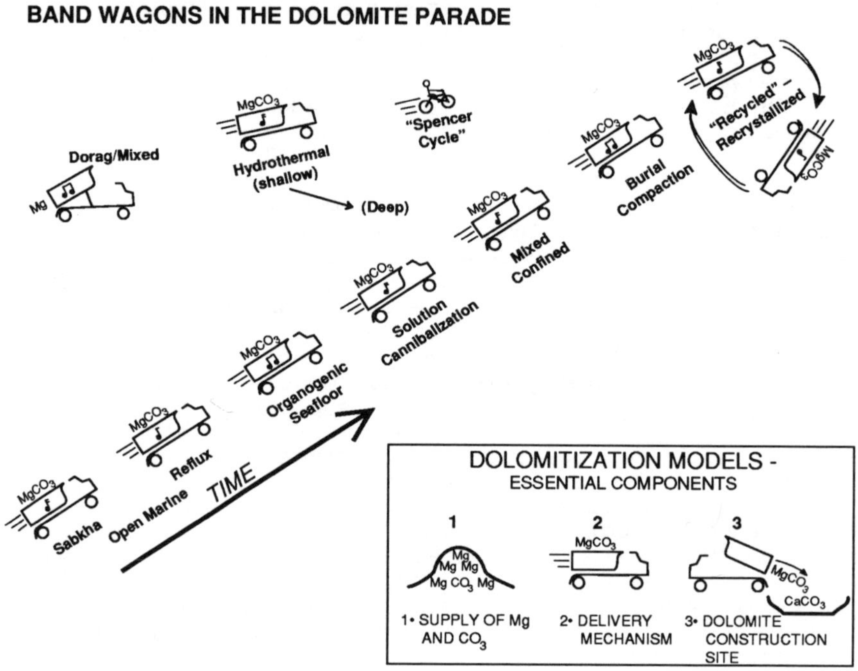

Figure 1. Essential components of (Morrow, 1990) and the different dolomitizing models or band wagons (after Eliuk, per. comm.).

The role of dolomitization with respect to the enhancement or reduction of porosity in dolostones is not well established. No general theories have recently been proposed for porosity evolution in dolostones (see recent poster session organized by at the 1993 AAPG meetings). The reason for this appears to be the many different ways in which dolomite may form, or be later modified during diagenesis and burial. Therefore it becomes very difficult to predict what processes really control porosity in dolostones, making it even more difficult to determine timing of these events.

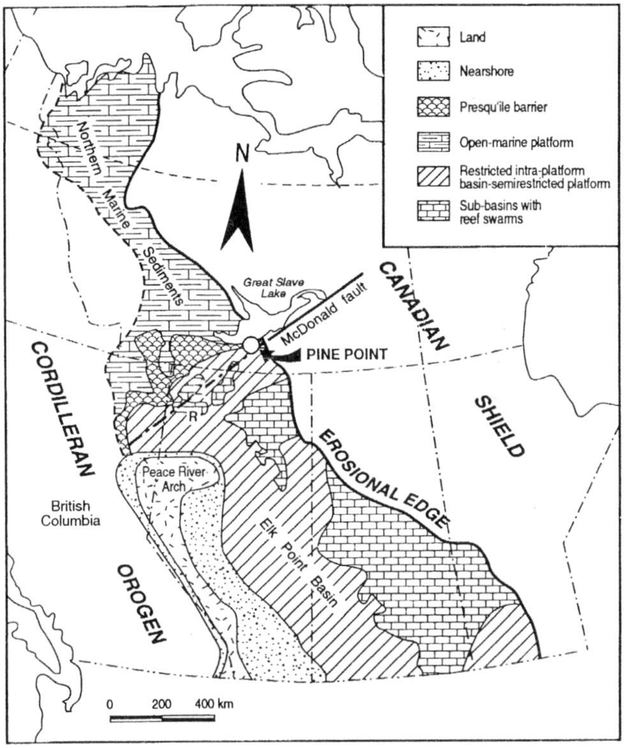

Figure 2. Simplified regional geological map of the Middle Devonian, Western Canada Sedimentary Basin (after Moore, 1988). the Presqu'ile barrier separates normal marine sediments to the north from evaporite and carbonate rocks to the south (from Qing and Mountjoy, 1992, with permission, Geology).

Important with respect to porosity are which chemical reactions were involved in the formation of a particular dolomite. The dolomitization equation can be written in different ways (Morrow, 1990b; Machel and Mountjoy, 1986). In addition to importing Mg ions one has to consider whether or not Ca ions were removed and whether CO_3^{2-} or HCO_3^- took part in the reaction. The reactions involved will govern whether or not porosity increases, remains about the same, or decreases (Machel and Mountjoy, 1986, 1987; see section on processes controlling porosity). In some modern environments dolomite appears to precipitate in pore spaces (Hardie, 1987) and only locally replaces some skeletal fragments. In many ancient examples secondary porosity appears to have developed either during dolomitization, or later by a separate dissolution event or events that dissolved some, or all, of the remaining calcite (Amthor et al., 1993; Drivet and Mountjoy, 1993). In addition there is increasing evidence that burial dissolution of earlier dolomites may have been locally extensive (Dravis and Muir, 1992; Mazzullo and Harris, 1992; Qing and Mountjoy, 1990; 1993 in press).

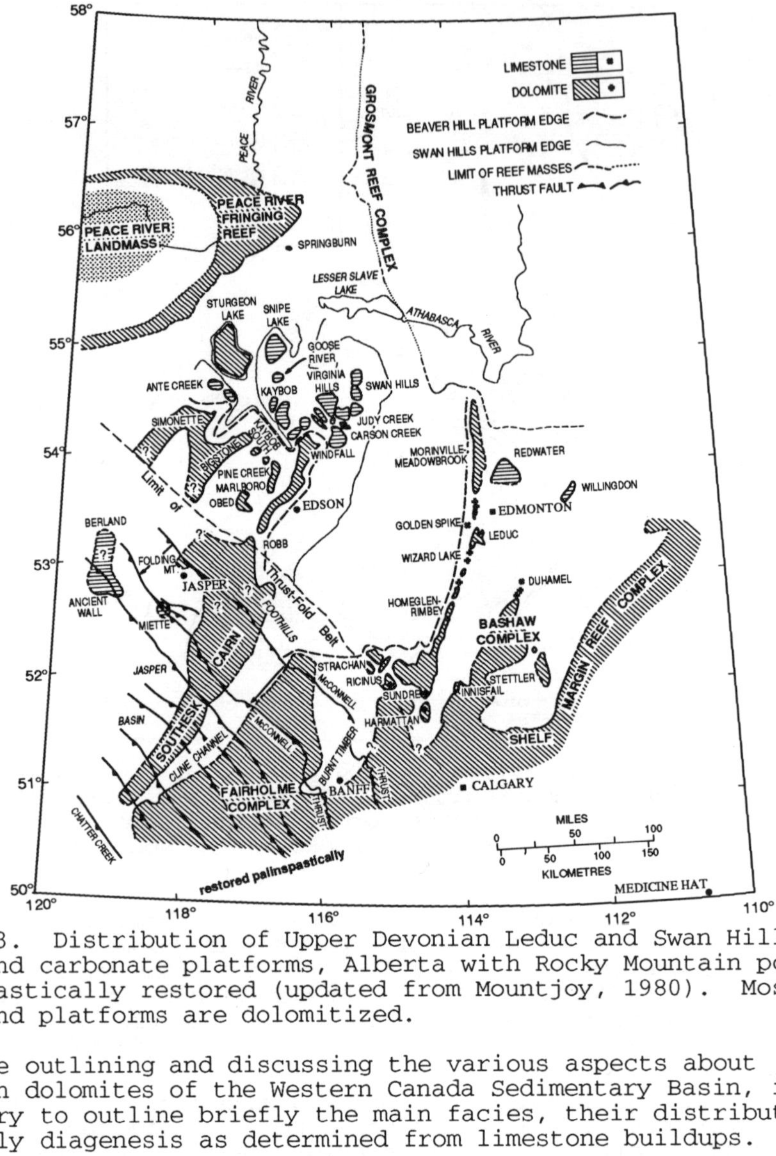

Figure 3. Distribution of Upper Devonian Leduc and Swan Hills reefs and carbonate platforms, Alberta with Rocky Mountain portion palinspastically restored (updated from Mountjoy, 1980). Most reefs and platforms are dolomitized.

Before outlining and discussing the various aspects about Devonian dolomites of the Western Canada Sedimentary Basin, it is necessary to outline briefly the main facies, their distribution and early diagenesis as determined from limestone buildups.

2. Nature of Carbonate Buildups

The facies of limestone buildups (reefs) of the Alberta subsurface (Figs. 2, 3 and 4) have been described in a series of papers, but only a few buildups have been thoroughly studied in terms of their diagenesis (Mountjoy and Krebs, 1983; Walls and Burrowes, 1985, 1990). These workers provided important data concerning the buildup facies, and early diagenesis that occurred prior to dolomitization which is so extensive in this basin. In general, cementation and solution are responsible for much of the observed variations in reservoir quality and heterogeneity in Devonian limestone reefs

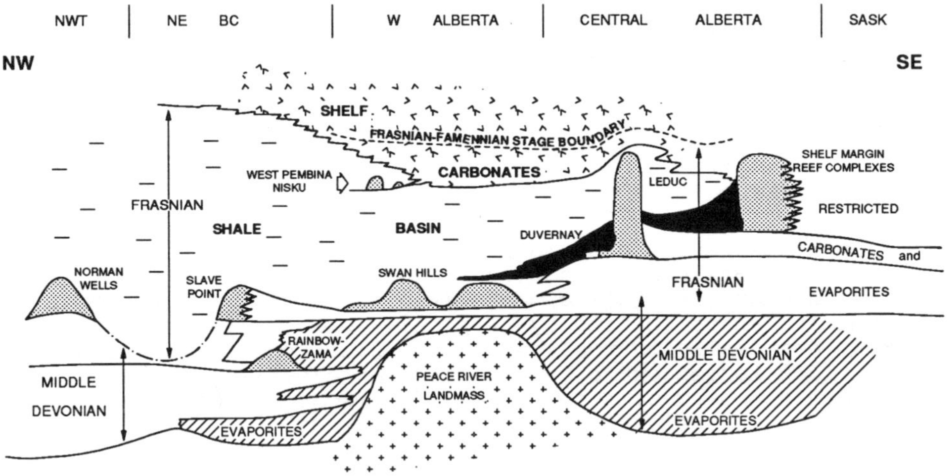

Figure 4. Schematic northwest-southeast cross section of the Middle and Upper Devonian from the Northwest Territories to southern Alberta and adjacent Saskatchewan, illustrating five levels of reef development (stippled), Keg River (Rainbow -Zama), Slave Point, Swan Hills, Leduc and Nisku (after Bassett and Stout, 1967). Hatchured pattern - evaporites, solid black represents Duvernay source beds.

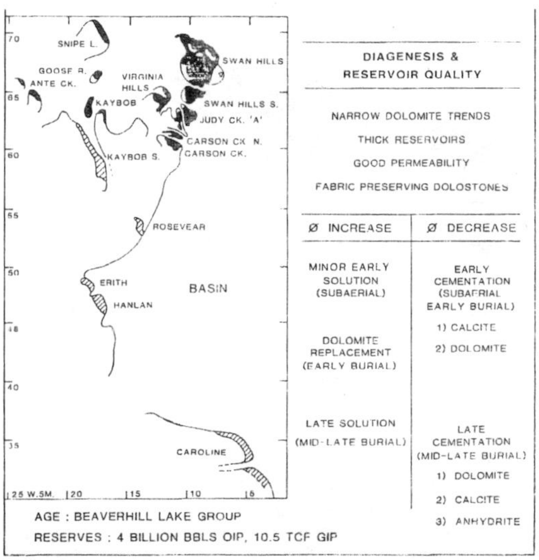

Figure 5. Distribution of Swan Hills reefs showing location of dolostone reservoirs (cross hatch pattern) and diagenesis and reservoir quality (from Walls and Burrowes, 1990).

In the submarine environment, cementation is far more important than solution, as for example in Golden Spike and Nevis buildups (Walls et al., 1979; Carpenter and Lohmann, 1989). The subaerial environment is characterized both by cementation and solution in response to vadose and meteoric waters, producing porous units interbedded with well cemented layers (Mountjoy, 1980). During subsurface diagenesis, pressure solution (stylolites) and associated cementation, and fracturing were the most important processes affecting reservoir character (Mossop, 1972; McGillivary and Mountjoy, 1975; Walls and Burrowes, 1990).

2.1. SWAN HILLS

Most Swan Hills reefs occur in linear patterns in the central part of the Alberta Basin with individual reefs typically separated by open marine channels or embayments (Fig. 5). They are up to 100 m thick and are characterized by vertically stratified facies patterns, especially in reef interiors which consist of subtidal and peritidal facies exhibiting considerable variation. Reef margins are composed of stromatoporoid-dominated framestones interbedded with skeletal grainstones and rudstones.

The diagenetic histories of different Swan Hills reefs are similar. Generally, reef or buildup margin facies contain better porosity and permeability than interior facies which commonly contain permeability barriers caused by layers of more muddy facies, and cementation. In Judy Creek for example, the interior has low porosity (5 to 10%) and low permeability (1 to 5 md), locally with some intervals with average porosity (8 to 12%, 10 to 50 md), and with much better reservoir quality at the reef margin (15%, up to 400 md, Wendte and Stoakes, 1982, Fig. 6).

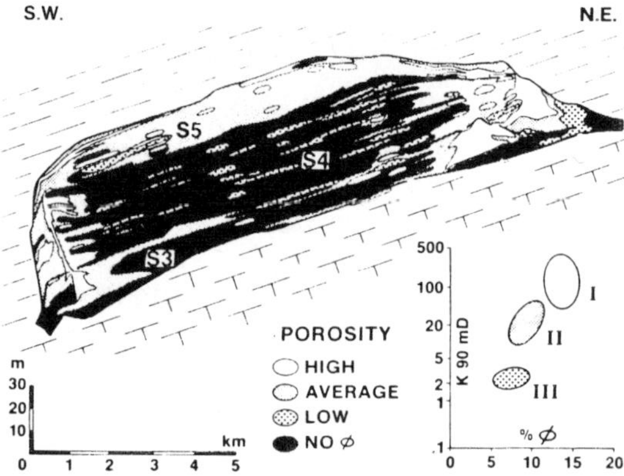

Figure 6. Cross section of Judy Creek oil field showing major porosity groups (from Jardine and Wilshart, 1982). Type 1 consist of stromatoporoid frameworks and grainstones, type 2 forereef and backreef grainstones, packstones and algal laminites, and type 3 packstones and wackestones of deeper forereef and interior lagoon. The major portion of the interior lagoonal facies in the S3 and S4 zones consists of non-porous lime mudstones (black).

The main processes that contribute to an increase or decrease in primary porosity are as follows. Solution, mainly resulting from subaerial exposure, enhances porosity beneath exposure surfaces in these reefs. Submarine cements are minor. Subsurface calcite and dolomite cements occur in Kaybob and appear to be directly related to stylolite formation (Wong and Oldershaw, 1981). They estimate a 20% reduction in bed thickness in Kaybob due to pressure solution. The marked reduction of porosity by calcite cements, associated with stylolites, is a common characteristic of most Swan Hills reefs. Although not recognized by these authors, some burial solution may also have taken place.

Dolomites generally make up less than 1% and occur near buildup margins as cements partly filling pores and fractures (Wong and Oldershaw, 1981; Viau and Oldershaw, 1984). Some fine crystalline dolomite occurs in, and up to a few centimetres below, green `shale' beds and appears to represent early sea floor and very shallow burial dolomite. Downdip, in the deeper parts of the basin, the Swan Hills platform margins are extensively dolomitized along NW-SE trends that appear to be related to faults (see section 3.4).

The diagenetic history, and porosity evolution, for Swan Hills limestone reefs indicates that porosity modification follows one of three pathways (Walls and Burrowes, 1990):

1. Extensive early cementation in submarine or subaerial environments that significantly reduced porosity and formed early permeability barriers (common in buildup interiors as thin beds and lenses below subaerial exposure horizons).
2. Minor surficial cementation but extensive burial compaction and stylolite formation (common in lime mud-rich lagoonal deposits, resulting in non-porous facies).
3. Minor surficial cementation followed by only moderate early through late burial cementation and compaction (retain porosities of 9 to 15%).

2.2. GOLDEN SPIKE

Golden Spike, one of the most thoroughly studied buildups, has facies similar to the large Redwater reef, except it is only 6 km^2 in area but has a similar thickness (170 m). The primary depositional porosity has been significantly modified by a complex diagenetic history which includes submarine and early-burial calcite cementation, subaerial calcite cementation and localized dissolution, and middle to late-burial calcite cementation (Fig. 7) (McGillivray and Mountjoy (1975), Reitzel and Callow (1976), Walls (1977), Walls et al. (1979), Walls and Burrowes (1985), and Carpenter and Lohmann (1989).

Porosity in fore reef and reef margin facies is consistently low due to occlusion by submarine cements and internal sediments of primary pores. Walls and Burrowes (1985) estimated that 70% of the cement volume is submarine in origin. Porosity in the reef interior is extremely variable due to the overlapping and sequential precipitation of submarine, subaerial and subsurface cements, as well as solution, that occurred mainly during subaerial exposure. Porosity occlusion by subsurface cements in the reef interior is slightly greater than on the margin, while diagenetic permeability barriers are common in the reef interior, inhibiting efficient hydrocarbon recovery (Table 1). Extensive stylolitization is also common in the reef interior (Mossop, 1972).

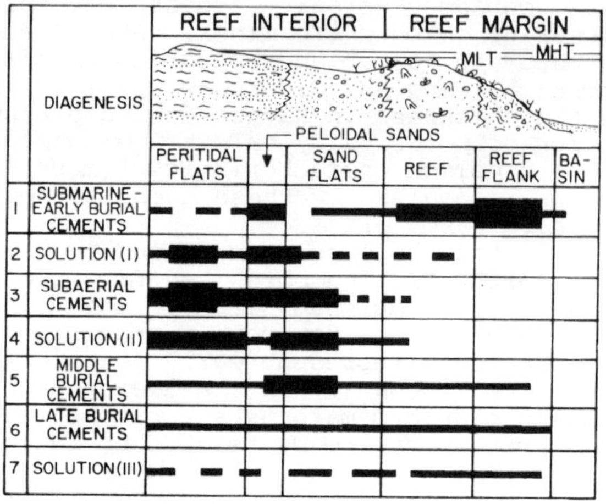

Figure 7. Distribution of main diagenetic features in Golden Spike buildup with respect to stages and facies (from Walls et al. 1979).

Table 1. Reservoir properties and major diagenetic processes effecting porosity and permeability in Golden Spike (from Walls and Burrowes, 1990).

		POROSITY (Ø)		PERMEABILITY (K)		MAJOR DIAGENETIC EFFECTS ON Ø, Kh	
		MEAN (%)	MODE (%)	MEAN (md)	MODE (md)	Ø, Kh INCREASE	Ø, Kh DECREASE
REEF MARGIN	DETRITAL CORAL	4	2	1	1	RARE SOLUTION	EXTENSIVE RADIAXIAL & FIBROUS CALCITE CEMENTATION.
	TABULAR STROMATOPOROID	5.5	4	10	5	RARE SOLUTION	RADIAXIAL & FIBROUS CALCITE CEMENTATION DEPOSITION OF INTERNAL SEDIMENT.
	MASSIVE STROMATOPOROID & STROMATOPOROID RUBBLE	10	8.5	100	10; 100	MOLDIC SOLUTION	CLEAR, NON-FERROAN & FERROAN CALCITE CEMENTS.
REEF INTERIOR	SKELETAL, PELOIDAL SANDS	12.5	10	50	20	MOLDIC SOLUTION	FIBROUS, NON-FERROAN & FERROAN CALCITE CEMENTS.
	ALGAL LAMINITES	7	6	60	10	SOLUTION ENLARGEMENT OF FENESTRAL PORES	MICROSTALACTITIC CEMENT, NON FERROAN & FERROAN CALCITE CEMENTATION

Very minor amounts of dolomite (as micrite replacement, microdolomite inclusions, and fracture filling) have been reported. Most of these dolomites are confined to green, argillaceous mudstones at the base of shallowing-upward cycles. Some minor dolomite also occurs along stylolites, as cement in fractures, and as a replacement of clear calcite cements (Walls, 1977). Walls (1977) speculated that rare fracture-filling dolomites might be related to periods of extensive dolomitization that occurred in the nearby Rimbey-Meadowbrook reef trend.

3. Examples of Dolomitized Reefs

Middle Devonian

Dolomitized buildups of Middle Devonian (Givetian) age occur throughout the Elk Point Basin in Alberta, Saskatchewan, Manitoba and North Dakota (Fig. 2). The Elk Point Basin is divided into several subbasins in which carbonate shelf and buildup sediments were deposited contemporaneously. To the northwest, hydrocarbon producing Keg River Formation carbonate buildups occur in three sub-basins: Rainbow, Zama and Shekilie (e.g. Schmidt et al., 1985, Qing and Mountjoy, 1989). Immediately to the northwest of these basins a shallow platform and series of reef complexes formed the Presqu'ile barrier (Skall, 1975; Qing and Mountjoy, 1994) that extends in a NE-SW direction for over 400 km from Great Slave Lake to the Peace River Arch. The Slave Point platform occurs immediately above the Presqu'ile barrier and is also extensively dolomitized (Gray and Kassube, 1963; Griffin, 1965).

3.1. RAINBOW DOLOMITIZED REEFS

More than 80 partially to completely dolomitized buildups of Middle Devonian Keg River Formation occur in the Rainbow sub-basin of northwestern Alberta. They vary in size from 0.6 km^2 to over 21 km^2 reaching thicknesses up to 250 m. Recoverable reserves are greater than 700 million barrels of oil and 850 billion cubic feet of natural gas. Data are summarized from Qing and Mountjoy (1989), Schmidt et al.(1985), and Walls and Burrowes (1985, 1990).

Rainbow limestone reefs contain abundant submarine cements that created rigid reef margins contemporaneously with reef growth. Primary porosity was reduced in the reef margin facies to less than 20% in many reefs. The reef interiors contain about equal amounts of submarine and subaerial cements, reducing porosity to about 20% by the end of subaerial diagenesis. Burial cements account for less than 15% of the total cement volume in most reefs, helping to preserve porosity following early diagenesis. Thus, reef interiors are generally more porous (13%) than reef margins (8%) (Fig. 8). Most porosity enhancement resulted during early exposure, except where buildups are dolomitized.

In six buildups (A, B, E, F, G, and Tehze), four petrographic types of dolomites were recognized by Qing and Mountjoy (1989): 1) gray, fine-crystalline dolomite; 2) isolated dolomite rhombs/patches; 3) matrix replacement dolomite; and 4) saddle dolomite. The petrographic and geochemical data indicate that these dolomites probably formed in three principle stages of dolomitization during progressive burial. Based on abundances estimated from cores studied by Qing and Mountjoy (1989), gray, fine-crystalline dolomite accounts for less than 1% of the total dolomite by volume. Isolated dolomite rhombs/patches are a major type of dolomite in the limestone buildups, and account for about 15% of the total dolomite in the Rainbow buildups.

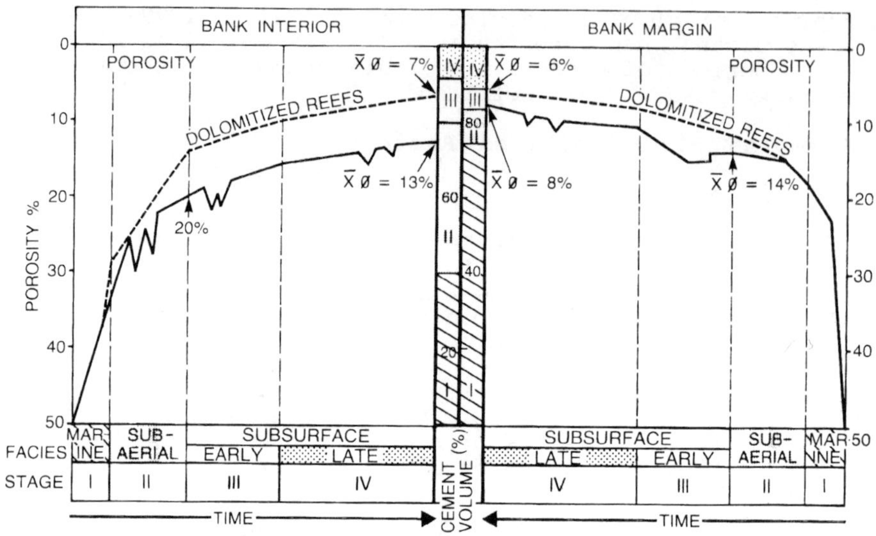

Figure 8. Summary of diagenesis and reservoir history for Rainbow reefs illustrating porosity reduction with time and the estimated cement volume (in percent) formed in each diagenetic facies and stage (from Walls and Burrowes, 1985).

Fine crystalline dolomite is interpreted to represent penecontemporaneously dolomitized lime muds. Dolomite rhombs and patches are similar to matrix-replacement dolomite. Matrix-replacement dolomite is fabric destructive and is most abundant in the dolostone buildups. These dolomites probably formed during shallow to intermediate burial because they: 1) postdate stylolites, 2) are widespread throughout the buildups and crosscut various facies, and 3) have light oxygen isotope compositions, and low Sr but high Fe and Mn concentrations. Dolomitizing fluids for dolomite rhombs/patches and matrix replacement dolomites possibly came from: 1) mechanical compaction; 2) dehydration of the adjacent Muskeg gypsum at relatively shallow burial depths; and 3) intermediate depth basinal fluids conducted updip along porous and permeable conduits. Either Devonian seawater or early compaction fluids were probably responsible for the early dolomite rhombs and patches. However, the main phase of matrix-replacement dolomitization took place later, possibly during early compaction (or even later) when the east side of the basin was intermittently tilted. These movements caused updip flow of intermediate depth basinal brines, during Late Mississippian to Jurassic time.

Saddle dolomite occurs as late-stage vug and fracture fillings. It is most depleted in $\delta^{18}O$ but most enriched in Mn and Fe, suggesting precipitation during deeper burial from warm basinal fluids. These fluids may have been expelled during Upper Cretaceous sedimentary and tectonic loading of the western side of the basin, and moved updip on the east side of the basin along existing platform, fault, and fracture conduits. The source of the Mg is uncertain. Some, or most, of the Mg may have resulted from chemical compaction of earlier matrix dolomites, or may have come from de-watering of primary carnallite during burial (Koehler and Kyser, 1991).

Matrix-replacement dolomite is either dense or porous. Dense matrix dolomites are the most abundant dolomite type in the dolostone buildups, forming 65% of the total dolomites by volume, while porous matrix dolomites (those with intercrystalline porosity) account for about 15%. Saddle dolomites are essentially restricted to dolostone buildups and have abundances of about 5%. The distribution of the dolomites in F buildup is shown in Figure 9.

In general, matrix-replacement dolomites are texturally destructive, and strongly modify (or obliterate) sedimentary and earlier diagenetic features. Why some matrix dolomites are porous and others are not is unclear, but it may be related to the primary facies (mudstone versus grainstones) and the amounts and rates of limestone dissolution during and/or after dolomitization. In F buildup, the spatial distribution of porous matrix replacement dolomites occur irregularly in all facies (Fig. 9).

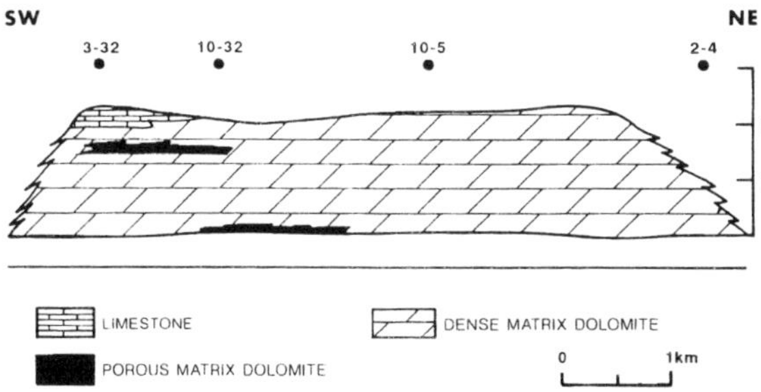

Figure 9. Dolomite distribution in F buildup. Dense matrix dolomites are the most abundant type and matrix dolomite occurs in all facies (from Qing and Mountjoy, 1989, with permission J.Sed. Pet).

The porosity of dolostone buildups (e.g. averaging 5% in F buildup) is generally half that of limestone buildups (e.g. 10% and 13% for Tehze and E respectively) as suggested by data summarized by Schmidt et al. (1985). Thus in these Rainbow reefs, matrix-replacement dolomitization appears to reduce porosity. The porosity in the dolostone buildups is irregularly distributed, resulting in a heterogeneous reservoir rock. The permeability of these dolostones, however, is much better than that of the limestones, because of fracturing, more uniform textures, and pore throat sizes (Qing and Mountjoy, 1989). In summary, the dolomitized Rainbow buildups have slightly less porosity but considerably higher permeability than limestone buildups, as noted by Walls and Burrowes (1985, 1990). In addition, dolomitized reefs show little porosity variation from reef interior to reef margin. However, in some Rainbow reefs the greatest variations in porosity occur between the porous and nonporous matrix dolomites (Qing and Mountjoy, 1989; Fig. 9).

3.2. PRESQU'ILE BARRIER (from Qing, 1991; Qing and Mountjoy, 1992, in press) (Fig. 10)

The Presqu'ile (Keg River Formation) barrier occurs in the northern part of the Western Canada Sedimentary Basin (Fig. 2). It is a 400 km long and 20 to 100 km wide region of carbonate shoals and reefs extending from outcrops in the Northwest Territories into the subsurface of northeastern British Columbia. It restricted seawater circulation in the southern part of the Western Canada Sedimentary Basin caused evaporites and carbonates to be deposited south of the barrier in the Elk Point Basin. North of the barrier subtidal shales and argillaceous carbonates were deposited (Fig. 2). Pine Point is located toward the present-day east end of the barrier, where various dolostones host more than 80 individual Mississippi Valley type (MVT) ore bodies (Skall, 1975; Kyle, 1981, 1983; Rhodes et al., 1984). The barrier is interpreted, on the basis of dolomite distribution and geochemistry, to have acted as a porous and permeable conduit system during burial. It focussed, and channeled, basin fluids that were responsible for extensive dolomitization, secondary migration of hydrocarbons, and local MVT mineralization.

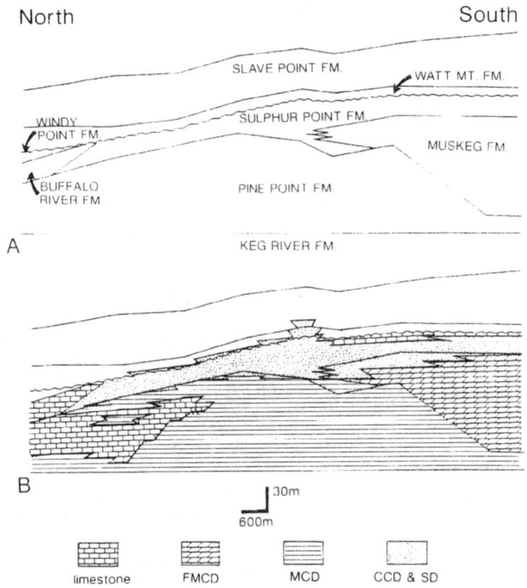

Figure 10. Stratigraphy and distribution of different dolomite types across the Presqu'ile barrier at Pine Point (from Qing and Mountjoy, 1990).

These carbonates have undergone diagenetic alteration in submarine, subaerial, and subsurface environments. Submarine diagenetic features include: micrite envelope, micrite, syntaxial, fine crystalline and fibrous cements. Local pendant cements and small-scale dissolution and brecciation are minor and interpreted to be associated with the sub-Watt Mountain subaerial exposure. Subsurface diagenesis resulted in the most important and extensive modifications in the Presqu'ile barrier, and include: blocky sparry calcite cement, compaction and stylolitization, dolomitization, hydrothermal dissolution, Pb-Zn sulphides at Pine Point, anhydrite, late-stage coarse-crystalline calcite, white calcite, elemental sulphur, fluorite, and bitumen (Qing, 1991; Qing and Mountjoy, 1992; 1994, in press).

Four types of dolomites are associated with the Presqu'ile barrier. In paragenetic sequence they are: fine-crystalline (FCD); medium-crystalline (MCD); coarse-crystalline (CCD); and saddle (SD). The origins and timing of these dolomites are established on the basis of their distinctive spatial distributions, cross cutting relationships, petrographic characteristics, diagenetic paragenesis, and geochemical signatures (Qing and Mountjoy, 1990; Qing, 1991).

3.2.1. *Fine-crystalline dolomite.* Fine-crystalline dolomite is restricted to the back-barrier facies and is interbedded with Muskeg evaporitic anhydrites. The heaviest $\delta^{18}O$ value of FCD is -1.58 ‰ PDB, which is within the range of calculated $\delta^{18}O$ values (-1.2 to -2.2 ‰ PDB) for dolomites that would precipitate from Middle Devonian seawater (assuming $\delta18$‰ of -4.0 PDB for marine limestones). The $^{87}Sr/^{86}Sr$ ratios of unaltered FCD range from 0.7079 to 0.7080, corresponding to the estimated $^{87}Sr/^{86}Sr$ ratio of Middle Devonian seawater (Burke *et al.* 1982).

FCD appears to have formed penecontemporaneously at or just below the sea floor from Middle Devonian seawater, because it was interbedded with anhydrite in the Elk Point Basin, and because its $\delta^{18}O$ values, $^{87}Sr/^{86}Sr$ ratios, REE patterns, and Sr concentrations are similar to those estimated and calculated for Middle Devonian seawater dolomites.

3.2.2. *Medium-crystalline Dolomite.* MCD occurs in the lower part of the Presqu'ile barrier in the Pine Point Formation and is restricted to the southern and central (lower) part of the barrier. The northern part of the barrier remains undolomitized. The heaviest $\delta^{18}O$ values for MCD is -3.7 ‰ PDB, slightly depleted compared with the estimated $\delta^{18}O$ signature of Middle Devonian seawater dolomites. However, the $\delta^{18}O$ values of most MCD fall between -5 to -7 ‰ PDB, and are much more depleted than the estimated $\delta^{18}O$ signature of Middle Devonian seawater dolomites. Three $^{87}Sr/^{86}Sr$ ratios, 0.7081, 0.7085 and 0.7087, range from the best estimated $^{87}Sr/^{86}Sr$ ratio of Middle Devonian seawater to being slightly more radiogenic (Fig. 11).

These data, combined with textural relationships, suggest that these dolomites formed during shallow burial at slightly elevated temperatures from fluids derived by means of compaction or loading of strata: a) from downdip in the western part of the basin, and/or b) from the Elk Point and Mackenzie basins adjacent to the Presqu'ile barrier. The restriction of MCD to the lower southern and central part of the barrier suggests dolomitizing fluids may have been derived from the Elk Point Basin, preferentially dolomitizing limestones adjacent to the Elk Point Basin. However, the primary O and Sr isotope signatures of MCD could have been modified during later neomorphism.

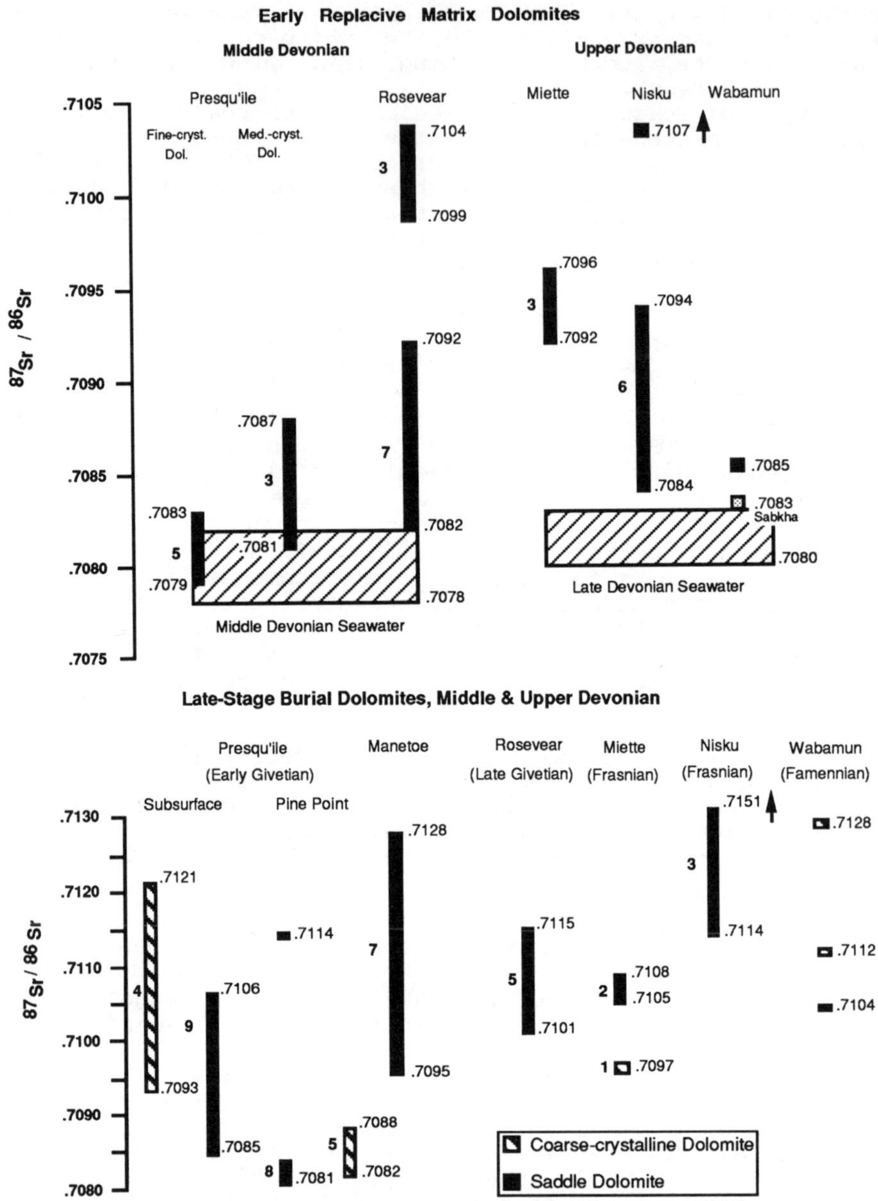

Figure 11. The $^{87}Sr/^{86}Sr$ values of early replacive matrix (upper diagram) and late-stage burial dolomites (lower diagram), Middle and Upper Devonian of western Canada, showing ranges for Middle and Upper Devonian seawater. Presqu'ile dolomites are divided into fine- and medium-crystalline types and into samples from Pine Point and the subsurface (from Mountjoy et al. 1992, with permission, Pergamon Press).

3.2.3. Coarse-crystalline and Saddle Dolomites (CCD and SD). CCD and SD nearly always occur together. They have similar $\delta^{18}O$ values, $^{87}Sr/^{86}Sr$ ratios, and trace element concentrations, and are interpreted to have formed from diagenetic fluids of similar chemical compositions. CCD represents dolomitized limestones, whereas SD postdates CCD occurring as dolomite cement filling vugs and fractures. CCD and SD postdate stylolites and locally extend above the sub-Watt Mountain unconformity, indicating that CCD and SD formed during burial. SD overlaps mineralization in the Pine Point area. Using various dating methods the timing of mineralization has been broadly constrained as occurring between Devonian to early Tertiary. Rb/Sr gives a late Devonian age whereas paleomagnetic data give a Cretaceous age (Symons et al., 1993). The homogenization temperatures of saddle dolomite fluid inclusions from 14 locations (Fig. 12) vary from 85 to 210°C, and gradually decrease from northeastern British Columbia (B.C.) (155 to 171°C) to Pine Point (95 to 106°C). The pressure-corrected homogenization temperatures exceed the maximum burial temperatures of the region, and therefore, the formation of CCD and SD probably represent regional hydrothermal event(s).

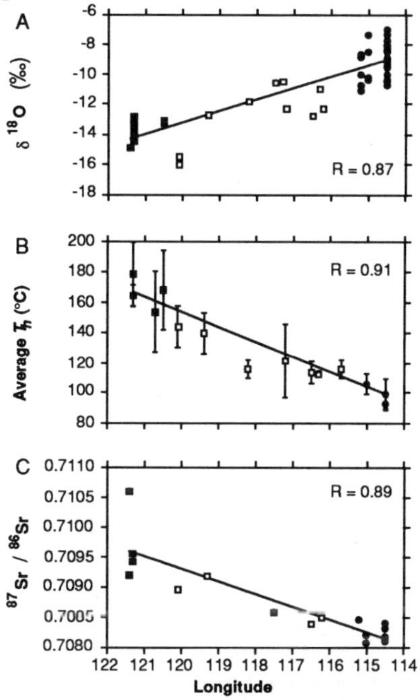

Figure 12. $\delta^{18}O$, homogenization temperatures (T_h), and $^{87}Sr/^{86}Sr$ of saddle dolomite cements vs. longitude. Dots are Pine Point samples, open squares are samples from Northwest Territories, and solid squares are samples from northeastern British Columbia (from Qing and Mountjoy, 1992, with permission, Geology).

The gradual decrease in $^{87}Sr/^{86}Sr$ ratios and homogenization temperatures (with corresponding increase in $\delta^{18}O$ values) from northeastern B.C. to Pine Point (Fig. 12) suggests a large-scale migration of hot and radiogenic dolomitizing fluids updip from west to east along the relatively permeable Presqu'ile barrier. The eastward updip flow of basin brines was probably established as a result of tectonic thrusting and compression, sedimentary loading, and uplift of the foreland basin. In the Western Canada Sedimentary Basin, the tectonic thrusting and compression occurred in two main episodes, the first during the Late Jurassic to Early Cretaceous, and the second during the Late Cretaceous to early Tertiary. The main stage of dolomitization and mineralization probably occurred during the second episode because: 1) fluid flow was likely to have been more active during the Late Cretaceous to early Tertiary as a result of a more extensive gravity-driven flow system that was formed by uplift of the foreland basin; 2) the Western Canada Sedimentary Basin reached maximum burial during the Late Cretaceous and early Tertiary, providing an adequate supply of hot brines in the deeper part of the basin for dolomitization; and 3) the presumed Cretaceous age of mineralization.

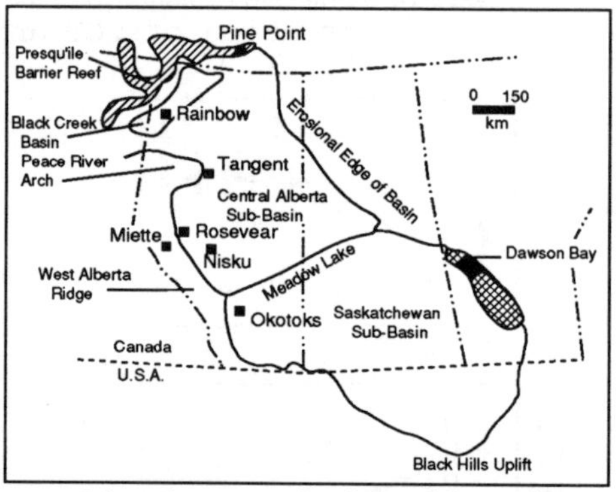

Figure 13. Generalized map of the Elk Point Basin showing Saskatchewan sub-basin, Winnipegosis Formation outcrops near Dawson Bay, and Presqu'ile barrier (from Mountjoy et al. 1992, with permission, Pergamon Press).

The Presqu'ile barrier (Fig. 13) appears to have acted as a deeply buried regional porous and permeable conduit system. When hot fluids with radiogenic Sr were expelled, tectonically, from deeply buried sediments into the Presqu'ile barrier, the temperatures in the conduit would increase, thus explaining why the homogenization temperatures of saddle dolomite fluid inclusions exceed the maximum burial temperatures of the Western Canada Sedimentary Basin. As hot fluids moved eastward updip along the Presqu'ile barrier, they would have gradually cooled, resulting in the eastward decrease in homogenization temperatures and a corresponding increase in $\delta^{18}O$ values of CCD and SD.

The radiogenic Sr could have come from deeply buried sediments and/or from the crystalline basement, during tectonic thrusting and compression. In support of this, SD in the western downdip part of the Presqu'ile barrier are the most radiogenic. As dolomitizing fluids moved updip eastward along the Presqu'ile barrier, the radiogenic Sr gradually mixed with less radiogenic formation waters, resulting in the $^{87}Sr/^{86}Sr$ ratios decreasing eastward along the Presqu'ile barrier (Mountjoy et al. 1992, Table 2).

Table 2. Potential sources of Sr isotopes in subsurface brines, Western Canada Sedimentary Basin (from Mountjoy et al., 1992, with permission, Pergamon Press).

		$^{87}Sr/^{86}Sr$
"Connate" seawater	Mid-Devonian	0.7080
	Silurian	0.7080–0.7088
	Ordovician	0.7077–0.7090
	Cambrian	0.7090+
Mid-Devonian evaporite brines		0.7080
Pressure solution of underlying strata	Ordovician	0.7070–0.7090
	Cambrian	0.7090+
	Late Proterozoic	0.7075
Adjacent illite shales		Slightly radiogenic? Higher in older rocks
Cretaceous clastic wedge		Volcanic and igneous detritus highly radiogenic
Metamorphic fluids		Variable—moderate to highly radiogenic?
Precambrian basement (McNutt *et al.*, 1990)		Radiogenic
		0.7112–0.7136 Yellowknife
		0.7188–0.7279 Thompson
		0.7105–0.7171 Sudbury I
		0.7250–0.7400 Sudbury II
		0.7080–0.7103 Norita
		0.7065–0.7278 Eye-Dashwa
Present-day western Canada subsurface fluids		
Edmonton area, Connolly *et al.* (1990) Tertiary and Upper Cretaceous		0.7057–0.7090

Few data have been published on the porosity and permeability of the Presqu'ile barrier carbonates, except for the earlier studies of the Clarke Lake gas field (Gray and Kassube, 1963; Griffin, 1965; Comby, 1975). The extensive dolomitization that occurred during burial and hydrothermal diagenesis, makes this trend very different from other Middle and Upper Devonian reefs to the south in Alberta. The late-stage dolomites are estimated to account for between 40% (reef interiors) to 55% (margins) of the total cement volume. In addition, pores are also filled with ferroan calcite, quartz and metallic sulphides (Walls and Burrowes, 1985). Walls and Burrowes estimate the porosity to be about 10% along the barrier and reef margins and 7% in the interiors. A recent study of the isolated Sierra reef complex west of the Presqu'ile barrier (Collins and Lake, 1989), indicates that dolomites average 10% porosity, whereas the overlying limestones mostly lack porosity or locally have up to 6%. The porosity in these dolomites consists of intercrystalline, solution enlarged vuggy, and fracture porosity.

3.3. WINNIPEGOSIS REEFS (Fig. 14)

Reef complexes crop out along the shores of Lake Winnipegosis and the islands of Dawson Bay in central Manitoba (Teare, 1990; Fig. 13) and occur in the subsurface of Manitoba and Saskatchewan (Martindale and MacDonald, 1990). Winnipegosis buildups are located in the Saskatchewan sub basin delimited by the Meadow Lake

Escarpment to the north and the Transcontinental Arch (Black Hills Uplift) to the south. The Dawson Bay exposures of the Winnipegosis buildups occur on the northeast flank of the preserved portion of this subbasin (Fig. 13). The outcrop reefs provide excellent 3-D exposures of the facies in individual buildups and some larger complexes. Although the sediments are dolomitized, textures and depositional structures are generally so well preserved that these rocks can be confused with limestones. No commercially significant quantities of hydrocarbons were produced from the Winnipegosis of Saskatchewan until 1986 when Home Oil and Canadian Hunter discovered oil in the Tableland area (Fig. 14).

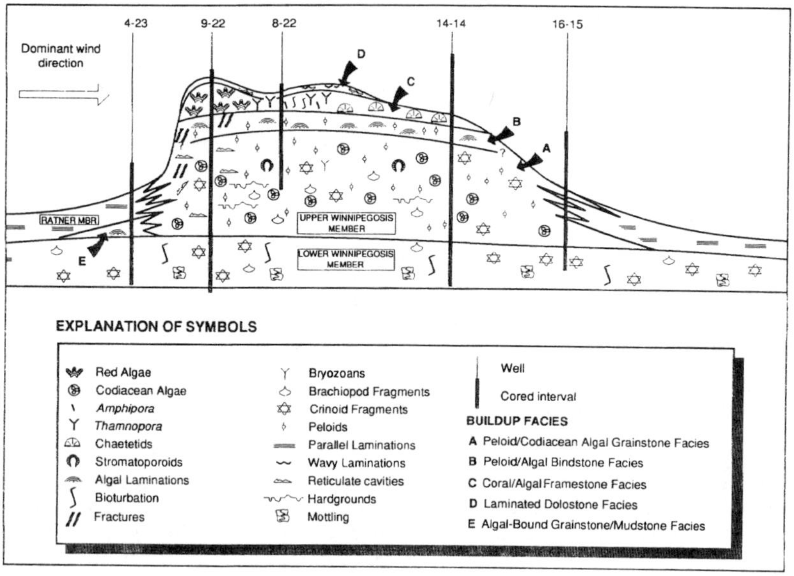

Figure 14. Schematic cross-section through the Tableland buildup showing stratigraphy and facies, subsurface Saskatchewan (from Martindale and MacDonald, 1990).

In both areas, the Lower Member of the Winnipegosis Formation (Upper Eifelian) rests unconformably on the Lower Elk Point Ashern Formation. Winnipegosis buildups (Givetian) directly overlie the Lower Member (platform) and commonly exceed 60 m in thickness (Fig. 14). In areas between buildups, the Lower Member is directly overlain by bituminous laminites of the Ratner Member, which are in turn overlain by anhydrites and halites of the Lower Prairie Evaporite (Wardlaw and Reinson, 1971). Upper Winnipegosis buildups consist of grainstones, packstones, framestones and bindstones. The lower two thirds of buildups in the Tableland area are composed of codiacean algal and peloidal grainstones in the form of a number of coalesced shoals. The upper one third of the buildups consist of branched corals, blue-green and red algae, forming a reef framework (Fig. 14). In Saskatchewan the Upper Member is completely dolomitized, whereas in North Dakota buildups retain part of their original limestone lithology. Buildups are capped by dolomitic laminites, deposited under subaerial conditions which terminated buildup deposition. The buildups are overlain by anhydrite which forms a seal for trapping hydrocarbons. The anhydrite thickens from 1 to 3 m above the buildups to 20 m in the adjacent basin forming a "halo" around the buildups (Martindale and MacDonald, 1990).

3.3.1. *Diagenesis.* The depositional facies are illustrated in the schematic cross-section (Fig. 14). The matrix, allochems and early cements in the buildups have been recrystallized and completely dolomitized. Cement precipitation (calcite, or high-Mg calcite) began during sedimentation with fibrous isopachous and radiaxial fibrous cements. Cementation continued until the buildups were in a deep burial environment. Blocky isopachous, blocky pore-filling, and syntaxial cements on crinoids were precipitated in a shallow burial environment. This was followed by replacement dolomitization. A limpid dolomite cement, precipitated in an intermediate to deep burial environment, occurs as overgrowths on pre-existing blocky cement crystals. Bladed anhydrite was precipitated in a deep burial environment after limpid dolomite precipitation and partial dissolution of both blocky calcite cements and dolomite rhombs.

Dolomitization

The timing and origin of replacement dolomitization in the Tableland buildup is uncertain, but probably occurred early, either as a result of mixed waters, or during shallow burial. In the Manitoba outcrops, mimetic dolomite replacement of radiaxial submarine and pendant vadose cements indicates that this generation of dolomitization postdated early cementation, and subsequent subaerial exposure of the buildup sediments.

Early replacement dolomites are widely distributed throughout the buildup, basin and platform sediments, and represent 60% of all Winnipegosis diagenetic products. White replacement dolomite is chalky, and constituent allochems are particularly poorly preserved. Conversely, light- to dark-brown replacement dolomite is more coherent, and the preservation of the cream coloured allochems and early diagenetic fabrics is improved. The platform sediments are characterized by yellow, chalky mottles surrounded by a distinctly different brown matrix. In the basin sediments, thin (<50), light to dark brown intervals of matrix dolomite are intercalated with the black, laminated, bituminous mudstones of the Brightholme Member. In the Ratner Member, 3.0 to 5.0 mm thick, brown intervals of dolomite are interbedded with laminated bituminous partings.

Replacement dolomite is divided into two categories based on crystal size: 1) microcrystalline dolomite (<20; commonly 5-10); and 2) finely crystalline dolomite (20-50; commonly 30-40). Both types of replacement dolomite are inclusion rich. Replacement dolomites neomorphically replace micritic sediments and mimetically dolomitized fibrous cements. These textural relationships suggest that replacement dolomitization postdated both early cementation, and mimetic dolomitization.

The fabric-destructive nature of finely crystalline dolomite affects allochems and early diagenetic fabrics throughout the buildups. In particular, allochem peripheries and marine cement intercrystalline boundaries are often the sites of neomorphic fine crystalline replacement dolomite.

The distribution of the two types of replacement dolomite does not appear to be facies controlled. However, the platform, basin and lowest buildup sediments are characterized by dense fine-crystalline dolomite, whereas most of the overlying buildup sediments contain both types of dolomites. These broad relationships between the occurrence of replacement dolomites and stratigraphic levels are particularly significant with respect to the distribution of porosity in the buildups.

In Manitoba, dolomitized cements exhibit a range of isotopic compositions for $\delta^{18}O$ (-4.9 to -6.35) and $\delta^{13}C$ (+0.27 to +2.12) (see below). Microcrystalline and finely crystalline replacement dolomites were selected from Dawson Bay area buildup, buildup flank, basin and platform sediments, and subsurface Saskatchewan buildup flank sediments. The means of the ranges of the isotopic values are as follows:

Environment	(n)	$\delta^{18}O$	$\delta^{13}C$
Buildup platform	2	-5.65	+0.42
Basin platform	1	-4.72	-0.03
Buildup	7	-5.29	+1.63
Buildup flank	2	-4.91	+2.77
Basin	4	-4.83	-5.06
Buildup (Sask.)	3	-5.73	+1.64
crinoids	2	-5.33	+2.25
brachiopod	1	-5.34	+1.26
calcrete nodule	1	-5.20	+1.88

The average $\delta^{18}O$ composition (-5.1) of Winnipegosis dolomites is lighter by one per mil with respect to non-dolomitized marine fossils (-3.9), and depleted by 4 per mil with respect to the value (-1.0) calculated from Land's (1980) equation. The difference between computed and predicted values may be attributed to several factors: 1) the isotopic composition of the dolomitizing fluids varied from that of the marine waters; 2) the isotopic composition of the dolomite may have been similar to that of marine waters, but was modified by subsequent diagenetic processes; 3) a combination of these two factors, and 4) neomorphic modification during later diagenesis.

Winnipegosis buildup dolomites buried at 2600 m in southeast Saskatchewan have mean values, $\delta^{18}O$ -5.73‰ and $\delta^{13}C$ +1.64‰, that closely correspond to the mean values for the Dawson Bay dolomites, although the Saskatchewan buildups were buried somewhat deeper.

If a surface temperature of 25°C is assumed, the calculated temperatures for these O isotopes correspond to burial depths of 300 to 600 m using a normal geothermal gradient of 30°C/km. If the dolomitizing fluids were marine waters modified by evaporation, then the $\delta^{18}O$ composition would be enriched, and the temperature range and burial depth for dolomitization would be greater. Therefore, the isotopic composition of the dolomitizing fluids appears to have been governed by elevated temperatures at shallow burial depths.

Kendall (1989) suggested a different interpretation with the buildups being dolomitized during drawdown, which occurred in the Elk Point Basin during deposition of the overlying evaporites. Drawdown would create a hydraulic system that would move basin margin fluids along with evaporated seawater basinward via the Winnipegosis platform, which presumably acted as a conduit. These fluids would have escaped through the porous buildups, dolomitizing them and precipitating a halo of gypsum around them, while salt was being precipitated further away from the buildups.

However, on the basis of C and O isotopes it seems unlikely that the Winnipegosis dolomitizing fluids were modified by the influence of meteoric waters, because these reefs were buried by a thick sequence of marine evaporites and carbonates. If the Dawson Bay dolomites were modified by recent meteoric waters, a considerably wider range of carbon isotopic values would be expected. Another possibility is that dolomitization occurred as a two step process involving evaporated fluids, and the observed isotopic composition may be explained by modification of an original evaporative signature by $\delta^{18}O$ depleted fluids during a second phase of dolomitization at elevated temperatures or during neomoprhism of these dolomites.

The $^{87}Sr/^{86}Sr$ values for Dawson Bay and Saskatchewan Winnipegosis limestones and dolomites (0.7080) fall on the Burke et al. (1982) seawater Sr isotope curve suggesting that: 1) the original limestones were precipitated in chemical equilibrium with Middle Devonian marine waters; and 2) the dolomitizing fluids were derived from modified Middle Devonian marine waters.

Reservoir characteristics

The reservoir in the dolomitized Tableland buildup is heterogeneous. The best reservoir rocks are the peloidal/codiacean algal grainstone shoals of the buildup interior and locally grainstones on the buildup flanks. Porosity in well 8A-22 can exceed 15% and permeabilities commonly exceed 50 md. Tighter zones can be attributed to fractures infilled by vadose silt and mud and to hardground and disconformity surfaces. Blocky isopachous cements of shallow burial origin have had only minor effects on porosity and permeability. In other cores the reservoir quality is much more variable. Other than the leaching of *Thamnopora* colonies in well 8A-22 (Fig. 14) which has increased porosity, reservoir characteristics were not significantly modified by subsequent diagenesis. In almost all cases, cavities and fractures in the upper facies are occluded by vadose silt and mud, resulting in highly variable but generally poor reservoir characteristics. A possible 2.5 m cave is filled with contorted, tight mudstones and shales. These tight zones may have a significant effect on vertical permeability in the buildup. Anhydrite cements are a minor component in leached portions of blocky isopachous cements, and in intercrystalline pore spaces in blocky mosaic cements that occur in fossil molds. It is also a minor component of shelter pores and occludes small solution vugs formed during deep burial. The effect of anhydrite cementation on reservoir quality is both minor and localised (Martindale and McDonald, 1990).

3.4. SWAN HILLS DOLOSTONE RESERVOIRS

Dolomitized Swan Hills reservoirs include Kaybob South, Rosevear, Erith, Hanlan and Caroline. The first four gas fields probably contain over 4.7 TCF of recoverable gas reserves and 450 million barrels of condensate. Caroline contains at least 1.7 TCF of recoverable gas, 200 million barrels of condensate, and 20 million tonnes of sulfur.

These reefs are similar to the Swan Hills limestone reefs in terms of their depositional facies and growth stages. However, instead of being isolated buildups they all occur within an aerially extensive carbonate bank (Fig. 5). They are divisible into 3 units; a basal platform of dark, well bedded Amphipora-rich biostromes; a middle stacked series of reef-fringed bank complexes, and upper stromatoporoid-rich biostromes.

3.4.1. *Kaybob South.* Gas and condensate are trapped against tight, argillaceous limestones that were deposited in an open marine embayment. Reservoir thickness ranges up to 75 m with an average gas pay of 30 m (Fig. 15A). The entire Swan Hills Formation is dolomitized and has an average porosity of 7.8%. Original gas-in-place was 3.7 TCF.

The dolomitized portion of the bank is 40 km long and extends westward from the bank margin for 3 km on the field's northwest end and 7 km in the southeast. Dolomitization was fabric preserving and original depositional facies can be recognized. Intercrystalline, vug, and moldic are the main porosity types. To the west the dolomite lenses are interbedded with nonporous bank interior limestones and eventually give way to well compacted, tight Swan Hills limestones (Fig. 15A).

3.4.2. *Rosevear.* The Rosevear gas field contains 447 BCF of gas-in-place, with average pay of 18 m and average porosity of 8.9%. It represents a classic diagenetic trap with a 2 km wide zone of porous dolomite localized along both sides of a narrow (1 km wide), deeper marine channel that is filled with tight limestone (Fig. 15B). The deeper marine channel and associated dolostones are restricted to the lower Swan Hills member and are overlain by tight shallow water limestones. Tight, bank interior limestones of the lower Swan Hills provide the updip trap for gas at Rosevear.

Dolomitization is fabric preserving and the predominant facies at the margins of the channel are branching stromatoporoid and Amphipora rudstones. These originally porous facies were subjected to several stages of subaerial exposure. The marine channel was later filled with dark lime muds and overlain by bank and biostromal limestones.

3.4.3. *Hanlan.* The Hanlan gas field is a dolomite reservoir with an average pay of 23 m with average porosity of 8.4%, and contains 1.4 TCF of gas-in-place. Dolomitization and porosity development at Hanlan are localized at the bank margin of the lower Swan Hills, with dolomitization rarely affecting the upper Swan Hills, similar to the Rosevear field (Fig. 16A). Updip basin facies of tight Waterways limestones and calcareous shales form the trap. The underlying Slave Point platform are also dolomitized.

Northwest-southeast oriented normal faults, mapped using subsurface data and seismic, subparallel the Swan Hills bank margin and the satellite Erith gas pool to the northwest in a NW-SE direction (Fig. 5). According to Walls and Burrowes (1990) these faults may have been responsible for the localization of the bank margins, and in turn may have played a significant role in subsequent dolomitization.

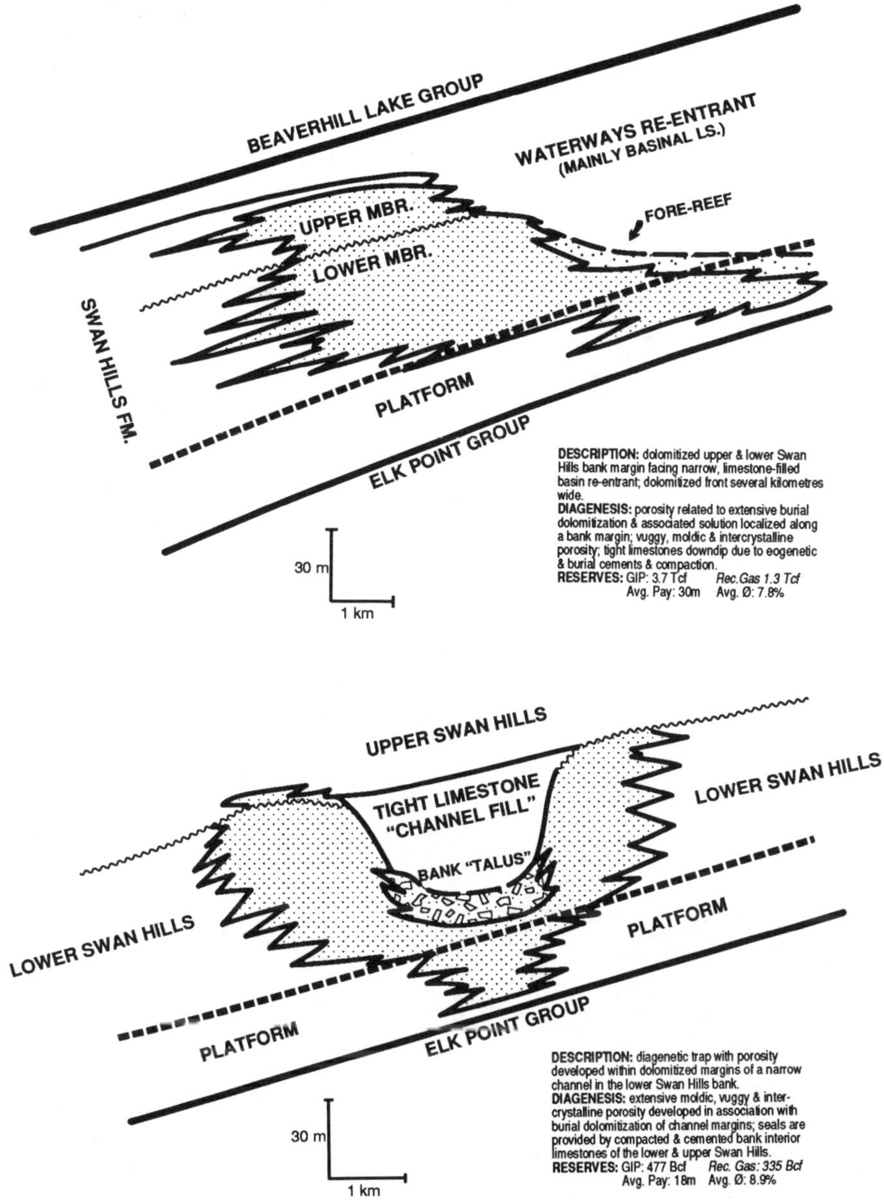

Figure 15A. Cross section of Kaybob south gas field showing dolomite (stipple). 15B. Cross section of Rosevear gas field showing dolostone reservoir rocks adjacent to tight limestone "channel" fill facies (after Walls and Burrowes, 1990).

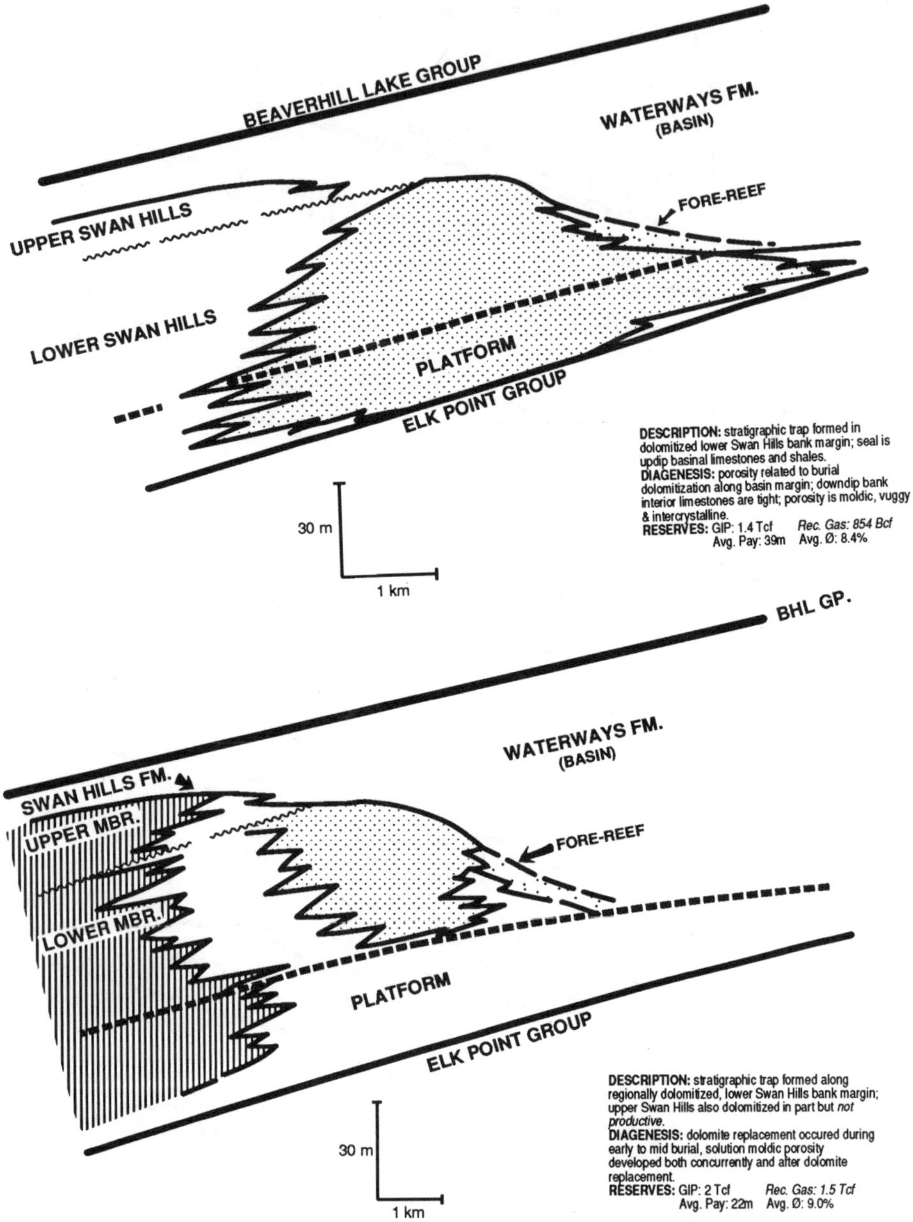

Figure 16. A. Cross section of Hanlan gas field showing that dolomitization occurs in both platform and lower Swan Hills member. B. Cross section of Caroline gas field. Stipple pattern bank margin dolostone reservoir and hatchured pattern downdip regional dolostones (after Walls and Burrowes, 1990).

3.4.4. *Caroline.* The Caroline gas field (discovered in 1986 and not yet in production) tested 25 mmcf/d, and 65 bbls. of condensate per mmcf in the 7-18 well, with H_2S levels as high as 35%. Shell's original estimate suggest 2 TCF of gas-in-place with 600 BCF of saleable gas, 200 million barrels of recoverable condensate, and 20 million tonnes of sulfur. Average pay thickness is 25 m and average porosity is 8%.

Dolomitization and porosity development are localized in a narrow 2 km belt along the updip Swan Hills bank margin, which extends for at least 25 km along the margin (Fig. 16B). Basinal Waterways tight limestones and calcareous shales form the trap. Downdip to the west the reservoir rocks interfinger with tight, bank interior limestones and about 7 km into the bank interior are tight dolostones. Further to the southwest the platform and lower and upper Swan Hills members have been subjected to dolomitization that is fabric destructive. These dolomites are much less porous and permeable than the bank margin dolostones and are non-productive.

Laflamme (1990) showed that the Caroline Swan Hills and Leduc dolomites are similar petrographically and geochemically, except for the following. Swan Hills dolomites are coarser (average 250 vs 140 um for Leduc), generally have greener fluorescence, have lighter $\delta^{18}O$ (-4.95‰ PDB vs -4.19‰ PDB for Leduc), but heavier $\delta^{13}C$ (+2.81‰ PDB vs 1.77‰ PDB for Leduc), and have slightly more radiogenic Sr (0.7086 vs 0.7082). These differences are not significant and thus the Swan Hills and Leduc carbonates could have been dolomitized by the same or similar dolomitizing fluids at about the same time.

3.4.5. *Characteristics of Swan Hills Dolostone Reservoirs*. Swan Hills dolostone reservoirs have several characteristics in common, which imply dolomitization in similar settings with comparable "plumbing systems" (Walls and Burrowes, 1990).

1) Dolomitization is concentrated along northwest trending updip bank edges facing deeper water, open marine basins (Hanlan, Caroline) or narrow channels or embayments (Rosevear, Kaybob South).
2) Dolostone reservoirs interfinger with, and laterally are replaced by, tight bank interior limestones.
3) Intercrystalline and moldic porosity dominate.
4) Reservoirs occur in narrow belts with excellent vertical permeability.
5) Dolomites are fine to medium crystalline, with euhedral to subhedral rhombs.
6) Dolomitization partially preserves primary textures.
7) Dolomitization is generally restricted to the platform and lower Swan Hills Member, except where associated with overlying Leduc dolostones.

The diagenetic sequence of these dolostones is summarized in Figure 17. Porosity destruction and/or preservation appears to have followed three possible pathways:

1) Extensive surficial cementation by calcite, followed by early dolomite replacement of these cements and the limestone matrix. Later dissolution is rare.

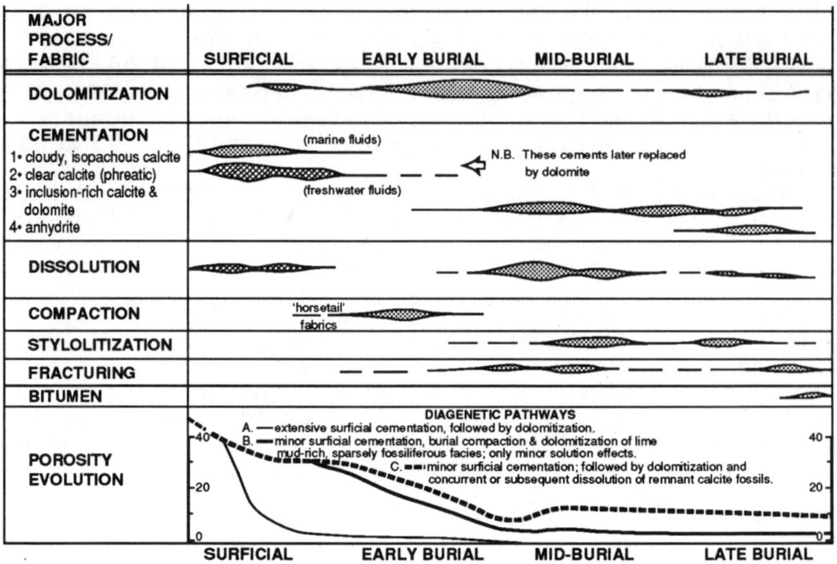

Figure 17. Diagenetic history and porosity evolution following three different diagenetic pathways (lower diagram) in Swan Hills dolostone reservoirs. Relative importance of processes indicated by width of bar: cross hatchured pattern indicates process occurs in selected intervals or areas (from Walls and Burrowes, 1990).

2) Minor surficial cementation mainly in mudstones and wackestones, followed by early burial compaction and dolomite replacement. Later solution is rare.
3) Minor surficial cementation, mainly in packstones, grainstones, and rudstones, followed by early burial compaction and dolomite replacement of limestone matrix; subsequent solution of calcite fossil fragments. These rocks are associated with a solution phase that postdates dolomitization. Significantly they retain the best porosity and permeability.

3.5. UPPER DEVONIAN MIETTE REEF COMPLEX - an outcrop example of a partially dolomitized buildup

The Miette buildup (Fig.18) provides important information about the nature and distribution of the different dolomites. As exposed in the Miette and McConnell thrust sheets, the Miette buildup is equidimensional with a width in a NW-SE direction of between 11 and 14 km, and in a NE-SW direction a width of at least 10 km and possibly 15 km depending on palinspastic reconstructions. The Arcs Member (uppermost Southesk Formation) extends an additional 5 to 10 kms further basinward (Fig. 18) giving the buildup at this stratigraphic level its greatest width of between 20 and 25 km (Mountjoy, 1965; 1989).

The thickness of the main central part of the buildup (Cairn and Peechee) is about 300 m. The Miette buildup occurs as an isolated mass about 50 km NW of the Southesk-Cairn complex and about 50 km southeast of the Ancient Wall buildup (Fig. 3, Mountjoy and Mackenzie, 1973; Mountjoy ,1987). The well layered shallow water

Cairn carbonates are locally fringed by stromatoporoid grainstones, packstones, and rudstones (in part organic framework) margins. Carbonate sedimentation kept pace with sea level rises and the depth of the surrounding basin gradually increased because of the much slower rates of basin sedimentation.

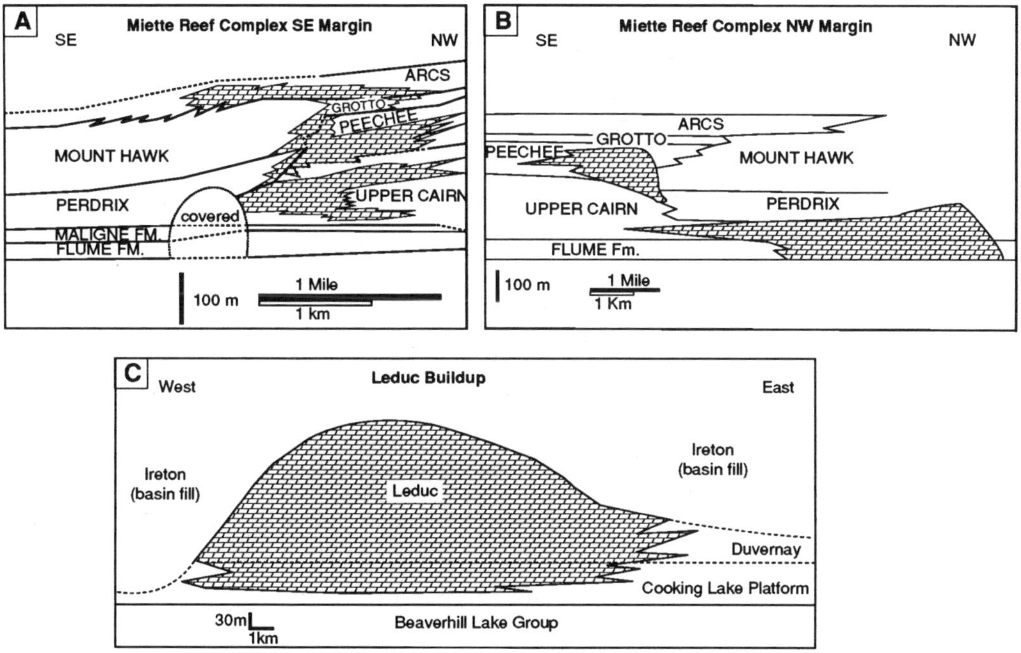

Figure 18. Schematic cross sections of Upper Devonian Leduc buildups. A, B Miette reef complex, Rocky Mountains. Dolomitization essentially restricted to the buildup margins (after Mattes and Mountjoy, 1980; Mountjoy, 1965, 1989). C. Leduc completely dolomitized buildup, subsurface central Alberta (from Amthor et al., 1993, with permission, Can. Soc. Petrol. Geologists).

The buildup interior is divided into two sequences: a lower biostromal sequence (Cairn Formation) and an upper, dominantly peritidal sequence (lower Peechee Member). The Miette buildup is fringed by a relatively narrow zone (300-1,000 m) with abundant stromatoporoids, ranging from biostromal layers which grade into lagoonal deposits, to small patch reefs of low relief that are discordant with surrounding strata (Fig. 18, Mountjoy, 1967; Noble, 1970). The stromatoporoid units consist of a variety of hemispherical, bulbous, irregular, tabular and encrusting forms with local Stachyodes thickets. Encrusting stromatoporoids, though uncommon, appear to be the only form unique to the margin facies.

Table 3. Characteristics and origin of dolomite types, Miette buildup, Miette thrust sheet (from Mattes and Mountjoy, 1980).

Dolomite Type	Fabric Components	Origin
1 Diffuse microcrystalline	Diffuse, micron sized crystals	Early diagenetic (pre-lithification?) first dolomitizing event
2 Mosaic	Coarse and fine bituminous matrix dolomite, dense and sucrosic textures, partially dolomitized matrix textures, dolomite spar replacing calcite spar-neomorphic and pseudomorphic varieties, non-luminescent, non-ferroan, precursor textures usually preserved	Late burial diagenesis
3 Pressure solution related	Fine to coarse crystalline dolomite, associated directly or indirectly with stylolites and solution seams	Pressure solution, either a reactate or cumulate product
4 White sparry	Coarse, relatively clear white dolomite, zoned, luminescent, precursor textures obliterated	Succeeded stages 1–3
5 "Baroque"	Coarse, clear scalenohedral cements, "Baroque" dolomite relatively iron rich, only in voids and fractures	Associated with brecciation

The Miette buildup has undergone an extensive and complex diagenetic history, summarized schematically in Figure 19 along with a burial history curve. Mattes and Mountjoy (1980) group 7 calcite cements into 5 diagenetic stages and recognize 5 different types of dolomite (Table 3), all of which are plotted relative to geologic time. Submarine cements are relatively minor in the buildup and have only been observed at one locality near the southeast margin. Dolomites types 2 and 4 are the most abundant and also the most porous rocks in the buildup. There is no evidence of extensive emergence and solution in the buildup. The white sparry (type 4) and saddle (type 5) dolomites were deposited during intermediate to deep burial.

The distribution of dolomites in the Miette buildup provides important information that can help geologists and reservoir engineers to visualize the complex variations that occur in the porosities of the different dolomites. The distribution of the dolomites also may help assess the nature of the fluid-flow systems responsible for dolomitization (Amthor et al., 1993). Dolomite is essentially confined to the buildup margin and adjacent basin-filling sediments. Within the dolomitized periphery, the degree of alteration varies from negligible to complete, with a spectrum of modification fabrics visible in individual beds over a few cms or meters laterally. The lateral extent of replacement dolomites vary in width from 250 to 500 meters in lower Cairn portion and 50 to 300 meters in upper Peechee portion. "Stringers" of replacement dolomite extend up to 2 km into the buildup and are closely controlled by bedding. Replacement dolomites are usually fabric selective, and many horizons consist of less than 60% dolomite because of the presence of unaltered or leached allochems (Mattes and Mountjoy, 1980). One can observe the progressive change from these partially dolomitized strata with matrix replacement (type 2), to progressive overprinting by more coarsely crystalline, texturally destructive dolomite (type 4) with a few calcite large fossils, to vuggy coarsely crystalline dolomites that have good intercrystalline porosity. These more coarsely crystalline dolomites are confined to within about 100 m of the reef margin (originally mapped as the reef margin dolomite of Mountjoy, 1965). Thus the later more crystalline phase of dolomitization was responsible for overprinting the earlier more finely crystalline matrix replacement dolomites, and also for dissolving the calcite components producing the vuggy porosity. This may well be the manner in which vuggy coarse-crystalline dolomites formed elsewhere in the basin, as for example in the Rimbey-Meadowbrook reef trend.

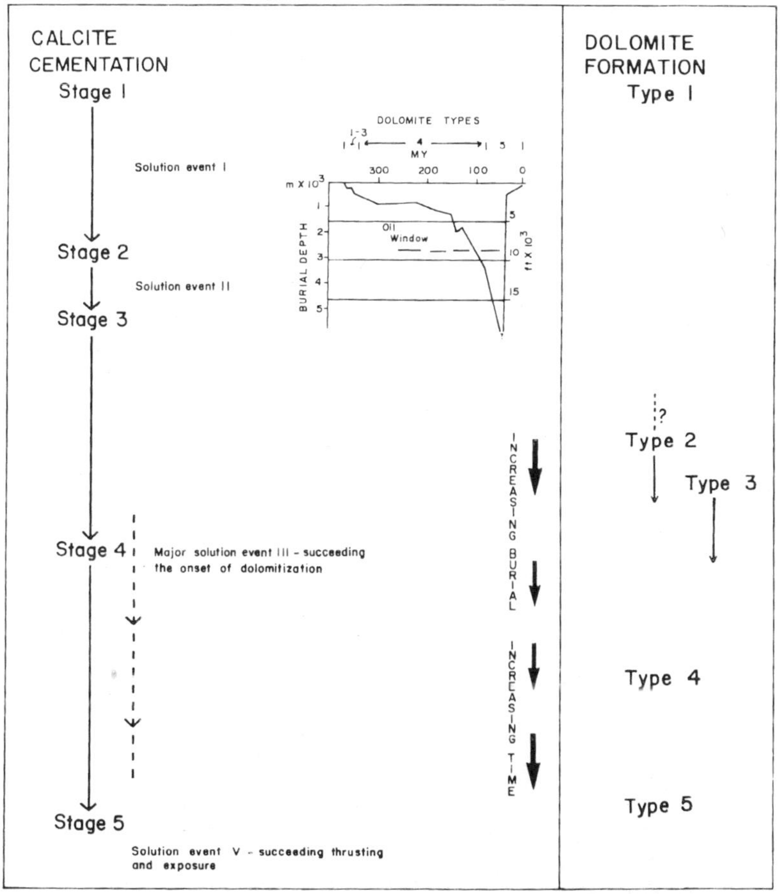

Figure 19. Paragenesis of different stages of calcite cementation and solution, and five types of dolomite, Miette reef complex, relative to burial history and oil window (from Mattes and Mountjoy, 1980).

The location of the conduit system that conducted the dolomitizing fluids to the Miette buildup margins is uncertain. It now appears that there would be insufficient Mg available for dolomitization utilizing a burial compaction model (see Amthor et al., 1993). Much of the underlying platform consists of cemented tight limestones that do not appear to have acted as a conduit system (Mountjoy, 1965, and work in progress; Cook, 1972). Possibly fault systems may have acted as a conduit system for these dolomitizing fluids, but dolomitized fillings in fractures and faults have not been observed beneath the Miette buildup in the field outcrops.

3.6. RIMBEY-MEADOWBROOK LEDUC REEFS

Subsurface Leduc Formation buildups extend for about 320 km along the linear Rimbey-Meadowbrook reef trend in central Alberta (Fig. 3). These buildups are extensively dolomitized where they are located above the dolomitized western margin of the underlying Cooking Lake platform, suggesting that dolomitization of the platform and buildups are related. The extent of dolomitization in the Cooking Lake platform ranges from scarce to complete. Complete dolomitization only occurs within about 5 kms of the reef trend and some dolomitization occurs throughout the East shale basin (Chouinard, 1993). Two major stages of dolomitization are recognized in the Cooking Lake and Leduc carbonates; 1) replacement dolomitization that formed 90 vol-% of the dolomite, and 2) later dolomite cementation. Each stage formed several dolomite types. Replacement dolomites comprise medium and coarse-crystalline, subhedral to euhedral planar mosaic dolomite (comprising up to 90 vol% in some buildups), and coarse-crystalline nonplanar anhedral dolomite (Fig. 20). Dolomite cements postdate all replacement dolomites and consist of white, coarse-crystalline dolomite that line voids and fractures (Amthor et al., 1993).

Stable isotope analyses of replacement dolomites from widely separated buildups range from $\delta^{13}C$ = 1.5 to 6.0 PDB, and $\delta^{18}O$ = -7.5 to -5.0 PDB. The planar and nonplanar dolomite cements are geochemically distinct from replacement dolomites, having the most negative $\delta^{18}O$ values (with means of -7.0 and -8.2, respectively). Coarse-crystalline planar dolomite cements show higher $CaCO_3$ and lower $MgCO_3$ concentrations relative to replacement dolomites. The massive replacement dolomites postdate limestone deposition and early-diagenetic features such as micritization, submarine cementation, and low-amplitude wispy stylolites. They cross-cut depositional facies boundaries and extend into overlying rocks, suggesting that pervasive replacement dolomitization postdates deposition of the Duvernay limestones and shales. These features are compatible only with subsurface dolomitization. Paragenesis, petrographic and geochemical data suggest that replacement dolomites formed in a shallow-burial environment (50-60°C or about 600 to 1200 m) during the Late Devonian to Early Mississippian, and the later stage dolomite cements formed at higher temperatures and greater depth, probably related to pressure-solution and/or late, warm basin fluids (Amthor et. al., 1993).

Mass-balance calculations indicate that burial compaction, which has been invoked to explain some Leduc Formation and other Devonian dolomites, cannot provide sufficient fluids for complete dolomitization of this part of the reef trend, this mechanism can only account for between 15 and 20% of these dolomites (Amthor et al., 1993). Additional fluid sources and fluid-flow mechanisms other than burial compaction are required to explain: 1) the amount of dolomite present, 2) the geometry of dolomites in completely and partially dolomitized buildups, and 3) the lack of dolomite in adjacent limestone buildups. These considerations suggest that an extensive regional fluid-flow system was necessary for complete dolomitization during early burial of the Rimbey-Meadowbrook reef trend. However, the sources of these fluids, the extent of the regional fluid-flow system, as well as a suitable driving mechanism are still uncertain.

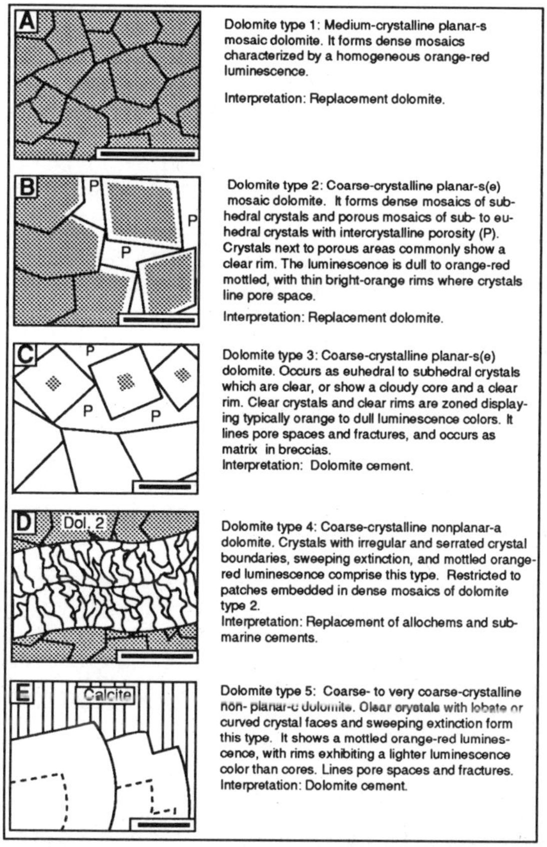

Figure 20. Five dolomite textures observed in Cooking Lake/Leduc Formation carbonates of the central Rimbey-Meadowbrook reef trend (Amthor et al., 1993, with permission, Can. Soc. Petrol. Geologists). Scale bars = 500 m.

Porosity

The most common porosity types are (Choquette and Pray, 1970): a) moldic and vuggy porosity, b) intercrystalline matrix porosity, c) breccia and fracture porosity, and d) interparticle porosity and solution-enlarged interparticle porosity (Amthor et al., 1993)

Molds and vugs are the most prevalent pore types, accounting for 50% to 100% of the studied cores. Moldic porosity is commonly after Amphipora and stromatoporoids. Vuggy porosity is more common, simply because there is every gradation from fabric selective moldic to non-fabric selective vuggy porosity, and most vuggy porosity is probably solution-enlarged moldic porosity. The shape of the vugs can commonly be related to the shape of the original skeletal components, e.g. bulbous or tabular stromatoporoids, Amphipora. In many instances irregular vugs are present that are very large (>6 cm), which do not give any indication as to their precursor fabric. Size and shape of moldic and vuggy porosity is highly variable. They range from 1 mm to more than 6 cm in diameter. The most common size range is about 2 to 30 mm.

Matrix porosity is either intercrystalline or solution-enlarged intercrystalline porosity (pinpoint of McNamara and Wardlaw, 1991), and rarely accounts for more than 10% of the pore types. Most of the Leduc dolostones form a compact rock of medium- to coarse-crystalline planar-subhedral dense mosaic dolomite. Intervals with effective intercrystalline porosity are confined to relatively narrow bands and patches commonly associated with vuggy porosity. What controls these zones is uncertain but may reflect areas of higher primary porosity.

Brecciation and fracturing is common in all studied buildups (Fig. 21), the intensity, however, varies ranging from isolated occurrences to several meter thick zones (e.g. Leduc buildup well 8-17-50-26 W4). Breccias consist of cm-sized angular dolomite clasts with point or sutured contacts, forming open framework breccia porosity. The dolomite clasts are identical to non-brecciated dolomite intervals, and display truncation of stylolites and fractures at the edges of clasts. Inclined stylolites within individual dolomite clasts indicate rotation and displacement. Clasts were sharply overgrown by minor amounts of compositionally zoned dolomite cements. Since these breccias postdate the replacement dolomites they reflect solution and collapse during burial (Amthor et al., 1993).

Interparticle porosity is restricted to Leduc/Cooking Lake limestones, where it most commonly occurs in peloidal grainstones. Solution-enlarged interparticle porosity is dominant in dolomitic limestones, where fossils are still calcite and the matrix is preferentially replaced and leached. Only with increasing dolomite content (around 75%) are fossils leached out, indicating a direct relationship between the amount of dolomite and vuggy and moldic porosity.

There is an overall decrease of porosity with increasing depth along the reef trend (Figs. 22 and 23, Drivet and Mountjoy, 1993; Amthor et.al., submitted). There are very poor correlations between porosity and permeability for individual wells. There appears to be a correlation between high K_h and K_v values and maximum and average porosity.

From the Rimbey-Homeglen buildup southwards, anhydrite constitutes the most pervasive and volumetrically important porosity occluding phase, filling vuggy, moldic, and fracture-related porosity (Fig. 23). Large vugs or "cavern-like" porosity

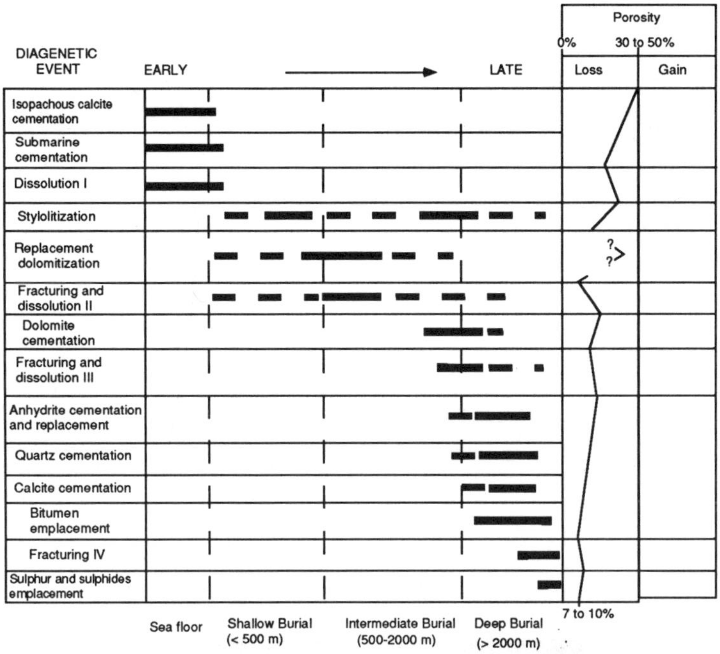

Figure 21 Inferred changes in porosity with respect to the paragenetic sequence including dissolution phases in the southern Rimbey-Meadowbrook reef trend. The present average porosity of the buildup interior and margins are 7 and 10% respectively (Drivet, 1993; Drivet and Mountjoy, 1993).

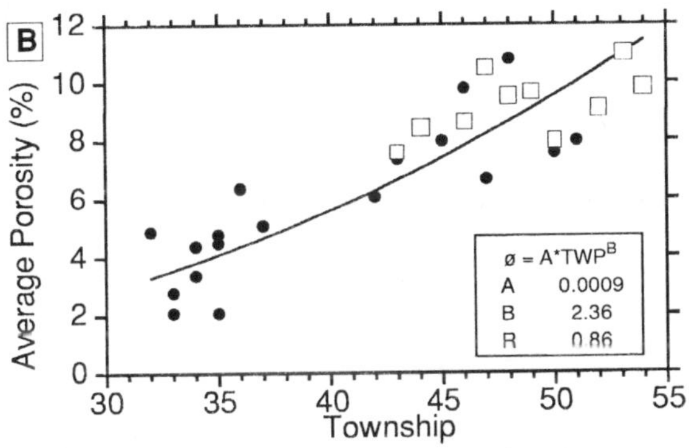

Figure 22 Average porosity plotted versus north-south townships along the Rimbey-Meadowbrook reef trend. Data (Energy Resources Conservation Board) obtained from averaging all core porosity analyses from the Leduc and Cooking Lake formation in a series of wells. Curve respresents best fit using a statistical computer program. Township 55 represents burial depths of about 2500m (max) and Township 30 burial depths >3500m (see Fig. 3 for township locations). Some buildup interior wells (filled circles) have the same porosity as buildup flank wells (open squares). Formula provides equation of best fit curve. r, correlation coefficient. From Mountjoy and Amthor (1994).

completely cemented by anhydrite is a common texture characterised by the collapse and displacement of dolomitized stromatoporoid layers. In addition to anhydrite cement, anhydrite replacement is also responsible locally for decreasing intercrystalline porosity when replacing porous matrix dolomite. The updip northward decrease in anhydrite coincides with an increase in average porosity values from 2% to 7% (data from ERCB core analyses). Thus, the most porous reservoir dolomites occur in the northern part of the Rimbey-Meadowbrook trend (from township 37 northward), where the burial depth is shallower, and the amount of anhydrite is lower (Fig. 23 , Drivet and Mountjoy, 1993).

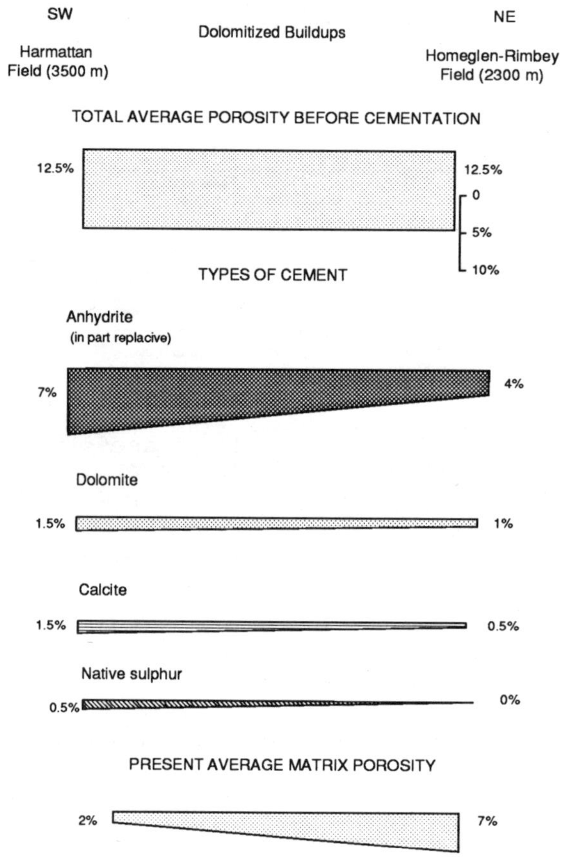

Figure 23. Porosity decrease along central and southern part of Rimbey-Meadowbrook reef trend caused by filling of pores by late-stage cements (mainly anhydrite) (Drivet and Mountjoy, 1993; and work in progress).

In general, porous dolostones can be attributed to solution-controlled vuggy and moldic porosity (where not filled with anhydrite), and to fracturing and brecciation, which is more intense in dolostones than in limestones. In addition, based on the presence of large vugs filled with replacement dolomite fragments and late-stage anhydrite, a second phase of extensive burial dissolution resulted in solution and collapse of vuggy replacement matrix dolomites. This solution may be responsible for the lack of core recovery and rubble zones in some cores (Drivet and Mountjoy, 1993). Thus porosity in dolostone reservoirs tends to decrease less rapidly with depth, a characteristic noted for South Florida dolomites (Schmoker and Halley, 1982). These features are currently being mapped and studied in the central and deeper part of the Rimbey-Meadowbrook reef trend. (see section 4).

Possible Solution resulting from TSR (thermal sulfate reduction)

Some authors attribute extensive subsurface solution in some dolomites to have resulted from acid solutions generated by H_2S and CO_2 produced during TSR (Eliuk, 1984; Machel 1987a; Krouse et al., 1988; Heydari and Moore, 1989; Machel, 1989; Reimer and Teare, 1992; Dravis and Muir, 1992; Mazzullo and Harris, 1992). Because TSR involves a series of different reactions (Machel, 1987a), it is difficult to determine the relative amounts of acid solutions that might be released. In addition, the release of HCO_3- and the reaction of S with methane or other HC gases (Heydari and Moore, 1989) favours the precipitation of late-stage calcites or dolomites (Machel, 1987b), occluding some, or all, of the porosity generated. On balance, there may be little or no increase of porosity associated with TSR reactions (Machel, personal communication, 1992).

Extensive late-stage solution (hydrothermal) has taken place along the Rimbey-Meadowbrook reef trend (Fig. 21, Amthor et al., 1993; Drivet and Mountjoy, 1993) and in other parts of the western Canada Sedimentary Basin (Eliuk, 1984; Machel, 1987a,b; Krouse et al., 1988; Reimer and Teare, 1992; Dravis and Muir, 1992; Qing and Mountjoy, 1994, in press). Whether or not this solution is related to TSR reactions or occurred earlier is not certain. For the Keg River Formation of the Comet platform, Dravis and Muir (1992) interpret that the dissolution resulted from both H_2S and CO_2 bearing brines and TSR. However most solution took place prior to the emplacement of bitumen. The extensive hydrothermal solution in the Presqu'ile barrier took place prior to, and prepared the way for, mineralization which appears to have been associated with TSR (Qing, 1992; Qing and Mountjoy, 1994, in press).

South of the Homeglen-Rimbey buildup in the Rimbey-Meadowbrook reef trend, TSR reactions are suggested by the presence of free sulphur and abundant pyrite occurring as geopetal sediment that postdates late-stage, white (coarsely crystalline and saddle) dolomites. Although it is tempting to associate the extensive subsurface solution in this reef trend to TSR reactions, it appears that much of the solution occurred prior to TSR. Thus, in the Rimbey-Meadowbrook reef trend, the Comet platform and the Presqu'ile barrier other fluids appear to have been involved in causing the porosity, possibly CO_2 or fluids charged with organic acids.

3.7. DEEP BASIN STRACHAN-RICINUS REEFS

The downdip buildups southwest of the Rimbey-Meadowbrook reef trend provide excellent examples in which the paragenetic relationships, and reservoir properties between limestone and dolostone reservoirs, can be studied. Reservoir rocks of Ricinus, Crimson and Chederville gas fields are completely dolomitized, whereas the Strachan and Phoenix fields vary from being partially dolomitized to undolomitized. Partially dolomitized reservoirs consist of skeletal framestones, floatstones and rudstones with grainstones and packstone matrices. Submarine cement is abundant along the margins of both Strachan and Phoenix. These dolomitized limestones have poor reservoir character. Pore types are mainly vugs, molds, intraskeletal, fractures, and minor intercrystalline. Porosity is very low and averages 1.5% and permeability ranges from 0.1 to 3.7 md.

In contrast, the dolomitized reservoirs have higher porosities, although they are similar in texture and geochemistry to dolomitized reservoirs in the updip portion of the Rimbey-Meadowbrook reef trend, noted above. They are characterized by replacement dolomites that comprise 95 vol-% of the reservoir, and late dolomite cements. Pore types are vugs, molds (*Amphipora*, fenestral-like), intercrystalline, breccia, and fracture. Porosities vary between 1.2 and 20.7% and average 8.4%, and permeabilities range between K_h 14 to 100 md, and K_v 8 to 40 md.

The Strachan gas field forms two separate pools. The eastern D3-A pool is partly to completely dolomitized. The western D3-B pool consists of partially dolomitized rocks. The gas compositions and the reservoir quality differs in each pool. The D3-A pool with more extensive dolomitization is the better reservoir with porosities ranging from 2 to 10% (avg. 4%), and permeabilities from 1 to 40 md. The gas contains 9.3% H_2S, 2.1% CO_2 and 20 barrels of condensate. The higher porosities in the D3-A pool appear to be due to less chemical compaction in the dolomites than in the limestones, and greater solution in the dolomites. The pore types in the dolomites consist of vugs, molds etc. as noted above for other dolomitized reservoirs in this reef trend. Pore types in the limestones are mainly intraskeletal (partly filled with bitumen), small vugs (solution of fossils), and fractures. Microfracturing (hairline fractures) is very abundant in the upper part of the reservoir (see below).

4. Processes Controlling Porosity in Dolomitized Carbonates

As noted earlier, the chemical reactions involved in dolomitization will control whether or not porosity of the initial limestones are changed. A general impression gained from modern and Pleistocene examples is that dolomitization does not increase porosity significantly, but that associated dissolution of aragonitic components causes extensive removal of material (with little or no dolomitization) to the extent that some carbonate coastlines are retreating as for example in the Yucatan peninsula of eastern Mexico (Back et al. 1986; Hanshaw and Back, 1980).

In the case of the Devonian of western Canada, during early burial it appears that strata with some primary porosity have been dolomitized with little change in porosity (e.g. Miette, Swan Hills). This is difficult to assess adequately in completely dolomitized reservoirs. Also one observes considerable variation in the amount of porosity locally within cores, and from place to place, in the buildups. The reasons for the local development of intercrystalline porosity next to dense low porosity dolomites is not clear, but may be related to the primary depositional facies. In the case of vugs and cavities there has been considerable solution of the larger fossil components (Walls and Burrowes, 1990).

Timing Whether or not dissolution happened at the time of dolomitization is not always clear, but the weight of the evidence, in my view, supports a later solution event(s) of calcite components that remained after early matrix dolomitization. This is especially evident in examples of the dolomitization and solution of stromatoporoids (e.g. Miette and Caroline). In many cases the concentric laminations of the outer part of the stromatoporoid have collapsed into the interior forming a vug with several broken laminations. Later these cavities are partially or completely filled with saddle dolomite, calcite, and/or anhydrite. In other cases *Amphipora* and some allochems were dissolved forming a modified moldic porosity (Mattes and Mountjoy, 1980; Amthor et al., 1993). Similar relationships have been noted by others in the Devonian of western Canada (Walls and Burrowes, 1985, 1990).

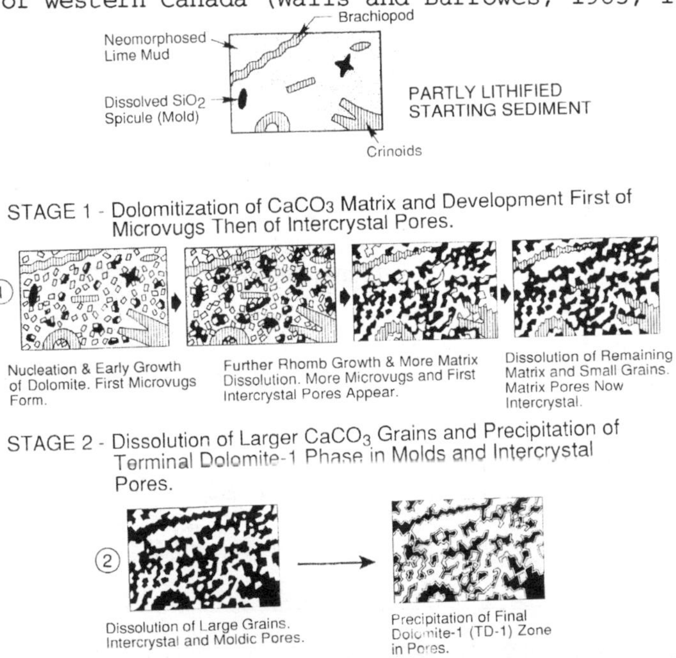

Figure 24. Diagrammatic model showing inferred steps in conversion of lime-mud sediment in the Carboniferous Burlington-Keokuk Formation to dolomite I, with accompanying development of pore space (from Choquette et al., 1992, with permission J. Sed. Pet.).

In Carboniferous strata of the U.S. mid-continent Choquette et al. (1992) have carefully documented, with SEM, that intercrystal and micro-intercrystal pores resulted from progressive dissolution of micrite during the first phase of dolomitization. This is interpreted to have occurred during shallow burial (30 to 100 m). Dolomite rhombs grew in the matrix and at the same time solution dissolved part of the muddy matrix forming microvugs. This was followed by more dolomitization and the formation of intercrystal pores, and enlargement of the microvugs, as more of the matrix was dissolved. Later the larger calcite grains were dissolved forming moldic pores and causing enlargement of the intercrystal pores (Fig. 24). CL zonation of the dolomites constrains the timing of the porosity development to the later stages of this first dolomitizing event, and implies a flow system of regional extent. Replacement of some of these early dolomites by a later dolomite had essentially no effect on the pore systems of the first dolomite.

4.1. LATE-STAGE SUBSURFACE DISSOLUTION

Recently subsurface solution has received considerable attention (see review by Mazzullo and Harris (1992) and short course "Subsurface Dissolution Porosity in Carbonates", Canadian Society of Petroleum Geologists (June, 1992)). Included here are brief summaries of the nature of mesogenetic pore types and the processes potentially responsible for them.

Mesogenetic pores are similar to, and thus easily confused with, eogenetic pore types. The pores are both fabric and nonfabric selective. They can be recognized by their cross-cutting relationships such as: postdating earlier cements, stylolites, fractures etc. These pores are often elongated, larger than adjacent grains or skeletal elements, and associated with stylolites and fractures (Table 4).

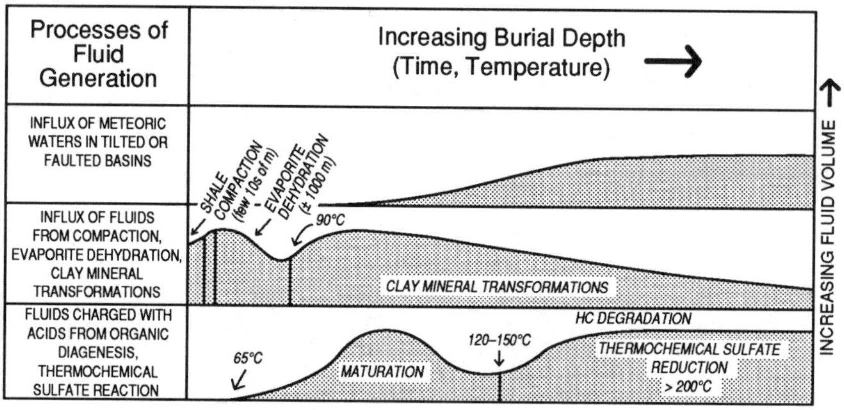

Figure 25. Geologic processes that modify chemistry of connate fluids in the burial environment. Shading shows times when process may release fluids and/or acids and gases. Thermochemical sulfate reduction can also occur at temperatures as low as 100 C (from Mazzullo and Harris, 1992).

Table 4. Subsurface (Mesogenetic) Dissolution. Criteria for Recognition. Modified after Mazzullo and Harris (1992).

- pores **similar to** and **easily confused** with eogenetic pores

- may occur in **low-porosity** rocks, by fluids following
  - fractures, - microporosity, - stylolites

- **pores are both fabric and nonfabric selective**

  - allomolds — micropores
  - interparticle and intraparticle — vugs
  - intercrystalline — stylolites
  - fractures
  - caverns and breccia associated with solution collapse

- **criteria for recognition - pores that cross cut earlier cements**

  - cements that postdate mechanical & chemical compaction
  - cements with hydrocarbon inclusions
  - pores associated with stylolites and fractures
  - elongated and large pores in grainstones.

Mesogenetic dissolution is a complex process that can occur during several stages and over a long period of time (Fig. 25). The processes that generate fluids that may dissolve carbonates are (Table 5):

1. Organic maturation - CO_2, CH_4, organic acids, H_2S
2. Thermochemical Sulphate Reduction reactions (TSR)
3. Upward cooling of CO_2-rich warm brines
4. Silicate hydrolysis of clay minerals during diagenesis

TABLE 5. Potential processes responsible for subsurface (mesogenetic) dissolution

<u>Potential processes responsible for dissolution</u>

A. <u>Organic Maturation generates</u>

- CO_2, - CH_4, - organic acids, - H_2S

B. <u>Thermal Sulphate Reduction (TSR)</u>

- can cause <u>dissolution</u>, and <u>precipitation</u> of

<u>Products</u> - carbonate cements (calcite and dolomite)
- solid bitumen
- sulphur
- anhydrite

(on balance seems to increase porosity locally in Western Can. Sedimentary Basin)

C. <u>Upward cooling of CO_2-rich warm brines</u>

- saturation of dissolved carbonate increases with decreasing temperature (Bakalowicz et al., 1987)

D. <u>Silicate hydrolysis of clay minerals during diagenesis in adjacent shales (Hutcheon and Abercrombie, 1990, and others</u>

Following are some examples of late-stage dissolution.

4.1.1. *Pine Point and Presqu'ile dissolution vugs and breccias.* Meteoric karst has been inferred to be responsible for extensive solution and brecciation at Pine Point and in many other MVT districts. In many cases, there is little or no direct evidence to support a meteoric origin for solution and the timing of dissolution is not sufficiently well constrained. The study of the Presqu'ile barrier indicates that dissolution related to the sub-Watt Mountain exposure differs from subsurface dissolution associated with the MVT deposits at Pine Point in at least three aspects (Qing and Mountjoy, 1990, 1994 in press; Qing, 1991):

1) dissolution vugs related to meteoric water during the sub-Watt Mountain exposure are small (usually 1 to 2 cm), while vugs that host MVT deposits at Pine Point are much larger, ranging from centimetres to metres in size, and are more extensive;

2) the dissolution vugs and breccias related to the sub-Watt Mountain unconformity are commonly filled with green shale, and locally with pendant calcite cements. Dissolution vugs at Pine Point are commonly filled with saddle dolomite, sulphide, and late-stage calcite cements;

3) dissolution features resulting from subaerial exposure only occur below the sub-Watt Mountain unconformity, while later-

stage vugs and breccias occur continuously across this unconformity.

Therefore, it appears that meteoric waters caused only minor dissolution in the upper part of the Presqu'ile barrier during the short period of exposure that formed the sub-Watt Mountain unconformity. Although meteoric dissolution could have enhanced porosity and permeability in the barrier and provided conduits for later hydrothermal fluids, the majority of dissolution vugs and breccias were produced by hydrothermal fluids during burial (Qing and Mountjoy, 1994 in press). These results have important implications for mineral exploration: 1) MVT deposits are not always related to a unconformity. Mineralization can occur in any available vugs and/or fractures whatever their origin (Qing, 1991; Qing and Mountjoy, 1994 in press; in press), and 2) Evidence of meteoric-karst for other MVT districts should be critically re-examined. In cases such as Pine Point, the influence of subaerial exposure has been over-emphasized and the possibilities of later extensive hydrothermal dissolution of carbonates overlooked.

4.1.2. *Keg River and Muskeg Formations, Comet Platform.* Recently Dravis and Muir (1992) in a careful and detailed petrographic and geochemical study have documented deep-burial dissolution of earlier dolomites in the Upper Keg River and Muskeg formations of the Comet Platform that borders the northwest side of the Rainbow Basin. They showed that this burial solution is the main reason for development of reservoir quality. The scale of dolomite dissolution varies from pin-point microporosity to moldic and large vugs, including 10 cm diameter or larger pores associated with breccias and zebra dolomites. Using enhanced petrographic techniques (including blue light fluorescence) they document that replacement dolomitization postdates earlier strongly stylolitized and grain sutured limestones, clearly indicating a burial origin.

Dissolution was coincident with, or post-dated, most pressure solution fabrics. It occurred, therefore during deep burial, as is shown by porosity which cross-cuts stylolites and the fact that dissolution has enlarged stylolites and etched late stage saddle dolomites. The anomalously high concentrations of carbon dioxide and hydrogen sulfide, metallic sulfides and ubiquitous late diagenetic anhydrites indicate the presence of corrosive burial fluids that promoted dolomite dissolution. Dravis and Muir (1992) interpret that thermochemical sulfate reduction, and the presence of Ca-rich fluids, were the likely mechanism for promoting dissolution.

4.2. LATE STAGE FRACTURING

Fractures occur in most dolostone buildups and are highly variable in their distribution. Included here are two examples of fracturing. The first is from the Wabamun Formation on the east flank of the Peace River arch, which has undergone at least three distinct stages of fracturing. The second example is one of late stage (very small hairline) fractures from the Strachan field in the deeper part of the Alberta basin.

4.2.1. *Tangent and Eaglesham fields, Wabamun, eastern Peace River.* Faults and associated larger scale fractures occur in different parts of the basin (Fig. 26). Faulting in this area is related to differential subsidence of the Peace River Arch. At least three phases of fracturing occurred and dolomitization appears to be associated with the last two phases. Matrix, medium to coarse-crystalline dolomites are the principal reservoir rock, but have highly variable porosity. They are interpreted to have formed at temperatures between 40 and 65 °C and depths of about 50 to 1500 m, probably from brines derived from adjacent and underlying strata that were expelled along fracture-fault conduits. Sucrosic, saddle and reworked dolomites fill pores and fractures. They postdate the second stage fractures and were precipitated during deep burial based on fluid inclusion temperatures of 100°C. A few dolomites have fluid inclusion temperatures near 200°C, suggesting that they probably formed from hot brines that ascended along faults and fractures from deeper parts of the basin and/or crystalline basement.

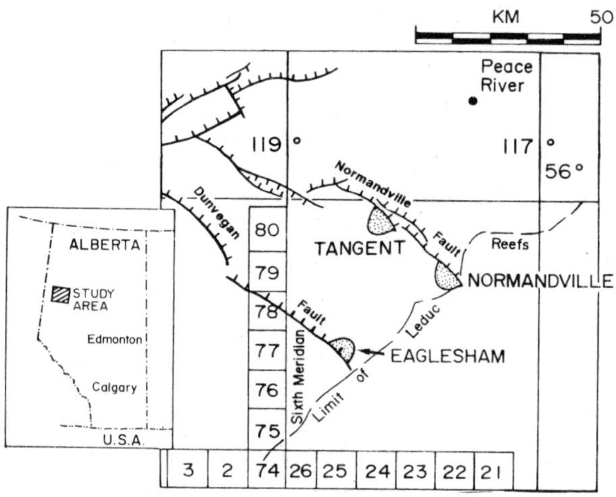

Figure 26. Location of Wabamun Normandville, Tangent and Eaglesham fields with respect to major Dunvegan and Normandville faults and the basinward limit of Leduc reefs around the Peace River Arch (from Mountjoy & Halim-Dihardja, 1991, permission, J. Sed. Pet.).

Exploring for Wabamun diagenetic dolomite reservoirs is difficult. Not all of the best developed reservoirs occur within fractured and dolomitized zones adjacent to faults since, locally, matrix dolomites lack porosity and some pores and fractures are completely filled with saddle dolomite. Faults and fractures are not always related to dolomitization. It is difficult to determine which fault and fracture systems have acted as fluid conduits. Most Wabamun dolomitization occurs in areas underlain by Leduc reef dolomitized carbonates, suggesting that the Leduc may have been part of a regional conduit for these dolomitizing fluids (also see Dix, 1993).

4.2.2. *Microfracturing in Strachan and other deep basin reservoirs.* The isolated Strachan buildup (Fig. 3) occurs in the deeper part of the Alberta Basin and contains major gas reserves.

Previously undescribed, abundant hairline microfractures (< 1 mm) occur in both partially and completely dolomitized rocks in the upper 100 m of the buildup. These microfractures are filled with bitumen and display three patterns: 1) semi-radial extending from vuggy and moldic pores, 2) random extending from irregular pores, and 3. subhorizontal extending from subvertical fractures. They crosscut all sedimentary and diagenetic products, including some late-stage anhydrite cements (Marquez et al., 1992).

Similar microfracturing patterns occur in the Phoenix and Harmattan fields, which are also isolated buildups. Minor microfractures are present in the Ricinus West, Crimson and Cheederville gas fields. Microfracturing of this type has not been reported from the Rimbey-Meadowbrook reef trend. Significant pressure differences were reported at the time of discovery between many of these buildups that contain hairline microfractures. In Strachan a 15,525 KPa pressure break was observed. The differences in initial pressures suggest that these buildups were isolated from the main Rimbey-Meadowbrook trend. These buildups also lack a water drive. In northern Alberta microfractures of this type are present in the Robb and Colt fields, both isolated from the main Simonette reef complex (Fig. 3).

Microfractures in the Strachan gas field are: 1) associated with all pore types, 2) always filled with bitumen, 3) occur as a late diagenetic event, and 4) appear to be restricted to isolated buildups in the deep (>4,000 m) part of the Alberta Basin.

Overpressuring caused by thermal cracking of oil to gas during increasing burial, explains many of the characteristics of the microfractures; their association with all pore types, bitumen filling, their late timing, and their restriction to isolated buildups in the deep part of the Alberta Basin (Marquez et al., 1992).

Porosity and permeability values from core analysis suggest that the presence of the bitumen filled microfractures in the upper part of the Strachan buildup considerably decreases the reservoir quality of the carbonate rock, reducing porosity to between 2 and 4%. This part of the buildup contains only minor amounts of late-stage calcite and dolomite cements.

5. Summary and potential dolomitizing models

Generally two phases of burial dolomitization have been recognized in many parts of the Devonian of the Western Canada Sedimentary Basin and at different stratigraphic levels: 1) a relatively early replacement, matrix dolomitization, and 2) a later coarse-crystalline and saddle dolomite - a cement that fills pores and fractures. There is little agreement on the depth and timing of these two phases of dolomite. Some workers consider the two phases to be part of one dolomitizing process because the two dolomite types overlap and grade into each other, suggesting that the later stages of a single dolomitizing process were more coarsely crystalline (Spencer, 1987; personal communication, 1992). Most workers have observed relatively minor amounts of overprinting by the later dolomite phase and generally see the late dolomites in vugs or in fractures that crosscut the earlier matrix replacement dolomites (Mattes and Mountjoy, 1980; Machel, 1985; Machel and

Anderson, 1989; Qing and Mountjoy, 1989; Kaufman et al., 1990; Mountjoy and Halim-Dihardja, 1991; Amthor et al., 1993; Dix, 1993). Although the C and O isotopes of these dolomites tend to overlap each other, the limited Sr isotopes show a clearer separation (Fig. 11; Mountjoy et al., 1992).

5.1. EARLY REPLACEMENT MATRIX DOLOMITES

Most workers report that replacement matrix dolomites postdate early near-surface calcite cements, wispy stylolites, and sutured grain contacts. These dolomites are cross-cut by rare high amplitude stylolites. In addition, the ^{18}O isotopes can be used to estimate temperatures of formation, provided these dolomites have not been modified later by neomorphism. Generally these data suggest temperatures of between 35 and 75°C, assuming Devonian seawater was -2.5‰ SMOW (determined from marine calcite cements and fossils), or somewhat higher temperatures (70 to 100°C) if one assumes seawater had the same isotopic composition as today's oceans (Fig. 27). The combined textural and isotopic data support a shallow burial origin at depths ranging between 500 and 1500 m.

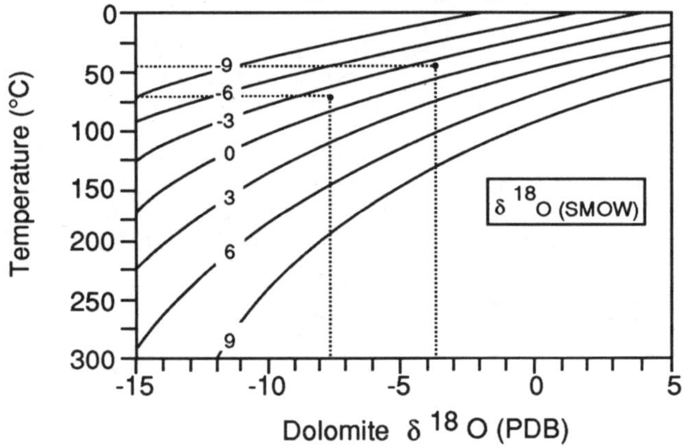

Figure 27. Plot of equilibrium relationships between temperature and isotope composition of water and dolomite (Land, 1985), assuming a $\delta^{18}O$ of the dolomitizing fluid of -2.5‰ SMOW, a range of $\delta^{18}O$ = -3.8 to -7.5 PDB for replacement dolomites. Replacement dolomites could have formed at minimum temperatures ranging from about 40 to 70°C.

If the above assessment is correct, the dolomitization problem becomes one of supplying sufficient Mg at these shallow depths over most of the basin and at different stratigraphic levels. This necessitates an extensive, interconnected conduit system throughout the basin. Earlier, it seemed plausible that fluids derived from compaction of the basin muds adjacent to reefs and platforms and from deeper in the basin might be adequate to explain the observed dolomitization (Illing, 1959; Mattes and Mountjoy, 1980; Machel and Anderson, 1989 among others). However, mass balance calculations showed there were serious difficulties with this model (Machel and Anderson, 1989; Amthor et al., 1993) and clearly it cannot produce more than about 20% of the dolomite, and can only explain dolomitization in relatively small buildups. Another serious difficulty against this model is that it can not explain those buildups that lack dolomitization (e.g. Golden Spike, Redwater, most Swan Hills buildups, etc., Fig. 3) because these buildups would have received similar amounts of compaction fluids from the adjacent basin. In addition computer modeling (Kaufman, personal communication) of compaction fluid flow also indicates that fluids could only be derived from a 10 km zone adjacent to the buildups. Evidently the amounts of compactionally-derived fluids are inadequate to explain any of the completely dolomitized reefs.

The juxtaposition of limestone and dolostone buildups in the Rimbey-Meadowbook reef trend led a number of workers (e.g., Illing, 1959; Machel and Mountjoy, 1987) to suggest that dolomitization only occurred where the dolomitizing fluids flowed along a major conduit system that followed the western margin of the Cooking Lake platform. This conduit system was later utilized during secondary oil migration (leaky pipeline of Stoakes and Creaney, 1984, 1985). Although there is reasonable evidence for a conduit system, the source of the Mg-bearing fluids, and how were they pumped through the vast conduit system, remains unclear. Although many suggestions have been made, there is little agreement and we will have to await additional information from several different parts of the basin.

5.2. LATE COARSE-CRYSTALLINE AND SADDLE DOLOMITES

Most authors consider saddle and associated coarse-crystalline dolomites to be deeper burial and to have formed late in the burial history of the basin, based on petrographic and fluid inclusion evidence. Fluid inclusions indicate temperatures between 90 and 230°C, with the higher temperatures occurring in the more deeply buried dolomites in the western part of the basin (Qing and Mountjoy, 1994, in press), except in the case of the Peace River Arch where high temperatures have been reported from fracture-filling dolomites (Packard et al., 1990; Mountjoy and Halim-Dihardja, 1991).

In most cases fluid inclusions yield temperatures that are often slightly above maximum burial temperatures, assuming normal geothermal gradients of 30°C/km. Many authors interpret these data to indicate that these dolomites probably precipitated during deep burial because this is the only time when hot fluids would be available in abundance in many parts of the basin. Circulating hotter fluids updip from somewhat deeper in the basin along conduits (Qing and Mountjoy, 1994, in press) or along faults and fractures (Packard et al., 1990; Mountjoy and Halim-Dihardja, 1991; Dix, 1993) would explain those temperatures that are slightly above maximum burial temperatures.

Bachu and Cao (1992) showed that in Alberta the present geothermal gradient is about 25°C/km south of 54° latitude and is more than 35°C/km to the northwest near 58° latitude, caused primarily by increased heat flow of crustal origin. They concluded that the present transport of terrestrial heat from the basement to the surface is by conduction and that the flow of formation waters is too slow to advect heat significantly because of the low permeability of the rocks. However their analysis did not take into account conduit systems with much higher permeabilities.

One intriguing approach has been suggested by Aulstead and Spencer (1985) and Spencer (1987; and personal communication; Aulstead et al., 1988; Lowenstein and Spencer, 1990), who on the basis of fluid inclusion data from Prairie Evaporite Formation halites and dolomites from various Devonian units, considered that most of the Devonian-hosted dolostones including saddle dolomite formed early (Devonian through early Mississippian), at shallow burial, from moderately hot (100 to 200°C), high salinity fluids (15 to 30 wt % salt). In this model Mg-rich residual evaporite brines sink and move downdip by gravity and are gradually heated until they reach the relatively impermeable basement where they are modified by interactions with basement rocks (losing Na and gaining Ca). After reacting with basement rocks it is inferred that these hot brines ascended along faults and other conduits until, in the shallow subsurface, they dolomitized limestones and precipitated dolomite and anhydrite cements. At the surface they mixed with the inflow of evaporated seawater to form potash salts (using the Qaidam Basin as a modern analog, Spencer et al., 1990; Casas et al., 1992).

There are several problems with Spencer's model. In the few basement cores available little alteration of basement rocks has been observed. In addition, there are no known heat sources to drive the fluids upwards along faults and fractures and into conduits. The model requires that these fluids flowed rapidly during Devonian or Carboniferous times so that they would be hot enough to form dolomites containing fluid inclusions 100°C or higher at very shallow burial depths. Although this model explains some aspects of these dolomites, it fails to explain the distribution of the two different stages of dolomites. Also there is a lack of evidence for widespread heat sources to drive fluids sufficiently quickly so that hot brines could emerge near the surface in several different parts of the basin at the same time. However, recent downhole organic maturation data in the Liard Basin of the Northwest Territories suggest that geothermal gradients and heat flows were probably higher in the Paleozoic than at present in that region, and that oil and gas maturation and migration occurred during the late Paleozoic to early Mesozoic (Morrow et al., 1993). This maturation pattern is very different from that of the southern Western Canada Sedimentary Basin where maturation of organic matter occurred much later, during the Cretaceous and early Tertiary (Deroo et al., 1977; Bachu and Cao, 1992).

On the other hand, Koehler and Kyser (1991) suggested that Mg-rich brines may have resulted from de-watering during intermediate burial of primary carnallite to sylvite or by the reaction of carnallites with undersaturated fluids (Koehler et al., 1990) in the potash deposits of the Elk Point Basin of Saskatchewan. Such a process would have provided considerable Mg to the formation waters and may have reacted with limestones to form dolomites, especially in the Dawson Bay Formation which overlies the Prairie Evaporite. Fluid inclusion and isotope data suggest that secondary carnallite formed from pre-existing sylvite (as intercalations and as pods) relatively recently and at low temperatures (25-40°C). They dated the carnallite isotopically in the Alwinsal Willowbrook well as 30 Ma (Baadsgaard, 1987). Data from fluid inclusions indicate three distinct events during burial for the formation of halite. None of these processes involved the presence of either pure seawater or evaporated seawater, but involve formation from complex mixtures of meteoric waters, evolved basin brines, and seawater (Chipley and Kyser, 1991).

5.3. DOLOMITIZATION MODELS

Swan Hills dolostones are related neither to seafloor processes nor to evaporites because textural evidence indicates that replacement occurred after the initial stages of cementation and early compaction. There are no directly associated evaporitic facies, and limestone buildups show no evidence of marine dolomites. [This does not rule out the involvement of evaporated fluids, if they could be moved from sites of evaporation often 100's of kms away; e.g. the Elk Point Basin to the east]. An early replacement origin is supported by the fact that most dolomitization is postdated by a solution stage, stylolites, and later dolomite and calcite cements.

Walls and Burrowes (1990) suggested that potential fluids available for dolomitization during early burial would be 1) phreatic, 2) seepage reflux, or 3) warm connate waters, but did not favour any one process. In the Swan Hills reservoirs, hydrodynamic studies by the Geological Survey of Canada suggest that formation waters are in pressure communication, suggesting that these reservoirs are part of a common "plumbing system" that may have persisted from the Devonian to the present.

Laflamme (1990) compared the replacement matrix dolomites of the Swan Hills Caroline field with those of the overlying Leduc dolomites and found them to be similar. He suggested that the Swan Hills dolomites could have been related to fluids driven by regional fresh water seawater mixing zones. This mechanism is not feasible for the overlying Leduc dolomites in the Caroline area. Laflamme (1990) suggested that for the Leduc, fluids were derived from downdip in the basin. If his interpretations were correct there should be some C and O isotope differences between the Swan Hills and Leduc dolomites, but there are none. It appears more likely that these dolomites probably formed from the same dolomitizing fluids, with the two reservoirs being connected via faults and fractures, and the distribution of the Swan Hills dolostones being controlled by the original facies and early diagenesis. The Swan Hills platform was less dolomitized because it was compacted and well cemented at the time of dolomitization. Most of the platform was not connected with a through-going conduit system, except along its updip eastern margin. The source of these dolomitizing fluids is uncertain, but may have come from downdip or laterally along the Cooking Lake platform that underlies the Leduc buildups.

6. General aspects and comparisons with Devonian carbonates in other parts of the world.

6.1 SEA LEVEL CONTROL

Clearly all Devonian buildups and platforms of the WCSB were formed during a series of shallowing upward sedimentary cycles ending with exposure, providing considerable potential for evaporative/sabkha dolomitization. Dolostones in the Lower Ordovician of Nevada (Dunham and Olson, 1980) and the Cambro-Ordovician of the Appalachians (Montanez and Read, 1992) have been interpreted to have resulted from peritidal dolomitization during high stand low stand systems tracts, similar to that interpreted for the Triassic Latemar platform by Goldhammer et al. (1990).

The limestone buildups from different parts and stratigraphic levels in the WCSB generally lack any dolomitization of the peritidal portions of the sedimentary cycles, as for example, in the Middle Devonian Winnipegosis buildups (Martindale and McDonald, 1990) and Rainbow (Schmidt et al., 1985; Qing and Mountjoy, 1989), and in the Upper Devonian - various Swan Hills buildups (Havard and Oldershaw, 1976; Wong and Oldershaw, 1981; Walls and Burrowes, 1986; Scaturo, et al., 1989), and Leduc buildups - Golden Spike (McGillivary and Mountjoy, 1975; Walls, 1978), Redwater (Wendte, 1974), Miette and Ancient Wall (Mountjoy, 1965; Mountjoy and Mackenzie, 1973; Mattes and Mountjoy, 1980; Mountjoy and Burrowes, 1982). One reason for this lack of peritidal dolomitization is that the periods of exposure may have been brief and insufficient to develop extensive evaporated brines and evaporites in the individual reefs (except in the platforms that bordered the eastern part of the basin, see Wilson, 1975), because the basin was subjected to considerable differential subsidence related to the Antler orogeny (Mountjoy and Savoy, 1992; Savoy and Mountjoy, 1994 in press; McLean and Mountjoy, 1993). The Middle Devonian to Lower Carboniferous carbonate development in general represents a long transgressive interval (Moore, 1988; Savoy and Mountjoy, 1994 in press).

6.2 COMPARISONS WITH OTHER DEVONIAN BASINS

When compared with other Devonian carbonate basins, the Western Canada Sedimentary Basin is extensively dolomitized. Mountjoy and Krebs (1983) suggested that the lack of mesogenetic (shallow to deep burial) dolomites and hydrocarbon deposits in Western Europe appeared to be because of the absence of suitable pore systems due to the earlier filling of pores by fibrous and sparry calcites. Similarly, the Devonian of Western Australia lacks significant dolostones (Playford et al., 1989; Hurley and Lohmann, 1989), except for some carbonate platform margins in the Geikie Range (Wallace, 1990; Wallace et. al., 1991) and in the back-reef lithologies of the Oscar Range (Kerans, 1985; Hurley, 1986). These authors suggested a burial compaction mechanism for dolomitization. Information available from China also suggests that most of the Devonian carbonates there are limestones.

A key difference between the Western Canada Sedimentary Basin and sedimentary basins elsewhere in the world is that the others lack extensive evaporites. Perhaps these evaporated brines (which formed in several different stratigraphic levels on the eastern side of the basin) may have seeped downwards into the basin. Provided such brines could have been pumped into the porous reefs and platforms, they may have caused extensive dolomitization during shallow burial of these Devonian strata. This possibility is currently being investigated.

7. Summary and Conclusions

It is difficult to quantify the characteristics of the Devonian dolostone reservoirs of western Canada, partly because so many diagenetic events have taken place in them. There are two main phases of dolomitization. First a major phase of dolomitization that replaced many limestone platform and reefs during shallow burial overlapping with incipient stylolitization and suturing of grain contacts (500 to 1500 m). This phase accounts for between 80 and 90% of all dolostones in the basin. Sr isotopes indicate that the fluids involved in dolomitization were derived either from Devonian seawater or modified seawater. It is unclear whether these fluids were derived from downdip in the basin or from updip evaporites. Direct precipitation from seawater can be ruled out, except for some Winnipegosis reefs. Also there is little evidence of dolomites being related to exposure during high stand and low stand systems tracts, except in the shallow water carbonate platforms that bordered the basin to the east and southeast.

A second phase of dolomitization took place from hydrothermal fluids (100°C or hotter) containing radiogenic strontium, presumably from formation waters similar to present day formation waters in deeper parts of the basin. This resulted in widespread but minor amounts of coarse-crystalline and saddle dolomite. In some of the deeper parts of the basin (> 3000 m present depths) very small amounts of late-stage dolomites were precipitated after oil generation and presumably resulted from TSR reactions.

The distribution of porosity in the original limestones had a primary control on replacement dolomitization. Where the original limestones were porous and permeable they were extensively dolomitized, but only if they were connected to regional conduit systems during shallow burial. The heterogeneous nature of carbonate porosity in most limestones has produced even more heterogeneous dolomite reservoirs, making it exceedingly difficult to predict the distribution of porosity and permeability in them. Generally where limestones have been changed to dolostones with intercrystalline porosity the rock textures are more uniform and the permeability has been improved.

Where extensive and complete dolomitization has occurred, it appears that these strata were part of, or connected to, extensive conduit systems that were oriented perpendicular to the flanks of the basin; e.g. the dolomitized Rimbey-Meadowbrook reef trend, and the Presqu'ile barrier. These conduit systems conducted dolomitizing fluids either downdip or updip. The sources of these dolomitizing fluids are still uncertain. Later, during burial, these conduit systems allowed other dolomitizing fluids to pass (forming saddle dolomites), and TSR-generated fluids and hydrocarbons to move updip.

The porosity and permeability of the replacement dolostones has been modified, and locally enhanced, by dissolution events associated with hydrothermal fluids and fluids related to TSR reactions. These fluids dissolved some dolomites and any remaining calcites increasing porosity in intercrystalline pores, vugs and along fractures improving the reservoir quality. Locally this resulted in solution collapse in some Leduc buildups in Alberta, parts of the Comet Platform (NE B.C.), and at Pine Point in the

N.W.T. Evidence for TSR reactions in much of the deeper parts of the Western Canada Sedimentary Basin (>3000 m present depth) is indicated by the presence updip of abundant anhydrite, late-stage calcite and dolomite cements, and downdip the occurrence of metallic sulphides, free sulphur, and an increase in porosity and the amounts of H_2S in reservoir gases. Locally these late-stage anhydrite and carbonate cements have considerably reduced porosity and permeability updip from where the TSR reactions took place.

In general, it seems that dolomitization and burial diagenesis have reduced the overall porosity of most buildups by 1 to 10% compared to precursor limestone buildups. However, limestone buildups buried deeper than 3000 m have lower porosities than dolomite reservoirs, apparently because of greater pressure solution of the limestones. Locally in dolostone buildups permeability has been increased, especially in dolomites with intercrystalline porosity.

More detailed studies of specific reservoirs are needed before quantitative models of the porosity and permeability and its distribution in dolomite reservoirs can be developed. The Western Canada Sedimentary Basin is unique in terms of the enormous amounts of dolomites present compared to Devonian basins in western Europe, Australia, and China. A key difference in the character of these other basins is that they lack extensive evaporites. In the WCSB a series of evaporite deposits formed along the eastern flank of the basin during much of the Middle and Upper Devonian. Mg-rich fluids from these deposits may have seeped downwards into the basin resulting in the early replacement dolomitization.

8. Acknowledgements

Over the years a number of geologists have freely shared and discussed information, problems and ideas concerning diagenesis and dolomitization. I am grateful to a number of graduate students with whom I have shared data, ideas and concepts, especially J. Amthor, H. Chouinard, M. Halim-Dihardja, E. Drivet, A. Laflamme, X. Marquez, H. G. Machel, B. Mattes, H. Qing, M. Teare, and S. Whittaker. J. Amthor and H. Machel have provided provocative and challenging discussions over the years. For permission to use some data and figures from their short course notes, I am grateful to Geoff Burrowes, Bill Martindale, and Dick Walls. Part of the text has been summarized from a review paper on burial dolomites (Mountjoy and Amthor, 1994, in press). I benefited from discussions at the Banff National Conference on Earth Science, September 1992, sponsored by the extension department of the University of Alberta; "Dolomite - from process and models to porosity and reservoirs". I appreciate the helpful and critical comments and suggestions of J. Amthor, E. Drivet, H. Machel, X. Marquez and S. Whittaker on various versions of the text, and of B. Sellwood on the final manuscript.

The Natural Sciences and Engineering Research Council of Canada has supported my research on carbonates and their diagenesis over a number of years, as have Chevron, Home, Norcen, PetroCanada and Shell Oil companies in Calgary.

References

Amthor, J.E., Mountjoy, E.W., Machel, H.G., (1993) 'Subsurface dolomites in Upper Devonian Leduc formation buildups, central part of Rimbey-Meadowbrook reef trend, Alberta, Canada', Bulletin of Canadian Petroleum Geology 41, 164-185.

Amthor, J.E., Mountjoy, E.W., and Machel, H.G. (submitted) 'Regional-scale porosity and permeability variations in Upper Devonian Leduc buildups: implications for origin and distribution of porosity in carbonates', American Association of Petroleum Geologists Bulletin.

Aulstead, K.L., and Spencer, R.J. (1985) 'Diagenesis of the Keg River Formation, Northwestern Alberta: fluid inclusion evidence', Bulletin of Canadian Petroleum Geology 33, 167-183.

Aulstead, K. L., Spencer, R. J. and Krouse, H. R. (1988) 'Fluid inclusion and isotopic evidence on dolomitization,Devonian of Western Canada', Geochimica et Cosmochimica Acta 52, 1027-1035.

Baadsgaard, (1987) 'Rb-Sr and K-Ca isotope systematics from potassium horizons in the Prairie Evaporite Formation, Saskatchewan, Canada', Chemical Geology 66, 1-15.

Bachu, S. and Cao, S. (1992) 'Present and past geothermal regimes and source rock maturation, Peace River Arch area, Canada' American Association of Petroleum Geologists Bulletin 76, 1533-1549.

Back, W., Hanshaw, B.B., Herman, J.S. and VanDriel, J.N. (1986) 'Differential dissolution of a Pleistocene reef in the ground-water mixing zone of coastal Yucatan, Mexico', Geology 14, 137-140.

Bassett, H.G. and Stout, J.G. (1967) ' Devonian of western Canada', Proc. Intern. Devonian System, Canadian Society of Petroleum Geologists 1, 717-725.

Burke, W.H., Denison, R.E., Hetherington, E.A., Koepnik, R.B., Nelson, H.F. and Otto, J.B. (1982)'Variations of seawater $^{87}Sr/^{86}Sr$ throughout Phanerozoic time', Geology 10, 516- 519.

Carpenter, S.J. and Lohmann, K.C., (1989) '$\delta^{18}O$ and $\delta^{13}C$ variations in Late Devonian marine cements from the Golden Spike and Nevis reefs, Alberta, Canada', Journal of Sedimentary Petrology 59, 792-814.

Casas, E., Lowenstein, T.K., Spencer, R.J.and Pengxi, Z. (1992) 'Carnallite mineralization in the nonmarine, Qaidam Basin, China: Evidence for the early diagenetic origin of potash evaporites', Journal of Sedimentary Petrology 62, 881-898.

Chipley, D. and Kyser, T.K. (1991) 'Large scale fluid movement in the Western Canadian Sedimentary Basin as recorded by fluid inclusions in evaporites', 6th International Williston Basin Symposium, Saskatchewan Geological Society Spec. Publ. 11, 265-269.

Chouinard, H., (1993) 'Lithofacies and diagenesis of the Cooking Lake platform carbonates, Alberta Basin subsurface, Canada', unpublished M.Sc.thesis, McGill University 120p.

Choquette P.W. and Pray, L.C. (1970) 'Geologic nomenclature and classification of porosity in sedimentary carbonates', American Association of Petroleum Geologists Bulletin 54, 207-250.

Choquette, P.W., Cox, A., and Meyers, W. J. (1992) 'Characteristics, distribution and origin of porosity in shelf dolostones: Burlington-Keokuk Formation (Mississippian), U.S. Mid-continent', Journal of Sedimentary Petrology 62, 167-189.

Collins, J.F. and Lake, J.H. (1989) 'Sierra reef complex, Middle Devonian, northeastern British Columbia', in H.H.J. Geldsetzer, N.P. James and G.E. Tebbutt (eds.), Reefs, Canada and adjacent areas, Canadian Society of Petroleum Geologists Memoir 13, 414-421.

Comby, O. (1975) 'Diagenesis et compaction mesogenestiques des roches carbonates application au complex recifal de Clarke Lake, Colombie Britannique, Canada', Ph.D. thesis, L'Universitie Paris, Paris.

Cook, H.E. (1972) 'Miette platform evolution and relation to overlying bank ('reef') localization, Upper Devonian, Alberta', Bulletin of Canadian Petroleum Geology 12, 779-807.

Deroo, G., Powell, T.G., Tissot, B., McCrossan, R.G. and Haquebard, P.A. (1977) 'The origin and migration of petroleum in the Western Canada Sedimentary Basin, Alberta - a geochemical and thermal maturation study' Geological Survey of Canada Bulletin 2262, 136p.

Dix, G.R. (1993) 'Patterns of burial- and tectonically controlled dolomitization in an Upper Devonian fringing-reef complex: Leduc Formation, Peace River Arch area, Alberta, Canada', Journal of Sedimentary Petrology 63, 628-641.

Dravis, J. J., and Muir, I. D. (1992) 'Case study of burial dissolution - Dolomites - Devonian Upper Elk Point Group, Western Canada, in G.R. Bloy (ed.), Submarine dissolution porosity in carbonates: recognition, causes and implications', Canadian Society of Petroleum Geologists, Short Course Notes, 55-117.

Drivet, E. (1993) 'Diagenesis and reservoir characteristics of Upper Devonian Leduc dolostones, southern Rimbey-Meadowbrook reef trend, Central Alberta', unpublished M.Sc. thesis, McGill University, 112p.

Drivet, E. and Mountjoy, E.W. (1993) 'Porosity variations in Upper Devonian Leduc dolomites, central Rimbey Meadowbrook reef trend, Alberta', American Association of Petroleum Geologists, annual meeting abstracts.

Dunham, J.B. and Olson, E.R. (1980) 'Shallow subsurface dolomitization of subtidally deposited carbonate sediments in the Hanson Creek Formation (Ordovician - Silurian) of central Nevada', in D.H. Zenger, J.B. Dunham and R.L. Ethington (eds.), Concepts and models of dolomitization, Society of Economic Paleontologists and Mineralogists Special Publication 28, 139-161.

Eliuk, L.S. (1984) 'A hypothesis for the origin of hydrogen sulfide in the Devonian Crossfield Member dolomite, Wabamun Formation, Alberta, Canada. Carbonates in Subsurface and

Outcrop: 1984', (Ed. by L.S. Eliuk), Core Conference Canadian Society of Petroleum Geologists, 245-290.
Goldhammer, R.K., Dunn, P.A., and Hardie, L.A. (1990) 'Depositional cycles, composite sea-level changes, cycle stacking patterns, and hierarchy of stratigraphic forcing: Examples from Alpine Triassic platform carbonates', Geological Society of America Bulletin 102, 535-562.
Gray, F.F. and Kassube, J.R. (1963) 'Geology and stratigraphy of Clarke Lake Gas Field, Northeastern Brithish Columbia', American Association of Petroleum Geologists Bulletin 47, 467-483.
Griffin, D.L. (1965) 'The Devonian Slave Point, Beaverhill Lake, and Muskwa Formations of northeastern British Columbia and adjacent areas', British Columbia Department of Mines and Petroleum Resources 50, 90 p.
Hanshaw, B.B. and Back, W. (1980) 'Chemical mass-wasting of the northern Yucatan Peninsula by groundwater dissolution', Geology 8, 222-224.
Hardie, L.A. (1987) 'Dolomitization: a critical view of some current views', Journal of Sedimentary Petrology 57, 166-183.
Havard, C.J. and Oldershaw, A.E. (1976) 'Sedimentary and early diagenetic cycles, Snipe Lake reef complex, Alberta', Bulletin of Canadian Petroleum Geology 24, 27-69.
Heydari, E. and Moore, C.H. (1989) 'Burial diagenesis and thermochemical sulphate reduction, Smackover Formation, southeastern Mississippi salt basin, Geology 17, 1080-1084.
Hurley, N.F. (1986) 'Geology of the Oscar Range Devonian reef complex, Canning basin, Western Australia', Ph.D. dissertation, Ann Arbor, Michegan, Univ. of Michegan, 269 p.
Hurley, N.F. and Lohmann, K.C. (1989) 'Diagenesis of Devonian reefal carbonates in the Oscar Range, Canning Basin, Western Australia', Journal of Sedimentary Petrology 59, 127-145.
Illing, L.V. (1959) 'Deposition and diagenesis of some Upper Paleozoic carbonate sediments in western Canada', Fifth World Petroleum Congress, New York, Proceedings Section 1, 23-52.
Jardine, D. and Wilshart, J.W. (1982) 'Carbonate reservoir description' Society of Petroleum Engineers, 10010, 13pp, Dallas, Texas.
Kaufman, J., Meyers, W.J. and Hanson, G.N. (1990) 'Burial cementation in the Swan Hills Formation (Devonian), Rosevear field, Alberta, Canada', Journal of Sedimentary Petrology 60, 918- 939.
Kendall, A.C. (1989) 'Brine mixing in the Middle Devonian of western Canada and its possible significance to regional dolomitization', Sedimentary Geology 64, 271-285.
Kerans, C. (1985) 'Petrology of Devonian and Carboniferous carbonates of the Canning and Bonaparte basins', Western Australian Mining and Petroleum Resources Institute Report 12, 203 p.
Koehler, G.D. and Kyser, T.K. (1991) 'Fluid events as recorded in the hydration water of carnallite from Elk Point Basin of western Canada', Summary of Investigations

1990, Saskatchewan Energy and Mines Miscellaneous Report 90-4, 270-274.

Koehler, G.D., Kyser, T.K. and Danyluk, T. (1990) 'Sr isotope evidence for the petrogenesis of carnallite in the Middle Devonian Prairie Evaporite Formation, Saskatchewan', Summary of Investigations 1990, Saskatchewan Energy and Mines Miscellaneous Report 90-4, 218-222.

Krouse, H.R., Viau, C.A., Eliuk, L.S., Ueda, A. and Halas, S. (1988) 'Chemical and isotopic evidence of thermochemical sulphate reduction by light hydrocarbon gases in deep carbonate reservoirs', Nature 333, 415-419.

Kupecz, J.A., Montanez, I.P. and Gun, G. (1992) 'Recrystallization of dolomite with time', in R. R. Rezak and P. Lavoie (eds.), Carbonate Microfacies: Frontiers in Sedimentary Geology, New York, Springer-Verlag, Ch. 14, 187-194.

Kyle, J.R. (1981) 'Geology of Pine Point lead-zinc district', in K.H. Wolf (ed.), Handbook of strata-bound and strataform ore deposits, Elsevier, New York, v.9, 643-741.

Kyle, J.R. (1983) 'Economic aspects of subaerial carbonates', in P.A. Scholle, D.G. Bebout and C.H. Moore (eds.), Carbonate depositional environments, American Association of Petroleum Geologists Memoir 33, 73-92.

Laflamme, A.K. (1990) 'Replacement dolomitization in the Upper Devonian Leduc and Swan Hills formations, Caroline area, Alberta, Canada', Unpublished M.Sc. thesis, McGill University, Montréal, 138 p.

Land, L.S., (1980) 'The isotopic and trace element geochemistry of dolomite: the state of the art', in D.H. Zenger, J.B. Dunham and R.L. Ethington (eds.), Society of Economic Paleontologists and Mineralogists, Special Publication no. 28, 87-110.

Land, L.S. (1985) 'The origin of massive dolomite', Journal of Geologic Education, 33, 112-125.

Land, L.S. (1991) 'Dolomitization of the Hope Gate Formation (north Jamaica) by seawater: Reassessment of mixing-zone dolomite', in H.P. Taylor, Jr., J.R. O'Neil, and I.R.Kaplan (eds.), Stable Isotope Geochemistry: A Tribute to Samuel Epstein, The Geochemical Society, Special Publication 13, 121-133.

Lowenstein, T.K. and Spencer, R.J. (1990) 'Syndepositional origin of potash evaporites: petrographic and fluid inclusion evidence, American Journal of Science 290, 1-42.

Machel, H.G. (1985) 'Facies and diagenesis of the Upper Devonian Nisku Formation in the subsurface of Central Alberta', unpublished Ph.D. diss: Montreal, McGill University, 392 p.

Machel, H.G. (1987a) 'Saddle dolomite as a by-product of chemical compaction and thermochemical sulfate reduction', Geology 15, 936-940.

Machel, H.G. (1987b) 'Some aspects of diagenetic sulphate-hydrocarbon redox reactions' in J.D. Marshall (ed.), The Diagenesis of Sedimentary Sequences, Geological Society of London, Special Publication 36, 15-28.

Machel, H.G. (1989) 'Relationships between sulfate reduction and oxidation of organic compounds to carbonate diagenesis,

hydrocarbon accumulations, salt domes, and metal sulfide deposits', Carbonates and Evaporites, 4, 137-151.
Machel, H.G. and Anderson, J.H. (1989) 'Pervasive subsurface dolomitization of the Nisku Formation in Central Alberta' Journal of Sedimentary Petrology 59, 891-911.
Machel, H.G., and Burton, E.A. (1991) 'Burial-diagenetic sabkha-like gypsum and anhydrite nodules', Journal of Sedimentary Petrology 61, 394-405.
Machel, H.G. and Mountjoy, E. W. (1986) 'Chemistry and environments of dolomitization - A reappraisal', Earth Science Review 23, 175-222.
Machel, H.G. and Mountjoy, E. W. (1987) 'General constraints on extensive pervasive dolomitization - and their application to the Devonian carbonates of western Canada', Bulletin of Canadian Petroleum Geology 35, 143-158.
Martindale, W. and MacDonald, R.W. (1990) 'Sedimentation and diagenesis of the Winnipegosis Formation Tableland area, southeast Saskatchewan' in G.R. Bloy & M.G. Hadley (eds.), The Development of Porosity in Carbonate Reservoirs, Canadian Society of Petroleum Geologists, Short Course Notes, Sect. 6, 6.1-6.19.
Marquez, X., Mountjoy, E.W., and Amthor, J. (1992) 'Microfracturing in deeply buried carbonate buildups: Upper Devonian Strachan and Ricinus reservoirs, Alberta', American Association of Petroleum Geologists-Canadian Society of Peptroleum Geologists annual meetings abstracts, 81.
Mattes, B. W. and Mountjoy, E. W. (1980) 'Burial dolomitization of the Upper Devonian Miette buildup, Jasper National Park, Alberta' in D.H. Zenger, J.B. Dunham, and R.L. Ethington (eds.), Concepts and models of dolomitization, Society of Economic Paleontologists and Mineralogists Special Publication 28, 259-297.
Mazzullo, S.J. (1992) 'Geochemical and neomorphic alteration of dolomite: a review', Carbonates and Evaporites 7, 21-37.
Mazzullo, S. J., & Harris, P. M. (1992) 'Mesogenetic dissolution: its role in porosity development in carbonate reservoirs' American Association of Petroleum Geologists Bulletin 76, 607-620.
McGillivary, J. G. and Mountjoy, E.W. (1975) 'Facies and related reservoir characteristics Golden Spike reef complex, Alberta', Bulletin of Canadian Petroleum Geology 23, 753-809.
McIlreath, I.A. and Morrow, D.W. (1990) 'Diagenesis', Geoscience Canada Report Series 4, 338 p.
McLean, D.J. and Mountjoy, E.W. (1993) 'Upper Devonian buildup margin and slope development, Canadian Rocky Mountains', Geological Society of America Bulletin 105, 1263-1283.
McNamara, L.B. and Wardlaw, N.C. (1991) 'Geological description of the Westerose reservoir, Alberta', Bulletin of Canadian Petroleum Geology 39, 332-351.
Montanez, I. P., and Read, J. F. (1992) 'Fluid-rock interaction history during stabilization of early dolomites, Upper Knox Group (Lower Ordovician), U.S. Appalachians' Journal of Sedimentary Petrology 62, 753-778.

Moore, P.F. (1988) 'Devonian Geohistory of the western interior of Canada', in N.J. McMillan, A.F. Embry, and D.J. Glass (eds.), Canadian Society of Petroleum Geologists, Devonian of the World 1, 67-83.

Morrow, D.W. (1990a) 'Dolomite - Part 1: The chemistry of dolomitization and dolomite precipitation', in I.A. McIlreath and D.W. Morrow (eds.), Diagenesis, Geoscience Canada reprint series 4, 113-123.

Morrow, D.W. (1990b) 'Dolomite - Part 2: Dolomitization models and ancient dolostones', in I.A. McIlreath and D.W. Morrow (eds.), Diagenesis, Geoscience Canada reprint series 4, 125-139.

Morrow, D.W., Potter, J., Richards, B. and Goodarzi, F. (1993) 'Paleozoic burial and organic maturation in the Liard Basin Region, northern Canada' Bulletin of Petroleum Geology 41, 17-31.

Mossop, G.D. (1972) 'Origin of the peripheral rim, Redwater reef, Alberta', Bulletin of Canadian Petroleum Geology 20, 238-280.

Mountjoy, E.W., (1965) 'Stratigraphy of the Devonian Miette reef complex and associated strata, eastern Jasper National Park, Alberta', Geological Survey of Canada, Bulletin 110, 132 p.

Mountjoy, E.W. (1967) 'Factors governing the development of the Frasnian, Miette and Ancient Wall reef complexes (banks and biostromes) Alberta', International Symposium Devonian System: Calgary, Alberta Society of Petroleum Geologists 2, 387-408.

Mountjoy, E.W. (1980) 'Some questions about the development of Upper Devonian carbonate (reefs), western Canada', Bulletin of Canadian Petroleum Geology 28, 315-340.

Mountjoy, E.W. (1987) 'Upper Devonian Ancient Wall Reef Complex, Jasper National Park, Alberta', Field Trip 5A Guidebook, Second International Devonian Symposium, Calgary, Canadian Society of Petroleum Geologists 65pp.

Mountjoy, E.W. (1989) 'Miette reef complex (Frasnian), Jasper National Park, Alberta', in H.J.J. Geldsetzer et al. (eds.), Reefs, Canada and adjacent areas, Canadian Society of Petroleum Geologists, Memoir 13, 497-505.

Mountjoy, E.W., Qing, H. and McNutt, R.H. (1992) 'Strontium isotopic composition of Devonian dolomites, Western Canada Sedimentary Basin: Significance of sources of dolomitizing fluids', Applied Geochemistry 7, 59-75.

Mountjoy, E.W. and Burrowes, G. (1982) 'Upper Devonian Miette Reef Complex, Jasper National Park, Alberta', International Association Sedimentologists, 11th Congress Excursion 27A, 57 p., Hamilton.

Mountjoy, E.W. and Halim-Dihardja, M. (1991) 'Multiple phase fracture and fault-controlled burial dolomitization, Upper Devonian Wabamun Group, Alberta', Journal of Sedimentary Petrology 61, 590-612.

Mountjoy, E.W. and Krebs, W. (1983) 'Diagenesis of Devonian reefs and buildups, Western Canada and Europe - a comparison', Zeitschrift Deutsche Geologisch Gesellschaft 134, 5-60.

Mountjoy, E.W. and Mackenzie, W.S. (1973) 'Stratigraphy of

the southern part of the Devonian Ancient Wall carbonate complex, Jasper National Park, Alberta', Geological Survey of Canada, Paper 72-20.

Mountjoy, E.W. and Savoy, L.E. (1992) 'Cratonic-margin and Antler foreland-basin strata in the Upper Devonian and Lower Carboniferous of the southern Canadian Rocky Mountains', American Association of Petroleum Geologists Annual Meeting Abstracts, 93.

Noble, J.P.A. (1970) 'Biofacies analysis, Cairn Formation of Miette Reef complex (Upper Devonian), Jasper National Park, Alberta', Bulletin of Canadian Petroleum Geology 18, 493-534.

Packard, J.J., Pellegrin, G.J., Al-Aasm I.S., Samson I., and Gagnon, J. (1990) 'Diagenesis and dolomitization associated with hydrothermal karst in Famennian upper Wabamun ramp sediments, northwestern Alberta', in G.R. Bloy G.R. and M.G. Hadley (eds.), The Development of porosity in carbonate reservoirs, Canadian Society of Petroleum Geologists, Continuing Education short course, section 9, 9.1-9.19.

Playford, P.E., Hurley, N.F., Kerans, C. and Middleton, M.F. (1989) 'Reefal platform development, Devonian of the Canning Basin, western Australia', in P.D. Crevello et al (eds.), Controls on carbonate platform and basin development, Society of Economic Palontologists and Mineralogists Special Publication 44, 187-202.

Qing, H. (1991) 'Diagenesis of Middle Devonian Presqu'ile dolomite, Pine Point NWT and adjacent subsurface', Ph.D. thesis, McGill University, 287 p.

Qing, H. and Mountjoy, E.W. (1989) 'Multistage dolomitization in Rainbow buildups, Middle Devonian Keg River Formation, Alberta, Canada', Journal of Sedimentary Petrology 59, 114-126.

Qing, H. and Mountjoy, E.W. (1990) 'Petrography and diagenesis of the Middle Devonian Presqu'ile barrier: implications on formation of dissolution vugs and breccias at Pine Point and adjacent subsurface, District of Mackenzie', Current Research, Part D, Geological Survey of Canada, Paper 90-1D, 37-45.

Qing, H. and Mountjoy, E.W. (1992) 'Large-scale fluid flow in the Middle Devonian Presqu'ile Barrier, Western Canada Sedimentary Basin' Geology 20, 903-906.

Qing, H. and Mountjoy, E.W. (1994 in press) 'Hydrothermal origin of dissolution vugs and breccias in Middle Devonian Presqu'ile Barrier, host of Pine Point MVT deposits' Economic Geology.

Qing, H. and Mountjoy, E.W. (in press) 'Petrography and geochemistry of coarse-crystalline and saddle dolomites in the Presqu'ile Barrier: host rocks of MVT deposits and hydrocarbon reservoirs in the Western Canada Sedimentary Basin', American Association of Petroleum Geologists Bulletin 78, no. 1.

Reimer, J.D. and Teare, M.R. (1992) 'Thermochemical sulphate reduction and hydrothermal dolomitization (TSR-HTD): a diagenetic process that created and modified Middle Devonian reservoirs in northeastern Brithish Columbia',

American Association of Petroleum Geologists 1992 Annual Convention Meeting (abstract), 130.

Reitzel, G.A. and Callow, G.O. (1976) 'Pool description and performance analysis leads to understanding Golden Spike's miscible flood (preprint)', Petroleum Section, Canadian Institute of Mining and Metallurgy, Paper 7622, 11 p.

Rhodes, D., Lantos, E.A., Lantos, J.A., Webb, R.J. and Owens, D.C. (1984) 'Pine Point orebodies and their relationship to the stratigraphy, structure, dolomitization, and karstification of the Middle Devonian barrier complex', Economic Geology 79, 991-1055.

Savoy, L.E. and Mountjoy, E.W. (1994 in press) 'Cratonic-Margin and Antler Foreland-basin strata (Middle Devonian to Lower Carboniferous) of the Southern Canadian Rocky Mountains and adjacent plains', in S.L. Dorobek and G.M. Ross (eds.), Stratigraphic Evolution of Foreland Basins, Society of Economic and Paleontologists and Mineralogists Special Publlication.

Scaturo, D.M. Strobel, J.S., Kendall, C.G.St.C., Wendte, J.C., Biswas, G., Bezdek, J., and Cannon, R., (1989) 'Judy Creek: A case study for a two-dimensional sediment deposition simulation', in P.D. crevello, J.L. Wilson, J.F. Sarg, and J.F. Read (eds.), Controls on Carbonate Platform and Basin development, Society of Economic Paleontologists and Mineralogists Special Publlication 44, 63-76.

Schmidt, V., McIlreath, I.A. and Budwill, A.E. (1985) 'Middle Devonian cementation reefs encased in evaporites, Rainbow field, Alberta', in P.O. Roehl and P. W. Choquette (eds.), Carbonate Petroleum Reservoirs, Berlin, Springer-Verlag, 141-160.

Schmoker, J. W. and Halley, R. B. (1982) 'Carbonate porosity versus depth: a predictable relation for South Florida', American Association of Petroleum Geologists Bulletin 66, 2561-2570.

Skall, H. (1975) 'The paleoenvironment of the Pine Point lead-zinc district', Economic Geology 70, 22-45.

Spencer, R.J. (1987) 'Origin of Ca-Cl brines in Devonian formations, Western Canada Sedimentary Basin', Applied Geochemistry 2, 373-384.

Spencer, R.J., and Lowenstein, T.K. (1990) 'Evaporites', in I.A. McIIreath and D.W. Morrow (eds.), Diagenesis, Geoscience Canada, Reprint Series no. 4, 141-163.

Stoakes, F.A. and Creaney, S,. (1984) 'Sedimentology of a carbonate source rock: the Duvernay Formation of Central Alberta', in L. Eliuk (ed.), Carbonates in Subsurface and Outcrop, Canadian Society of Petroleum Geologists, Core Conference 1984, 132-147.

Stoakes, F.A. and Creaney, S. (1985) 'Sedimentology of a carbonate source rock: the Duvernay Formation of Alberta, Canada', in L.Eliuk (ed.), Carbonates in Subsurface and Outcrop, Society of Economic Paleontologists and Mineralogists, Core Workshop no. 7, 343-375.

Teare, M. (1990) 'Sedimentology and diagenesis of Middle Devonian Winnipegosis Reef complexes, Dawson Bay, Manitoba', M.Sc. thesis, McGill University, Montreal, 150 p.

American Association of Petroleum Geologists 1992 Annual Convention Meeting (abstract), 130.

Reitzel, G.A. and Callow, G.O. (1976) 'Pool description and performance analysis leads to understanding Golden Spike's miscible flood (preprint)', Petroleum Section, Canadian Institute of Mining and Metallurgy, Paper 7622, 11 p.

Rhodes, D., Lantos, E.A., Lantos, J.A., Webb, R.J. and Owens, D.C. (1984) 'Pine Point orebodies and their relationship to the stratigraphy, structure, dolomitization, and karstification of the Middle Devonian barrier complex', Economic Geology 79, 991-1055.

Savoy, L.E. and Mountjoy, E.W. (1994 in press) 'Cratonic-Margin and Antler Foreland-basin strata (Middle Devonian to Lower Carboniferous) of the Southern Canadian Rocky Mountains and adjacent plains', in S.L. Dorobek and G.M. Ross (eds.), Stratigraphic Evolution of Foreland Basins, Society of Economic and Paleontologists and Mineralogists Special Publlication.

Scaturo, D.M. Strobel, J.S., Kendall, C.G.St.C., Wendte, J.C., Biswas, G., Bezdek, J., and Cannon, R., (1989) 'Judy Creek: A case study for a two-dimensional sediment deposition simulation', in P.D. crevello, J.L. Wilson, J.F. Sarg, and J.F. Read (eds.), Controls on Carbonate Platform and Basin development, Society of Economic Paleontologists and Mineralogists Special Publlication 44, 63-76.

Schmidt, V., McIlreath, I.A. and Budwill, A.E. (1985) 'Middle Devonian cementation reefs encased in evaporites, Rainbow field, Alberta', in P.O. Roehl and P. W. Choquette (eds.), Carbonate Petroleum Reservoirs, Berlin, Springer-Verlag, 141-160.

Schmoker, J. W. and Halley, R. B. (1982) 'Carbonate porosity versus depth: a predictable relation for South Florida', American Association of Petroleum Geologists Bulletin 66, 2561-2570.

Skall, H. (1975) 'The paleoenvironment of the Pine Point lead-zinc district', Economic Geology 70, 22-45.

Spencer, R.J. (1987) 'Origin of Ca-Cl brines in Devonian formations, Western Canada Sedimentary Basin', Applied Geochemistry 2, 373-384.

Spencer, R.J., and Lowenstein, T.K. (1990) 'Evaporites', in I.A. McIIreath and D.W. Morrow (eds.), Diagenesis, Geoscience Canada, Reprint Series no. 4, 141-163.

Stoakes, F.A. and Creaney, S,. (1984) 'Sedimentology of a carbonate source rock: the Duvernay Formation of Central Alberta', in L. Eliuk (ed.), Carbonates in Subsurface and Outcrop, Canadian Society of Petroleum Geologists, Core Conference 1984, 132-147.

Stoakes, F.A. and Creaney, S. (1985) 'Sedimentology of a carbonate source rock: the Duvernay Formation of Alberta, Canada', in L.Eliuk (ed.), Carbonates in Subsurface and Outcrop, Society of Economic Paleontologists and Mineralogists, Core Workshop no. 7, 343-375.

Teare, M. (1990) 'Sedimentology and diagenesis of Middle Devonian Winnipegosis Reef complexes, Dawson Bay, Manitoba', M.Sc. thesis, McGill University, Montreal, 150 p.

Tucker, M.E. and Wright, V.P. (1990) Carbonate Sedimentology, Blackwell, London.
Viau, C.A. and Oldershaw, A.E. (1984) 'Structural controls on sedimentation and dolomite cementation in the Swan Hills field, central Alberta', in F.F. Krause and G. Burrowes (eds.), Devonian lithofacies and reservoir styles in Alberta, 13th Canadian Society of Petroleum Geologists Core Conference, 201-239.
Wallace, M.W. (1990) 'Origin of dolomitization in the Devonian carbonates on the Barbwire Terrace, Canning basin, Western Australia', Sedimentology 37, 105-122.
Wallace, M.W., Kerans, C., Playford, P.E. and McManus, A. (1991) 'Burial diagenesis in the Upper Devonian reef complexes of the Geikie Gorge region, Canning Basin, western Australia', American Association of Petroleum Geologists Bulletin 75, 1018-1038.
Walls, R.A., (1978) 'Cementation history and porosity development, Golden Spike (Devonian) Reef Complex, Alberta', Unpublished Ph.D. thesis, McGill University, Montréal, 307 pp.
Walls, R. A. and Burrowes, G. (1985) 'The role of cementation in the diagenetic history of Devonian reefs, Western Canada, in N. Schneidermann and P.M. Harris (eds.), Carbonate Cements: Society of Economic Paleontologists and Mineralogists, Special Publication 36, 185-219.
Walls, R.A. and Burrowes, G. (1990) 'Diagenesis and reservoir development in Devonian limestone and dolostone reefs of Western Canada', Canadian Society of Petroleum Geologists, Short Course Notes, Section 5, 5.1-5.17.
Walls, R.A., Mountjoy, E.W. and Fritz, P. (1979) 'Isotopic composition and diagenetic history of carbonate cements in Devonian Golden Spike reef, Alberta', Geological Society of America Bulletin 90, 963-982.
Wardlaw, N.C. and Reinson, G.E. (1971), 'Carbonate and evaporite deposition and diagenesis, Middle Devonian Winnipegosis and Prairie Evaporite Formations of Saskatchewan', American Association of Petroleum Geologists Bulletin 55, 1759-1786.
Wendte, J.C. (1974) 'Sedimentation and diagenesis of the Cooking Lake Platform and Lower Leduc reef facies, Upper Devonian, Redwater, Alberta', Unpublished Ph.D. Diss., University of California, Santa Cruz, 237 p.
Wendte, J.C. and Stoakes, F.A. (1982) 'Evolution and corresponding porosity distribution of the Judy Creek reef complex, Upper Devonian, Central Alberta', in W.G. Cutler (ed.), Canada's great hydrocarbon resevoirs, Canadian Society of Petroleum Geologists Core Conference, 63-81.
Wilson, J.L. (1975) 'Carbonate facies in geologic history', Springer-Verlag, Berlin-Heidelberg, Germany, 471 p.
Wong, P.K. and Oldershaw, A. (1981) 'Burial cementation in the Kaybob reef complex, Alberta, Canada', Journal of Sedimentary Petrology 51, 507-520.

LOSSES AND GAINS IN WEATHERING PROFILES AND DURIPANS

V. PAUL WRIGHT
Postgraduate Research Institute for Sedimentology, The University,
P.O. Box 227, Whiteknights,
Reading, RG6 2AB, U.K.

ABSTRACT

The weathering profile is a site of major diagenesis and sediment production. Material is both lost in solution and also transported in particulate form, but significant concentrations (relative and absolute accumulations) of carbonate, sulphate, silica and Fe and A1 oxides and hydroxides can also occur. Quantitative techniques for identifying ancient weathering profiles and their diagenetic alteration are reviewed. Ancient weathering profiles displaying changes from humid to arid climates appear well represented in the geological record, reflecting the higher preservation potential for such sequences. Weathering profiles and sediments can contain a variety of groundwater-related absolute accumulations of carbonates, sulphates, etc. and criteria for their recognition are reviewed.

1. **Introduction**

The weathering profile is a key component of the geological cycle. It is the zone wehere the biosphere, atmosphere and hydrosphere interact with the lithosphere. It is arguably the most complex geochemical system yet most geologists receive little or no training in the study of weathering processes and how weathering profiles develop. There are a number of very good reasons why geologists should be aware of the importance of the weathering profile.

(i) Most siliclastic materials are created in weathering profiles and, in many cases, are further modified during transit from their site of formation to the site of deposition, by being incorporated in soil profiles as stored river sediment. The nature of the prevailing climate, length of time in the initial weathering profile and the length of time spent in soil profiles en route to the final site of deposition all affect the final sediment composition (e.g. see Johnsson & Meade, 1990). Clays which comprise the volumetrically most important sedimentary materials, are largely created in weathering profiles.

(ii) The weathering profile is one of the most widespread of all geological environments and forms an essential component of almost all outcrops, and is an integral part of continental environments.

(iii) Weathering profiles are the sites of formation many types of mineral concentrations including various ores and construction materials.

(iv) The nature of the weathering profile is very closely related to the climatic and geomorphic regime. Thus careful analysis of palaeo-weathering profiles can provide valuable informa-

A. Parker and B. W. Sellwood (eds.),
Quantitative Diagenesis: Recent Developments and Applications to Reservoir Geology, 95–123.

tion on palaeoatmospheric composition, palaeoclimates and palaeolandscapes, as well as clues to the nature of biological activity.

(v) Weathering profiles represent subaerial exposure surfaces and so the ability to recognize palaeo-weathering profiles is essential in sequence stratigraphic analysis, not only for recognizing different low stand systems but also for distinguishing minor (for example parasequence boundaries) from more significant 'breaks' in deposition.

The aim of this paper is to briefly review aspects of weathering profile formation and to show how quantitative techniques can be used to understand how such profiles develop, geochemically. To provide a geological dimension, the recognition and stratigraphic aspects of ancient profiles are also reviewed. The weathering profile is not only the site of the loss of materials but significant accumulations of various oxides, carbonates and sulphates also develop, particularly related to shallow groundwater activity. As such they represent a distinctive diagenetic facies which has largely been overlooked by geologists.

2. Composition of the Weathering Profile

Terminological disputes are rife in this field of study. Many workers distinguish an upper zone of biochemical weathering from a lower one of geochemical alteration (e.g. Macias & Chesworth, 1992). The difficulties in applying such a division to the pre-mid Palaeozoic examples would limit the use of such a division. There is now a concerted attempt to clear up the terminological problems via the 'Saprolite Newsletter' edited by S. Buol (North Carolina State University).

Figure 1 shows the basic division of the weathering profile used in this paper. The nature of the profile will ultimately depend on the climate, length of exposure and the composition and structure of the parent material. We can subdivide the alteration zone into two parts: soil and saprolite. The zone of soil formation takes place in the upper few metres of this profile, in the particularly biologically active part, and in general the rates of these processes are rapid and biochemical weathering is the dominant process causing the breakdown of the parent material. The rate of weathering is greatest in this zone as a result of high CO_2 content and organic acids, but much of the ions released are recycled by vegetation. Physical erosion of the soil also takes place by run off. As a consequence of pedoturbation (physical reorganization by soil processes) the original structure of the parent material is destroyed.

Beneath the soil is the saprolite, a weathering residuum which retains the structure (for example stratification) and fabric of the underlying host. The weathering in this zone leads to isovolumetric changes such that little or no physical disruption takes place. There is a transition zone between the saprolite and the soil where the products of weathering, for example clays, begin to exert physical changes, through wetting and drying, on the original rock or sediment structure and fabric.

As the parent material is weathered, in addition to chemical and physical changes, the density decreases as a result of the chemical removal of components and because of volume expansion (Fig. 1). A geochemical classification of weathering is discussed in a later section (see Fig. 3).

The main process involved in weathering is hydrolysis and hence the degree of chemical weathering is directly related to the amount of water entering the profile (ultimately controlled by climate) and the degree of biochemical activity (also climatically controlled). Recent reviews of chemical weathering have been given by Nahon (1991) and Chesworth (1992).

Meteoric leaching is commonly quoted as a means of creating significant porosity in ancient sandstones particularly at shallow depths (Bjørlykke, 1984; Shanmugam and Higgins, 1988), with

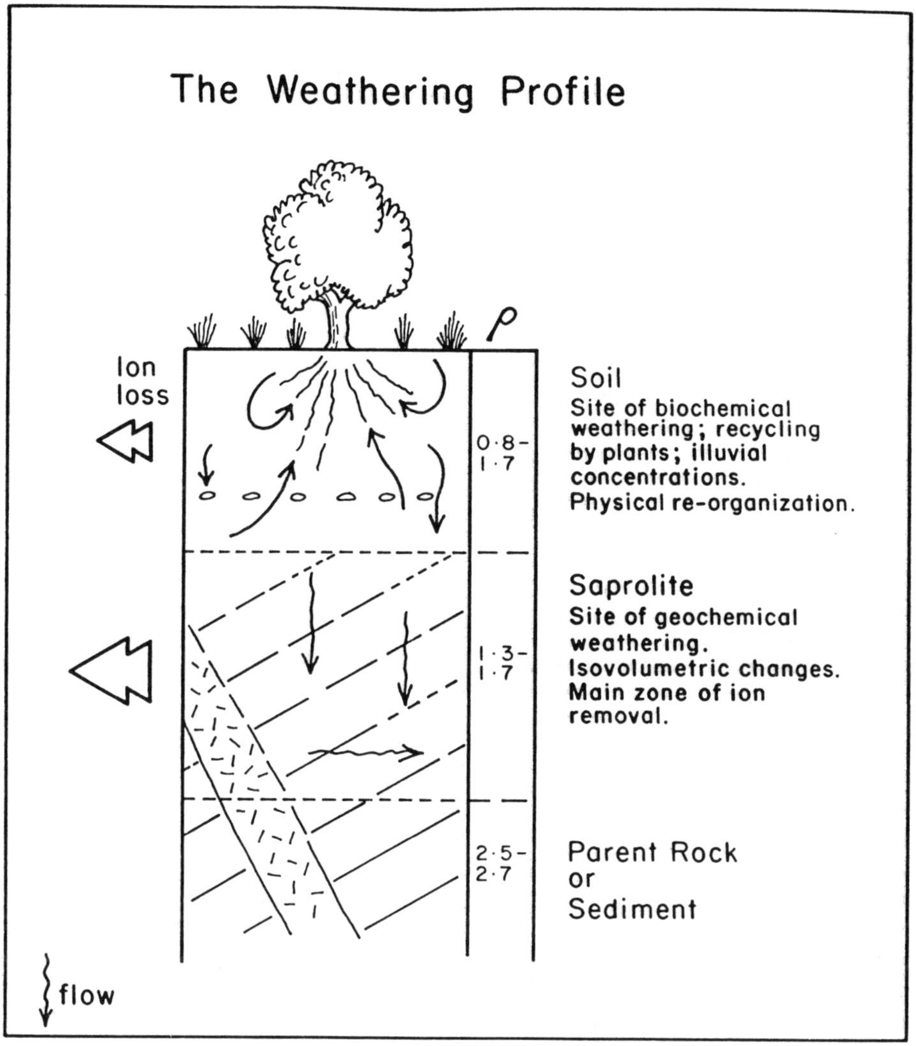

Figure 1. The weathering zone consists of two parts: the soil and the saprolite (see text). Marked density contrasts occur between zones with a general decrease upwards; volume expansion occurs from the parent material to the saprolite and soil. The term saprolite refers to a less indurated material which still retains relict macro-structure and fabric. It corresponds to the *coarse saprolite* of some authors (e.g. Nahon, 1991).

obvious economic implications. We might regard such zones as saprolites. Felspar dissolution is a major feature in such occurrences and typically creates dissolved grains in which kaolinite has precipitated (incongruent dissolution, see below). The kaolinite produced by the leaching of potassium felspar occupies, approximately, two-fifths of the volume of the original felspar grain (Bjørlykke, 1984) and the pore systems are very small. Recently Colin *et al.* (1992) have provided a porosity profile for a saprolite, although developed on Archaean gneiss from Gabon. Porosities of over 10% are encountered at depths of 50-60m below the surface, largely composed of intracrystalline porosity (10-30 m) within kaolinite masses produced by chlorite, biotite and felspar dissolution.

3. Dynamics of Weathering Profiles

Soils form more rapidly than saprolites, this is the basic reason why soils lacking prominent sub-soil weathering profiles are more widepsread than soils with thick saprolites. This is also the reason that while many continental sequences in the geological record contain numerous paleosols, palaeo-saprolites are generally restricted to major unconformity surfaces where the residence time of the material in the weathering zone may have exceeded millions of years.

The relative thicknesses and degrees of development of the components of the weathering profile depend on complex relationships between the rates of surface denudation, soil development and saprolite growth (Dethier, 1986; Pavich, 1986). Figure 2 shows that understanding the dynamics of profile growth involves appreciating the relationships between these various rates. The rate of surface denudation is typically greater than the rate of saprolite propagation, but the change in volume of material from the saprolite to the soil accounts for part of the reason that these rates appear to be in balance in many settings.

If the rate of surface denudation (L3, Fig. 2) is much higher than the rate of saprolite growth, bare, unweathered rock will result but where L3 is low, thick saprolites can develop. Such situations develop on low relief landscapes in ancient peneplain cratonic regions, such as much of central and western Australia and southern Africa. In these areas weathering has continued in some cases, since Cambrian times, while many areas have undergone extensive late Mesozoic and Tertiary weathering (Beckmann, 1983a; Bird & Chivas, 1988; Idnurm & Senior, 1978). Saprolites reach thicknesses of from tens to hundreds of metres (Beckmann, 1983a,b; Senior & Mabbut, 1979), and in the case of Australia are associated with rates of Phanerozoic craton lowering as low as 1.5-5 mm/kyr (Van der Graaff, 1981), as compared to worldwide, short-term average rates of 50-80 mm/kyr (Ritter, 1967). Tectonic, eustatic and climatic changes, of course, interupt these profoundly long periods and these old landscapes are dissected with surfaces of different ages. The saprolite history of each of the southern continents is quite different, reflecting, especially, tectonic behaviour (Beckmann, 1983b).

In summary very thick weathering profiles can only develop where rates of erosion are very low. Since rates of erosion are governed by relief, this typically occurs in regions with very low relief. However, thinner profiles can develop during less substained periods of subaerial exposure.

An additional consideration is the rate of soil thickening (L2 on Fig. 2). If the soil thickens continuously it may reach a point where it reduces infiltration of water into the saprolite to such an extent as to retard any further advance of the saprolite-substrate front.

4. Rates of Weathering Profile Formation

The rates of saprolite formation are generally low and most of the figures quoted fall in the range of 1-50 mm/Kyr. (Chesworth, 1992; Nahon, 1991, and case studies in Colman & Dethier, 1986). However, such rates are mainly short-term ones derived from geochemical studies of drainage

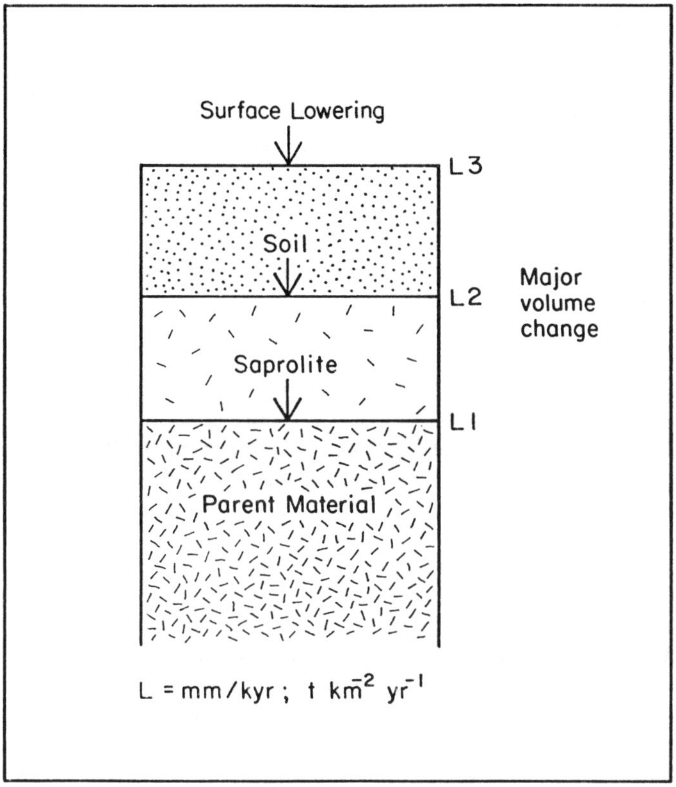

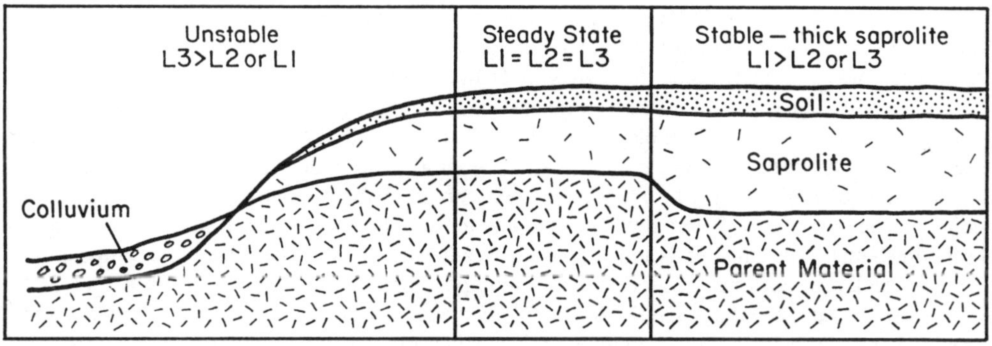

Figure 2. Components of landscape denudation and rates of erosion, soil and saprolite propagation. L1 - rate of transformation of parent to saprolite. L2 - rate of transformation of saprolite to soil. L3 - rate of surface erosion. For thick saprolites to develop the rates L1 must exceed both L2 and L3. Where the landscape is undergoing relatively rapid erosion (where L3 exceeds L2 or L1) the soil and saprolite cover will be ultimately removed.

basins. They are unlikely to be representative of long term rates where environmental changes over such long periods will have affected the processes operating. For example, the thick saprolites of southwest Queensland developed during the mid Cretaceous to early Cenozoic, a period of some 30 my. The kaolinised sediments reach a maximum thickness of 100 m giving a crude long term value of saprolite formation at 3.3 mm/Kyr.

An appreciation of saprolite development is more than just academic. As well as aiding in understanding the occurrence of laterite and bauxite deposits, it has become relevant to interpreting the occurrence of zones of extensive felspar dissolution in hydrocarbon reservoirs (Bjørkum *et al.*, 1990; Emery *et al.*, 1990).

If a zone of secondary porosity formation is related to subaerial exposure and saprolite formation, it may be predicted in the subsurface by its association with a seismically-resolvable unconformity. Such surfaces have traditionally been targets for hydrocarbon exploration. However, while some porous zones in the northern North Sea can be attributed to palaeo-saprolites (Emery *et al.*, 1990), others cannot. Bjørkum *et al.* (1990) have argued, using calculations for felspar dissolution and flow rates through the Jurassic Rannock Formation, also in the northern North Sea, that the rate of dissolution propagation and kaolinite formation would have been only a few tens of millimetres per thousand years, and much less than the estimated rates of erosion for the capping unconformity, where hundreds of metres of Jurassic sediments were eroded off during a period of a few thousand to a few tens of millions of years. They argue that to achieve the loss of sequence with the unconformity, the rate of erosion must have exceeded the rate of propogation of the dissolution front.

Zones of felspar dissolution, if due to saprolite formation, should possess a number of features. The degree of alteration should, of course, decrease with depth. The saprolite should be capped by a soil profile but truncation of weathering profiles is common and its absence is not a reliable criteria for dismissing a saprolite origin for an altered zone.

5. Quantitative Analysis of Weathering and Palaeo-Weathering Profiles

A large body of knowledge exists on the chemistry and mineralogy of weathering processes. Major reviews include the volumes edited by Colman & Dethier (1986), Lerman & Meybeck (1988), Martini & Chesworth (1992) and the review by Nahon (1991).

One particularly useful method of quantifying and displaying bulk chemical trends during weathering is the use of triangular plots of major oxides developed by Nesbitt & Young (1984, 1989), (see also Nesbitt, 1992; Rainbird *et al.*, 1990). The plots display only parts of the bulk chemical compositions of rocks (Fig. 3). They have been compiled for a range of igneous rocks; many sediments having highly heterogeneous compositions. The technique is an excellent way of displaying both weathering trends and, for palaeo-weathering profiles, also diagenetic compositional changes. Molar proportions are used to plot the compositions. Two plot types are used: the A-CN-K (felspar) diagram, and the A-CNK-FM (mafic) diagram. In the case of the former, since felspars constitute over 70% of the upper mantle, their weathering is a critical control on compositional changes (Nesbitt & Young, 1984). Nesbitt & Young (1984) used thermodynamic, mass balance and kinetic data to predict weathering trends (Fig. 3) and compared these with ones from actual weathering profiles (Fig. 4). The fit of actual and predicted evolutionary weathering trends is impressive.

One particularly interesting result from the comparison of the actual and predicted trends is that the Al of the felspars has been retained within the profile in weathering phases. This indicates that

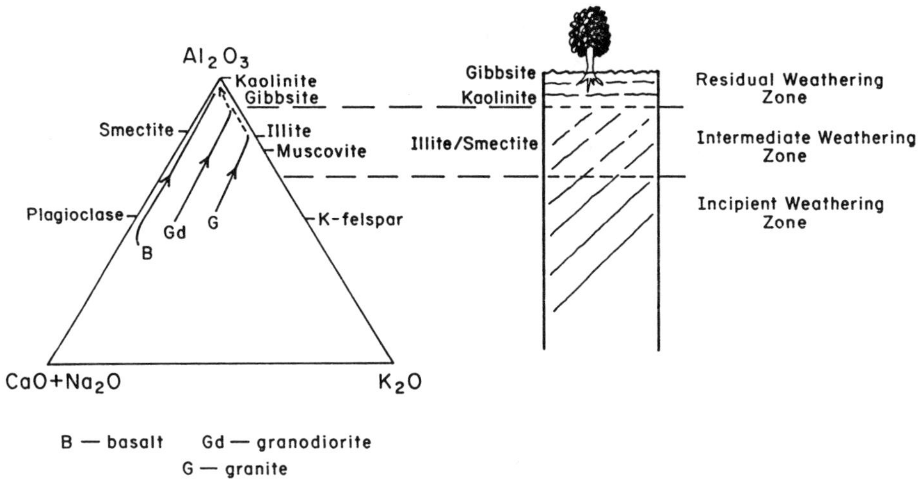

Figure 3. Geochemical classification of the weathering profile showing weathering trends for basalt, granodiorite and granite plotted on an A-CN-K (felspar) diagram. The solid arrowed lines show the compositional trends. Various phases, both primary (felspars) and secondary (smectite, etc.) are included. The weathering trend lines indicate that Ca and Na are removed from the host relative to K. This continues until most of the Ca and Na is removed from the solid phase, representing the loss of plagioclase. K-felspar, if present, is then broken down releasing K but its incongruent dissolution results in Al being retained.

Each weathering zone is produced from previously altered materials and the composition of the new mineral phases reflects the composition of weathering fluids. Near the surface, in the *residual weathering zone,* the soil solutions are acidic (reflecting local high P CO_2 levels and organic acids). The lack of solid bases in the weathered parent material results in little if any neutralisation. These fluids are called primitive weathering solutions and their low solute activity results in the formation of gibbsite and kaolinite.

The *intermediate weathering zone* has less weathered parent materials and reactions between weathering solutions and solid bases results in less aggressive solutions. The higher solute activity is associated with stability fields of such phases as smectites, illites and vermicullites. Deeper in the profile weathering solutions achieve equilibrium with respect to the host-rock minerals and little alteration takes place. For details see Nesbitt & Young, 1984.

weathering alters the felspars by incongruent not congruent (porosity-forming) dissolution (Nesbitt, 1992). Incongruent dissolution (for details see Nahon, 1991, p.53) is the replacement process where the new weathering phase develops at the site of dissolution and commonly leads to the pseudomorphing of the original mineral grain. Congruent dissolution leads to the removal in solution of the original mineral, hence creating porosity.

The weathering trends for the mafic minerals using the A-CNK-FM plot are difficult to predict but systematic trends in actual profiles do occur (Fig. 5); more Ca, Na and, K are removed than Fe and Mg for all igneous rocks, creating a trend subparallel to the left hand boundary (Fig. 5b).

This quantitative view of weathering trends allows the assessment of ancient weathering profiles (Figs. 4, 5). If a depth trend through a presumed ancient profile is comparable to that seen in a predicted or recent profile, the origin of the profile by weathering can be confirmed. However, few ancient profiles have not been altered during diagenesis and the effects of metasomatism or even of polygenetic soil development, can be readily observed from the trends (Figs. 4 & 5).

6. Stratigraphic Considerations

Thick weathering profiles require profoundly long periods of time for their formation, with at least hundreds of thousands of years being required to produce profiles a few metres thick. Surface conditions do not remain constant for such periods as a result of climatic and tectonic changes. As a consequence many weathering profiles, currently active and ancient ones, have undergone complex histories. The complexities of major weathering profile development are well illustrated by the early Tertiary forms of southern Britain and the Paris Basin (Issac, 1983a, b; Thiry, 1989).

Issac (1983a, b) has provided a picture of the polyphase weathering history of southwest England during Palaeocene to Oligocene times (Fig. 6). During the Palaeocene, deep weathering of the Cretaceous chalks and sands (Upper Greensand) resulted in the formation of thick lateritic kaolinite and flint nodule residua. These residual flints were reworked into fluvial gravels. Silcretes occur both within the deeply weathered Upper Greensand and the residual flint gravels. Some of these silcretes represent pedogenic forms, while others are groundwater in origin.

Following the Palaeocene, tectonism took place which was followed by extensive erosion which, locally, removed the Cretaceous cover and the saprolite, exposing a range of deformed Upper Palaeozoic siliciclastic sediments and granite. Some of the earlier flint nodule saprolites were reworked into fluvial gravels at the base of the sequence (e.g. the Buller's Hill and Aller Gravels). Deep strike-slip basins developed, such as the Bovey and Petrockstow basins, which became filled by Eoc-Oligocene lacustrine-swamp kaolinites and lignites. These deposits contain no reworked flint-bearing saprolites which must have been completely removed from nearby source areas. The clays were derived from deeply weathered Upper Palaeozoic sediments and granites, representing a second phase of saprolite formation. There has been much debate as to whether the kaolinite (both well ordered and disordered) may represent hydrothermally-altered granite and not saprolite, but a weathering origin seems more likely and much of the kaolinite was derived from the siliciclastic country rock as well as from the granite.

The Palaeocene weathering residues are composed of a relatively mature assemblage of quartz, kaolinite (disordered and ordered), meta-halloysite and goethite. The Eo-Oligocene residual and lake deposits contain quartz, disordered kaolinite, illite and mixed-layer clays. Issac (1982a) has interpreted these differences as reflecting the duration of weathering. The Palaeocene saprolites may represent periods of 1.5-2.0 x 10^7 years. The Eo-Oligocene weathering phase, whose assemblage may

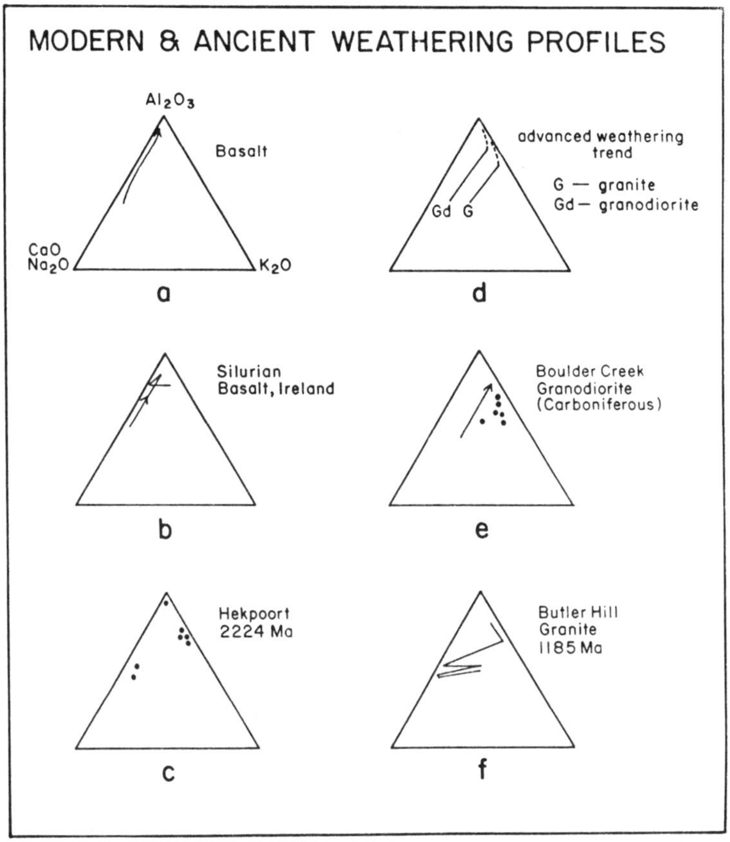

Partly based on Nesbitt & Young, 1989.

Figure 4. Felspar (A-CN-K) plot for basalts and granite/granodiorite weathering profiles. a - weathering trend for basalt. b - trend for well preserved Silurian weathering profile from Dingle, Ireland, which shows minor departure from normal trend (see Fig. 5). c - Heckport Basalt (Early Proterozoic, 2224 Ma); this profile contains illite, chlorite and quartz. The plot shows no weathering trend and more K_2O than in typical weathering profiles. The basalt has been buried to 15 km with a minimum burial temperature of 300°C. Any kaolinite or gibbsite has been lost. K has clearly been added as well as Mg and SiO_2, probably during early diagenesis. d - trends for granodiorite and granite, showing advanced trends in which all the Ca and Na-bearing minerals have been lost, followed by K-felspar, and illite. e - Boulder Creek granodiorite showing a theoretical weathering trend based on the original composition of the rock, and plots of the actual weathering profile compositions; K-enrichment has taken place and the thick profile (over 25 m) shows little kaolinite with illite as the most abundant mineral. Fresh felspar occurs even in the altered zone. The K-enrichment probably resulted in kaolinite being converted to illite and produced secondary K-felspar from kaolinite and degraded felspars. Metasomatism also took place (addition of Mg and Fe not shown here) but the lack of any enrichment in Na might suggest the fluids were continental in origin. f - Butler Hill Granite (Precambrian) showing enrichment of Na, but here not associated with elevated temperatures but possibly due to sea water alteration as the sequence is overlain by marine sediments. a, c-f from Nesbitt & Young, 1989; b - from Wright *et al.*, 1991. The plots are as molar proportions.

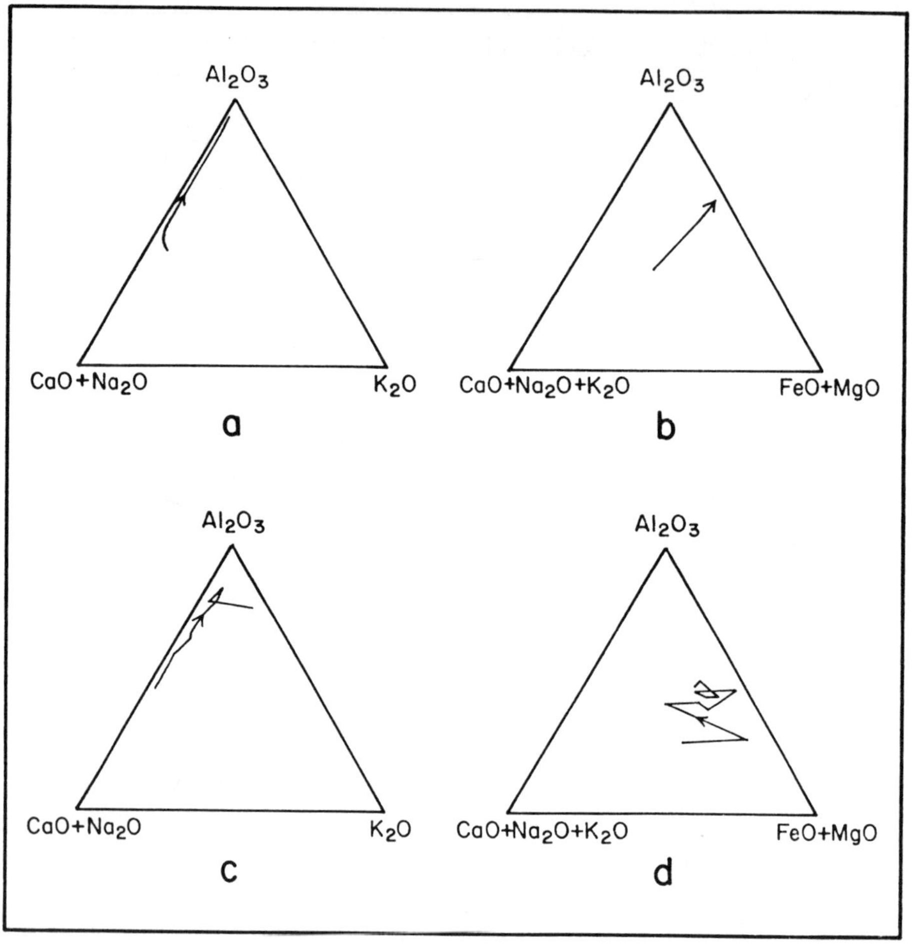

Figure 5. Comparison of weathering trends on modern and ancient basalts. a and b are A-CN-K (felspar) and A-CNK-FM (mafic) plots for the Baynton Basalt, after Nesbitt & Young, 1984. c-d are from a Silurian palaeo-weathering profile from Dingle, southwest Ireland, from data in Wright *et al.*, 1991. All plots are as molar proportions. a - shows the typical trend as Ca and Na are initially lost from the breakdown of plagioclase to produce smectite, which is later broken down to kaolinite or gibbsite. b - shows loss of felspar-bearing components, but A-CNK-FM trends show more variability than A-CN-K trends, resulting largely from the effects of oxidation states on the mobility of Fe. If weathering conditions are oxidative, ferric iron will be fixed in the profile while reducing conditions will result in ferrous iron being removed with a shift in the trend towards the Al_2O_3 apex. c - from an intra-Silurian weathering profile shows a clear weathering trend. The profile contains smectite even though the sequence has undergone actinolite-grade metamorphism. The diversion of the trend in the uppermost part of the profile reflects enrichment of K (a very common feature of most of Precambrian and Lower Palaeozoic palaeo-weathering profiles) interpreted as reflecting the addition of detrital illite due to bioturbation and mixing with overlying illitic sediments. d - the A-CNK-FM plots show no trend and is interpreted as reflecting Fe mobilisation in the original soil which probably underwent some hydromorphism and possible pseudo-gley formation.

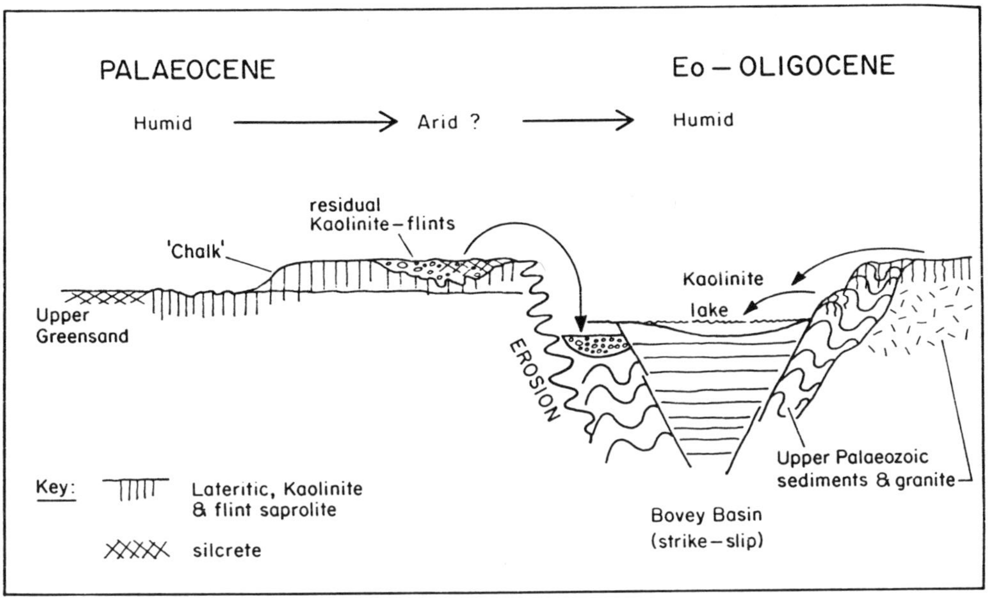

Figure 6. Schematic history of weathering profile development of southwest England (see text).

be less mature, took less, probably nearer 10^6 years. Climatic change also took place, with major cooling at some time during the Eocene. The phase of silcrete formation may represent a period of increased aridity.

There is other evidence for extensive saprolite formation during this interval. The sarsen stones of southern England are remnants of the early Tertiary silcretes, and represent both pedogenic and groundwater forms (Summerfield & Goudie, 1980). Saprolites of possible Palaeogene age have also been recorded from South Wales by Battiau-Queney (1984), including spectacular rotted sandstones and conglomerates in the Pennsylvanian 'Millstone Grit'. Thick lateritic and kaolinitic saprolites, up to 60 m. thick, occur in Britanny and are believed to represent Cretaceous and Eocene hot, humid exposure phases (Esteoule-Choux, 1983). These are also associated with silicrete horizons.

Comparisons can also be made between the saprolite history of south west Britain and the Paris Basin (Thiry, 1989). During the early Eocene deep weathering, under an alternating wet-dry regime, led to the formation of kaolinitic profiles which were reworked into floodplain and lake settings. By the middle Eocene the climate had become drier, (perhaps with a rainfall of 400 mm/yr, half that of the early Eocene, with a dry season of eight months). Silcretes developed, both pedogenic and groundwater forms, as well as calcretes and Mg-rich clays. By late Eocene times conditions had become more arid and evaporites formed in the lakes. This wet to dry trend may be comparable to that documented by Issac as the Palaeocene lateritic profiles became silcretised, but exact correlations are not possible at present.

A strikingly similar pattern is shown in the Cainozoic weathering history of Australia. From the Palaeocene to the Miocene the climate was predominantly warm and wet and resulted in deep saprolite formation (Beckmann, 1983a). Extensive pedogenic laterites formed in the Oligocene to Miocene although groundwater laterites, typically much younger, complicate the picture (see below). The alumino-silicate-rich substrates produced by deep weathering are ideal parent materials for later silcrete formation, either as pedogenic or groundwater forms. In Australia some, at least, of this extensive silcretization occurred in the Middle to Late Miocene when the climate became drier (Bird *et al.*, 1990; Milnes & Thiry, 1991). However, the actual history of saprolite/silcrete formation in Australia is highly complex, reflecting the vast time periods over which they developed (Milnes & Thiry, 1991; Ollier *et al.*, 1988).

These examples serve to illustrate the complexity of saprolite 'stratigraphy' which can arise as landscapes become dissected and as climates change. Groundwater cementation by Fe-oxides, aluminosilicates and silica can also further complicate the saprolite sequence.

Remarkably this type of polygenetic saprolite stratigraphy can even be seen in Precambrian deposits. Complex saprolites and silcretes have been described from the undeformed and unmetamorphosed Proterozoic Dubawnt and Hornby Bay Groups of the northwestern Canadian Shield by Ross and Chiarenzelli (1985). The weathering profiles occur on Archean granitoid gneisses and in lavas and volcaniclastics, representing a series of weathering events spread over a period from 1760-1666 Ma. Saprolites, representing incipient laterites, developed at a number of levels in the succession (Fig. 7). Silcretes overprinted some of these profiles but also developed in the overlying volcanic sequence, as 1-2 m lenses. Other silcretes developed in the Thelon Formation (Fig. 7), a thick (at least 2 km) sequence of sandstones and conglomerates. Here the silcretes reach thicknesses of 30

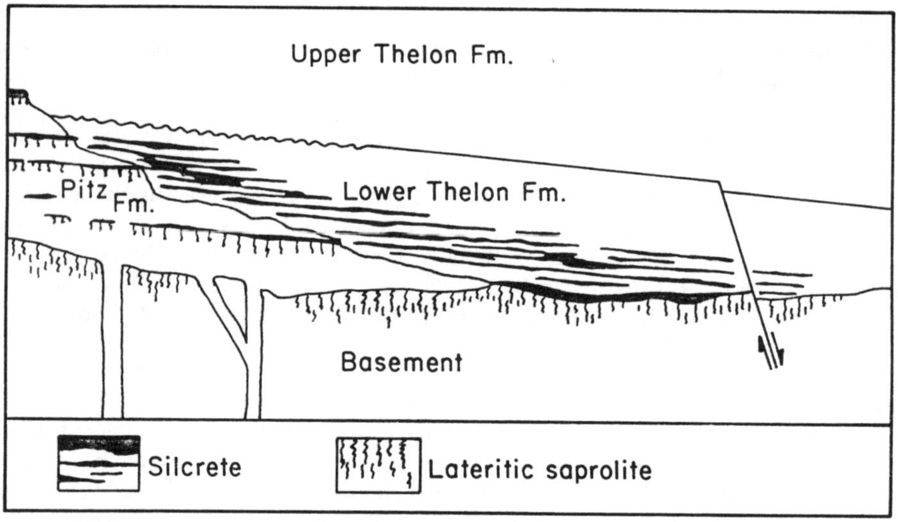

Figure 7. Weathering profile stratigraphy from the Proterozoic Dubawnt and Hornby Bay Groups of the northwestern Canadian Shield. After Ross & Chiarenzelli, 1985.

m and are polygenetic. The Hornby Bay Group represents an alluvial aeolian sequence formed in an intermontane basin. The silcretes formed both in colluvium, brecciated basement fractures, and in lacustrine deposits.

The stratigraphic distribution and form of many of the silcretes suggests a groundwater origin. The general climatic history appears to be one of initial deep saprolite formation (at least 50 m deep on the sub-Dubawnt surface), probably under a more humid climate producing lateritic profiles, followed by a later more arid phase with silcrete and evaporite genesis. Once again, a major unconformity reveals a similar set of saprolite-soil-groundwater products.

The examples given here all exhibit a humid phase overprinted by a more arid phase of weathering. This 'sequence' probably reflects the preservation potential of the profiles. Humid phases create profiles which are able to host the products of later 'arid' phases. If the reverse climatic sequence occurs the earlier arid phase may not create a significant saprolite, and the later, more humid phase, may cause erosion and replacement of any earlier materials.

7. Duripans

Mineral *concentrations* are also a major feature of weathering profiles and near surface diagenetic settings. The previous sections have concentrated on the decomposition of material in the upper part of the lithosphere but weathering and related phenomena also lead to both relative and absolute accumulations of various minerals. Such zones of partly to well indurated accumulation are termed duripans and include laterites and bauxites, calcretes, silcretes and gypcretes. Not only are these prominent features of many present day weathering profiles but they are widespread in the geological record.

A simple division can be made regarding the nature of these accumulations:

Relative accumulations are concentrations of minerals in the weathering zone associated with the corresponding loss of other, more labile components. Such minerals are the products of weathering whereby they are produced and fixed at sites while coexisting phases in the parent material have been removed. Laterites (see below) associated with weathering profiles are perhaps the best known example. Some silcretes are also produced in this way.

Absolute accumulations are concentrations of minerals resulting from the addition to pre-existing materials in near surface settings. Calcretes of pedogenic origin, for example, result largely from the concentration of calcium carbonate derived from dust. This carbonate is put into solution in the upper part of the profile and precipitated at some depth in the soil. However, extensive concentrations of Fe and Al oxides and hydroxides, various carbonates, silica and gypsum form in the zone of water-table fluctuations to create groundwater accumulates. The nature and importance of these groundwater-related accumulations are emphasized in this section because they represent a set of early diagenetic facies which appear to be more important in ancient continental deposits (particularly in red-beds) than has previously been appreciated. The most common types, laterites, calcretes and silcretes will be reviewed.

The contrasts in the origins of these two types of accumulations are best illustrated by looking at laterites.

7.1 LATERITES

Laterites are near surface concentrations of haematite and goethite, typically with Al hydrox-

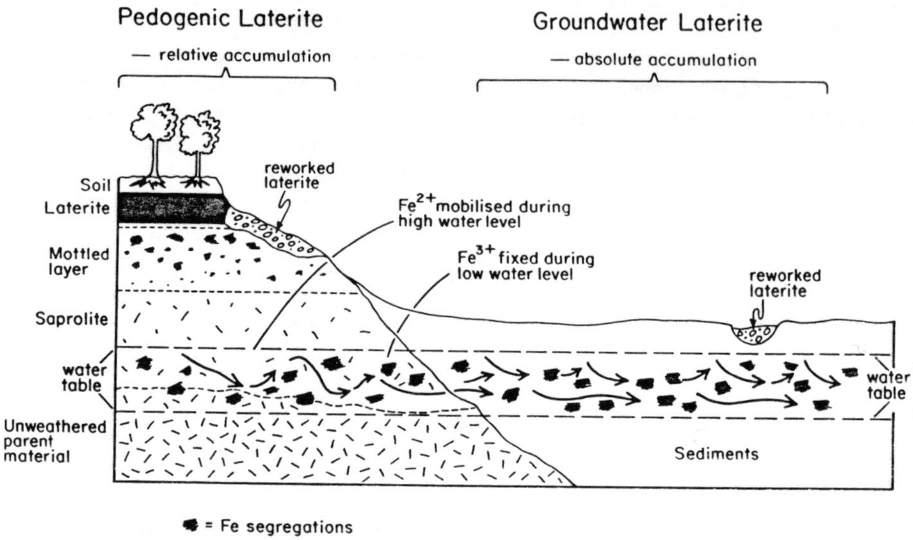

Figure 8. Major types of laterites. Pedogenic laterites form on stable landsurfaces as a result of deep weathering and the segregation of Fe oxides and oxyhydroxides. They are associated with distinctive geochemical profiles. Groundwater laterites develop around fluctuating watertables with reducing groundwaters. They usually represent the net addition of Fe to the host material or to localised segregations of finely disseminated iron oxides.
Modified from Fitzpatrick, 1988. See also Valeton, 1983.

ides, kaolinite and other materials. The term is effectively synonymous with ferricrete. Laterites in which the Al content is high enough to be economic, are called bauxites. Schellman (1982) has provided a discussion of laterite terminology.

The concentration of Fe can arise by two ways (Fig. 8): by the addition of Fe to the host material ("absolute accumulation"), typically transported in solution by groundwater, or by the 'relative accumulation' by the local re-organization of Fe due to weathering associated with extensive desilication. The former process is important in the formation of groundwater laterites, while the latter with laterites capping thick weathering profiles (pedogenic laterites).

The Fe is mainly released from the weathering of silicate minerals in rock, sediment or soil and is either transported some distance or only undergoes short distance segregations, to be precipitated as oxides, hydroxides or oxyhydroxides as a result of Eh changes.

From the nature of the source of the Fe a major distinction can be made between the pedogenic and groundwater laterites (McFarlane, 1976), although some groundwater laterites also result from local Fe segregation.

7.1.1 *Pedogenic Laterites*. As stated above the Fe is a relative accumulation, that is the Fe is a residuum and little or none has been added (Fig. 9). The Fe is concentrated by the weathering process and is associated with a characteristic weathering profile (Nahon, 1986, 1991; Herbillon & Nahon, 1988; McFarlane, 1983; Tardy, 1992; Tardy & Roquin, 1992), which may be tens of metres thick (Fig. 10). Above the unaltered parent (Fig. 10A) material is the saprolite zone in which the original lithologic structure, e.g. bedding, is retained but the rock is less indurated. No significant physical re-organisation has taken place. Iron oxyhydroxides have been released by weathering, either directly or mixed with newly formed clays (Fig. 10B).

The overlying unit is intermediate between the physically unaltered, weathered bedrock and the overlying soil. This zone, where extensive clay mineral formation has occurred, especially of

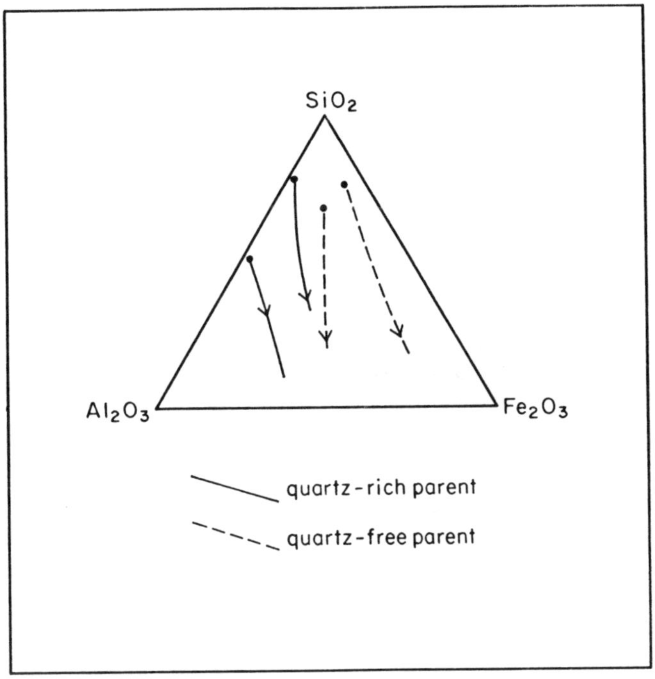

Figure 9. Weathering trends for laterite profiles. The loss of SiO_2 leads to a relative increase in the Fe and Al oxide contents. Based on McFarlane, 1987.

kaolinite, undergoes physical reorganization caused by the wetting and drying of the clays. Parent materials with a basic composition may have smectite-rich lower layers which can generate considerable deformation. However, these smectites are also converted to kaolinite with its much lower capability for expansion.

The main components here are Fe oxyhydroxides, kaolinite and relict quartz grains (Fig. 10C), and progressive mineral changes occur upwards through this zone. A key point in understanding how laterite profiles form is to appreciate that the profile is polyphase with each part having developed from the preceding one by weathering. *This first, lower part of the saprolite has been the source for overlying horizons, which show progressive changes in composition.* The Fe oxyhydroxides become progressively purer and better crystallized up-profile. The kaolinite crystals however become smaller and more poorly crystallized. This reflects the substitution of Fe in the octahedral sheets (Muller & Bocquier, 1986; Nahon, 1986). The Fe segregations become larger up the profiles while the relict quartz is commonly strongly corroded.

The segregations form as a result of minute scale movements of Fe related to localised Eh and pH changes affecting the solubility of Fe: ferrous iron is mobile, ferric iron is effectively immobile.

The segregation results in mottles and nodules set in 'bleached' zones which have lost their Fe and are composed mainly of kaolinite and quartz. According to Nahon (1986) the bleached areas are prone to translocation of material in particulate form. This results in an increase in microporosity and the resulting pores become filled with authigenic kaolinite and quartz (Fig. 10D), precipitated as void coatings and fillings. This material is then progressively replaced by iron hydroxides, typically Al-rich, and finally by haematite (Fig. 10E). The actual process of replacement has been discussed by

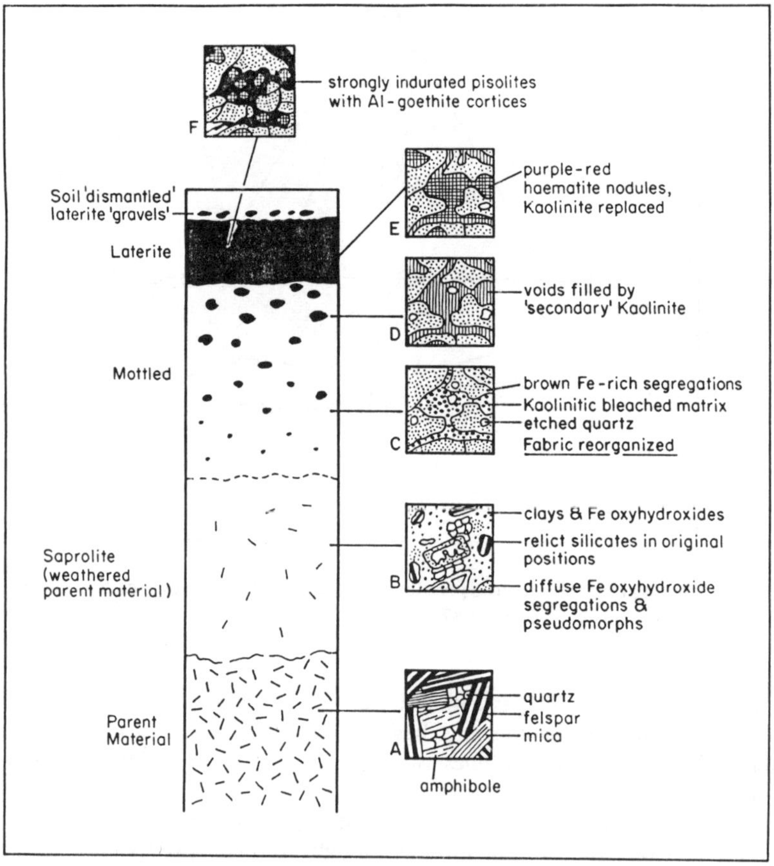

Figure 10. The laterite profile. Thicknesses range from a few metres to tens of metres. See text for details. Based on Nahon, 1986. See also Tardy, 1992.

Ambrosi *et al.* (1986) and Muller & Bocquier (1986, 1989). The Fe-rich kaolinites are destabilized in the presence of haematite and destroyed by hydrogen and hydroxyl ion activity related to reduction/oxidation reactions of Fe (a process known as ferrolysis). McFarlane (1987) has suggested that the kaolinite breaks down as a consequence of microbial dissolution.

The upper, most Fe-rich parts of the profile are typically pisolitic. The morphology and origin of these has been discussed by McFarlane (1983) and Tardy & Nahon (1985). The pisoids consist of a cortex of Al-goethite around the purple-red nodules (Fig. 10F), themselves having replaced kaolinite fills and coatings. The cortices replace the purple-red haematite nodules. The Al is probably derived from the replacement of kaolinite by haematite. This layer is highly indurated and is the most Al-rich part of the laterite profile. Details of this process are to be found in Nahon *et al.*, 1977 and Tardy & Nahon 1985.

Many laterite profiles are capped by laterite 'gravels' set in red clay matrix resulting from *in situ* chemical weathering (Nahon, 1976; Nahon *et al.*, 1977). The process involves the dissolution of pisoids of Al-goethite and the precipitation of kaolinite and pure goethite.

7.1.2 *Laterite and Bauxite (Alcretes).* Many laterites are Al-rich haematite and goethite and the further breakdown of such residual accumulations with the desilication of kaolinite, can lead to the formation of Al-oxides, hydroxides and oxyhydroxides such as gibbsite, boehmite and diaspore. These Al-rich materials are termed bauxites or alcretes and most are associated directly with laterites. However, not all bauxites occur with laterites, some 10% being associated with karst or palaeokarst (Bardossy, 1989). Such karst bauxites develop from the *in situ* weathering of previously weathered kaolonitic materials derived from lateritized areas, trapped in karst depressions.

7.1.3 *Rates of Laterite Formation.* Most pedogenic laterites are found on old, stable landsurfaces such as the cratonic areas of Africa and Australia. Many workers have stressed the long time periods required to form these profiles (McFarlane, 1976, 1983). Fritz (1975), argued, from a theoretical basis, that time periods of 1 Ma would be required to form a fully developed profile. However, many factors may act to retard or accelerate the rate of profile development (Nahon, 1986). Climate, a critical control, in particular is not stable over such extended time spans and changes in climate would invariably act to extend the time periods required. It appears that time periods of 1-10 Ma are typical for forming a full profile (Kronberg *et al.*, 1982; McFarlane, 1983; Nahon, 1986). Another measure of how slow the process is comes from assessments of the rates of weathering of the parent rocks to form the saprolite. Such rates vary from 5-50 mm/10^3yr depending mainly on rock type, the rate being about three times more rapid on basic or ultrabasic rocks than on acidic ones (Nahon, 1986).

7.1.4 *Climate.* Pedogenic laterites form under humid tropical climates with a long dry season, whereas groundwater forms develop under a much wider climatic range. In addition, many laterites are relict, particularly those of Australia (Fitzpatrick, 1988). The climate needs to be sufficiently wet to promote the breakdown and leaching of silicates but a dry season is required to *fix* the iron oxyhydroxides and oxides mobilised during the weather phases. This wet/dry alternation is also required for groundwater laterites but in this case it is the fluctuation of the watertable which acieves the change in oxidation state (see below).

7.1.5 *Groundwater Laterites.* As stated above not all laterites are pedogenic and occur with deep weathering profiles. Many, including bauxitic forms, occur not with stable landsurfaces but in lowland, commonly coastal, areas, in aggrading sedimentary sequences including deltaic and fluvial

Table 1
Differences between pedogenic and groundwater ferricretes

Pedogenic	Groundwater
Associated with distinctive vertical profile with upward increase in Fe.	Little or no "profile" development.
Associated with deeply weathered parent (saprolite).	Host may be unaltered other than Fe-oxide cemented.
Associated with kaolinitic clays.	May be associated with a wide range of clay minerals.
Exhibits variety of Fe and Al oxides and oxyhydroxides including geothite, maghemite, gibbsite.	Simple mineralogy with haematite, goethite and rare gibbsite.
Complex fabrics indicating polyphase derivation. Pisoids, with Al-goethite cortices, increase in frequency down profile.	Simple, typically single-phase fabrics. Pisoids, if present, apparently decrease in frequency downwards.
Profile may be capped by "dismantled" (brecciated) horizon with altered Al-goethite pisoids.	Top of ferricrete horizon not altered unless exhumed.
Associated with stable land surfaces.	May occur in any substrate including aggrading sediments.
Associated mainly with humid, tropical climates with long dry season but may be relict.	Found under a variety of climates including arid or semi-arid.

deposits. Such settings, with relatively high watertables, have, for example, been described from the coastal lowlands of India and South America (Patel, 1987; Valeton, 1983). The Fe is an *absolute accumulation*, having been brought in to the site by groundwater in the reduced, ferrous state.

Iron in the shallow groundwater can become fixed by oxidation during falls in the watertable (McFarlane, 1976), a seasonal climate assisting the variable movement of the watertable (Valeton, 1983). Unlike pedogenic laterites, groundwater forms can develop under a much wider range of climates, even under hypersaline, evaporitic settings (see references in Wright *et al.*, 1992).

Little information is available on the rates of formation of groundwater laterites but it seems reasonable to assume that the rate of Fe-accumulation is likely to be much more rapid than that in pedogenic forms.

These laterites are *not* associated with weathering profiles (saprolites or soils) but by virtue of forming relatively near the surface, may develop within pre-existing weathering profiles (Milnes *et al.*, 1985).

A range of occurrences of groundwater laterites has been documented from Australia and Southern Africa (Bourman *et al.*, 1987; Milnes *et al.*, 1985; Milnes *et al.*, 1987). These authors describe ferruginised bedrock and Fe-impregnated and indurated sediments, typically with relatively simple Fe-segregations and weakly altered hosts. The segregations are set in 'bleached' horizons up to a few metres thick, lacking kaolinite but with smectite, micas and felspars. The Fe concentrations appear to have formed by the segregation of finely disseminated Fe-oxides due to simple movement of Fe by groundwater under reducing conditions. Arguably if no Fe has been added by lateral flow, such segregations are not absolute accumulates.

Table 1 shows a list of criteria which can be used to distinguish the two types of laterite discussed in this chapter.

7.1.6 *Palaeo-Laterites.* Many ancient pedogenic lateritic profiles have been recorded from the geological record (McFarlane, 1983). The middle Cretaceous was a period of extensive laterite and bauxite formation (Bardossy, 1982; Nicholas & Bildgen, 1979; Sigleo & Reinhardt, 1988). Relatively few descriptions of ancient groundwater laterites have been given. Goldbery & Beyth (1984) have recorded both groundwater and pedogenic ferricretes and bauxites from the Triassic Budra Formation of southwestern Sinai. These formed under a relatively humid climate. Recently, Wright *et al* (1992) have described problematic haematite concentrations from Silurian and Permian red bed sequences from Ireland and the Spanish Pyrenees respectively. These occur as small, millimetre to centimetre-sized mottles within distal alluvial fan to playa margin sandstones and silty mudstones. Both sequences contain large volumes of readily weatherable volcaniclastics and the haematite concentrations have been interpreted as probably having formed around fluctuating watertables in lake margin settings; similar concentrations occur around some Australian acidic lakes. In this case even though the haematite concentrations are not associated with paleosols or identifiable weathering profiles, they do not appear to represent absolute accumulations of Fe-oxides but merely segregations from Fe-rich parent materials. These incipient lateritic concentrations formed under a semi-arid climate in both sequences, as evidenced by calcretes and various desiccation-related features in both sequences.

7.2 CALCRETES (AND DOLOCRETES)

Calcretes are relatively common in the geological record but the vast majority of examples documented, both Quaternary and earlier, are relatively thin, one to two metres thick nodular to massive carbonate accumulations, either calcitic (calcretes) or dolomitic (dolocrete). Wright & Tucker (1991) have provided a review of the terminology and profile characteristics of such carbonates. These represent carbonate accumulations in soils, that is, pedogenic calcretes or dolocretes. The carbonate is sourced from aeolian dust or less commonly from lithogenic carbonate in the host sediment. They are a form of absolute accumulation in most cases.

Whereas such pedogenic carbonates have been the focus of exhaustive treatment, there is

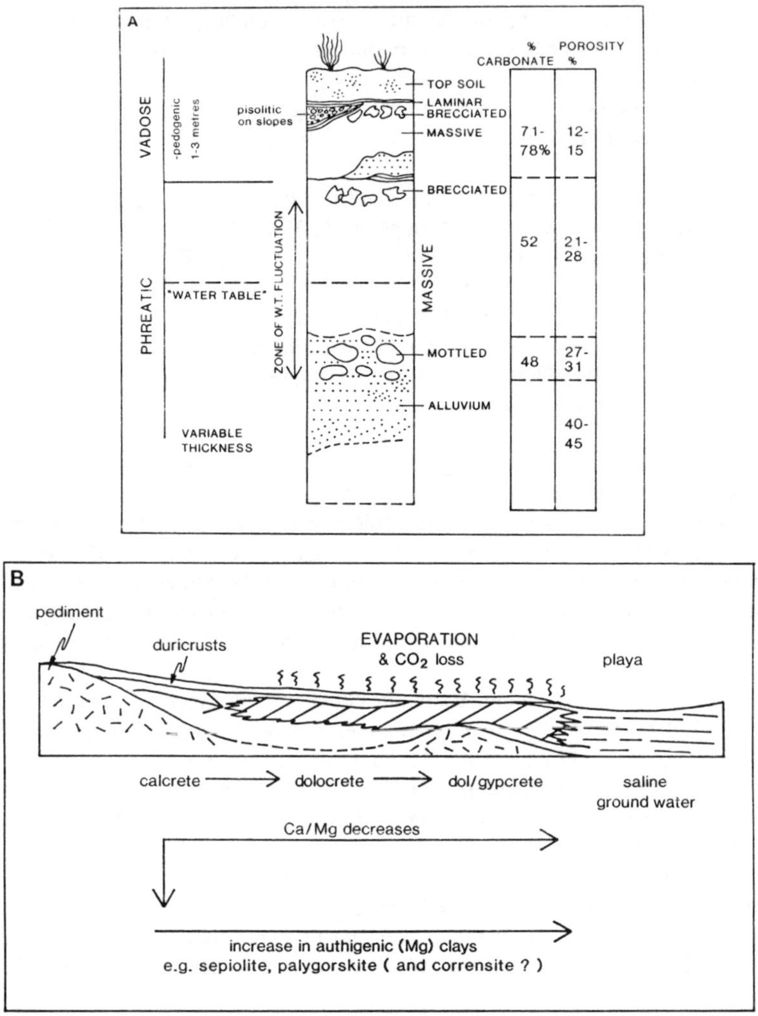

Figure 11. From Wright and Tucker 1991.

(A) Groundwater calcrete profile (based on Arakel, 1986 and other sources). Some workers (e.g. Carlisle, 1980) recognise two zones in the massive phreatic unit; an upper earthy zone with remnant soil and alluvium, and a lower 'porcellaneous', dense zone with abundant cracks and cavities.

(B) Evolution of groundwaters and their precipitates across an arid alluvial basin (see text). Based on Arakel (1986).

relatively little awareness amongst most geologists of the more spectacular forms of groundwater calcretes and dolocretes. Extensive zones of carbonate precipitation occur in some arid and semi-arid alluvial basins in Australia, as calcite and dolomite. These are closely related to zones of gypsum precipitation (gypcretes).

These zones result from the passive cementation, displacive and *replacive* growth of secondary carbonates. This result in some cases is the near total replacement of the host material including extensive dissolution of detrital silicates. The carbonate zones can reach dimensions of over 100 km long by 10 km wide with thicknesses exceeding 10 m (Carlisle, 1983; Arakel, 1986; Arakel & McConchie, 1982). Maizels (1987) records the presence of cemented Plio-Pleistocene alluvial gravels in Oman apparently over 200 m thick.

The carbonate in these zones is typically finely crystalline and locally highly porous (21-28% in some massive zones) (Fig. 11). A range of features suggesting displacive and replacive growth are present but obvious vadose features, such as meniscus or gravitational cements, are absent. Authigenic silica is commonly present (Arakel *et al.*, 1989) as well as a range of clays, especially Mg-rich forms such as sepiolite and palygorskite, and gypsum. On a local scale the carbonate zones may be lensoid in form, showing thickening to form domes which may break surface.

The carbonate zones are hosted in permeable alluvium, and typically show gradational upper and lower contacts, but some forms are capped by pedogenic carbonates. Many develop karstic features if directly subaerially exposed and their conduits become important recharge areas for groundwater (Arakel, 1991; Arakel *et al.*, 1990).

The carbonates precipitate from shallow mobile groundwaters which become progressively concentrated (particularly by evaporation) during down-dip flow. Initial groundwaters, recharged through the surrounding basement or alluvial-colluvial cover, or through exposed carbonate, is initially fresh in nature but during flow there is a progressive increase in the Mg/Ca ratio (Fig. 11), as calcite is precipitated, followed by dolomite then silica/silicates and evaporites. Evaporation is not the only cause of precipitation, with degassing of CO_2 and the common ion effect also playing a role.

Dolomite commonly forms an important component on these carbonates and is typically a primary pore-filling and fracture-lining cement or has replaced calcite, but developed where the carbonate zone is at shallow depth, in the more distal parts of the drainage lines (Arakel *et al.*, 1990; Jacobson *et al.*, 1988).

Extensive groundwater calcretes and dolocretes have also been recorded from Kuwait by Khalaf (1990) and El Sayed *et al.* (1991). Details of related gypcrete 'facies' in Australia have been given by Arakel & McConchie (1982).

Despite this occurrence of extensive groundwater calcretes and dolocretes in Quaternary drainage systems, there have been woefully few records of ancient analogues, with Tandon & Narayan (1981) and Purvis & Wright (1991) briefly describing minor possible equivalents. However, Spötl & Wright (in press) have described dolomites, up to 16 m thick, from the Upper Triassic arid alluvial deposits of the Paris Basin, which they have interpreted as groundwater dolocretes. These finely crystalline dolomites possess many textures attributable to displacive and replacive growth but only exhibit definite pedogenic textures in their uppermost one or two metres. They have high porosities (in this case bearing oil) which reflect probable early karstic solution, both in vadose and phreatic locations. Many of the pores are lined with early silica cement and silification of the dolomite also occurred. Isotopic and trace element analyses of the silica and carbonate supports an origin for both from evolved, slightly saline, non-marine groundwater. The dolocrete contains spheroidal dolomite, a highly unusual form which has also been recorded from groundwater dolocretes in Kuwait (Khalaf, 1990; El-Sayed *et al.*, 1991). It may be a characteristic feature of such dolomite

Table 2
Differences between pedogenic and groundwater calcretes and dolocretes

Pedogenic	Groundwater
Generally thin, 1-2 m. Thicker forms exhibit complex brecciated, pisolitic and peloidal fabrics.	May reach thicknesses of 10 m or more.
Typically display profile with orderly set of horizons; sharp topped.	Massive; gradational top.
Macrostructures include nodular, massive, laminar, pisolitic, prismatic.	Rarely display laminar horizons; never pisolitic or prismatic.
Rhizocretions may be common.	Rare rhizocretions.
Microfabric may exhibit beta-fabrics (biogenic microbial/root-related carbonates such as alveolar septal structure, needle-fibre calcite, *Microcodium*).	Typically alpha-fabrics (densely crystalline).
Vadose fabrics such as meniscus and pendant cements.	Absent.
Typically finely crystalline.	May show wider range of crystal sizes, including spheroidal dolomite.
Porosities are low.	Porosities locally high (>25%), typically horizontally elongate.
Rarely display lateral changes such as calcrete to dolocrete to gypcrete.	May show regular down-dip mineralogical changes.
Typically associated with more stable surfaces on floodplains.	Associated with drainage channels and playas.

zones.

Whereas examples of such well preserved groundwater dolocretes and calcretes may be rare in the geological record, early groundwater-related carbonate precipitation is a widespread phenomena in arid alluvial basins and it is likely that much of the early cementation, so commonly seen in many red-bed sequences, is attributable to the activities of shallow, mobile groundwaters. Criteria for distinguishing pedogenic from groundwater calcretes/dolocretes are given in Table 2.

7.3 SILCRETES

Silcretes are indurated zones of unconsolidated materials (sediments, saprolite, soils) cemented by various forms of secondary silica including opal, cryptocrystalline or well crystallized quartz (after Milnes & Thiry, 1992, who provide a thorough review of these materials).

Silcrete has been one of the less well understood types of precipitate related to weathering profiles, but recently, through the work of Thiry and co-workers, a clearer picture has emerged of its formation. As in the case of laterites and calcretes two broad types occur, pedogenic and groundwater silcretes. Passive cementation and replacement are the main processes operating. This is the case for laterites also and both these duripan types lack the common displacive fabrics seen in calcretes because silica and iron, being in covalently bonded compounds, favour adhesion with silicate-rich host rocks. The inability of calcite to form such bonds, reflecting the ionic bonding of calcite, results in calcite crystals preferentially forming cohesive bonds (Chadwick & Nettleton, 1990).

Milnes & Thiry (1992) have stressed that two main processes occur in silcrete formation. Silica (as opal) can form by incongruent dissolution with the transformation of clays to opal. This requires an acidic environment in which major cations including Al are more soluble from Si. In addition, congruent dissolution of silicates (detrital and authigenic) occurs which can reprecipitate as opal. The former process is characteristic of groundwater silcretes and the deeper horizons of pedogenic forms. The acidic conditions in some such silcretes may relate to local factors such as the oxidation of pyrite-rich shales. In pedogenic silcretes the dissolution is achieved by meteoric water, which can produce solutions with up to 6 ppm SiO_2. As such solutions become evaporated, silica can precipitate, its solubility decreased also by the concentration of various salts. Quartz in the main precipitate, reflecting lower concentrations of silica and other cations. In contrast silica in the lower parts of soil profiles and in the groundwater zone consists of opal, which requires higher levels of silica saturation, higher rates of precipitation and the presence of impurities (Milnes & Thiry, 1991). These differences in chemical environment are also reflected in the crystal forms observed with larger quartz crystals (but still only microquartz) and overgrowths occurring in the pedogenic forms, and high precipitation rates and impurities in the groundwater zone favouring abundant nucleii with low and hindered growth rates (with more complex fabrics including chalcedony). Thiry & Millot (1987) have discussed the distribution of silica fabrics in silcretes.

The conditions required for the formation of silcretes has been the subject of some disagreement. This arises partly because pedogenic silcretes appear not to be forming at the present time; silica does precipitate in some contemporary soils but does not appear to be forming silcretes. Another cause of the confusion may lie in the fact that silcrete can form in both the soil zone and around the watertable. Evaporation is required for pedogenic silcrete formation but exact climatic requirements are still unclear. Groundwater silicification has taken place relatively recently in many settings. Arakel *et al.* (1989) provide a detailed account of the silicification of groundwater calcretes in central Australia. The replacement of the carbonate, silica cementation and carbonate precipitation in their systems appears to take place without any significant change in the chemical environment,

Table 3
Differences between pedogenic and groundwater silcretes

Pedogenic	Groundwater
Generally 1-2 m thick.	May be slightly thicker.
May exhibit set of ordered profile features.	Lacks profile.
Exhibits range of macrofeatures - columnar, prismatic, pseudo-brecciated.	Massive, nodular.
Quartz with some opal; quartz typically microcrystalline, or as overgrowths, euhedral quartz.	Silica chiefly as opal and chalcedony.
Dissolution features present on primary and authigenic silica.	Silica has replaced clays with preservation of earlier host fabric.
Range of vadose features such as pendant and stalactitic masses. Range of illuvial features such as geopetals, grading, nodule cappings (cutans).	Cements are isopachous.
Typically as single profiles at landsurface.	May occur as multiple profiles with gradational margins

and the silicification is most marked in the zone of groundwater fluctuations. Arakel *et al.* (1989) noted that the silica lined irregular, horizontally elongated solution cavities in the calcrete, were a distinctive feature. Such cavities are an important component of the local aquifers and interestingly many of the oil-filled pores in the Upper Triassic Paris Basin groundwater dolocretes described by Spotl & Wright (in press) had the same morphologies and linings.

Table 3 provides a list of criteria for distinguishing pedogenic and groundwater silcretes.

It seems likely that pedogenic silcretes form very slowly but groundwater forms may develop more rapidly (Arakel *et al.*, 1989). In some cases groundwater forms become *younger* at depth, reflecting progressive falls in the watertable as landscapes become increasingly dissected. When this happens, earlier, higher groundwater silcretes may be affected by dissolution or even become exposed at the surface and undergo dissolution. There are very few records of pre-Cenozoic silcretes, and apparently none of groundwater forms.

8. Summary

Sediments are produced in weathering profiles, which represent one of the key diagenetic environments. The formation of these profiles can be studied in simple quantitative geochemical techniques which can also be used to identify and interpret palaeo-weathering profiles.

Duripans, indurated horizons within weathering zones or sediments, are prominent features of the near surface diagenetic realm. In the case of the three most well documented varieties, ferricretes (laterites), calcretes (and dolocretes) and silcretes, both pedogenic and groundwater forms occur. The latter represent a particular type of diagenetic facies which is not widely appreciated by geologists.

References

Ambrose,, J.P., Nahon, D. and Herbillon, A.J. (1986) 'The epigenetic replacement of kaolinite by hematite in laterite-petrographic evidence and the mechanisms involved', Geoderma, 37, 283-294.

Arakel, A.V. (1986) Evolution of calcrete in paleodrainages of Lake Napperby area, central Australia, Palaeogeogr., Palaeoclimatol., Palaeoecol., 54, 283-303.

Arakel, A.V. (1991) 'Evolution of Quaternary duricrusts in Karinga Creek drainage system, central Australian groundwater discharge zone', Austral. J. Earth Sc.i., 38, 334-347.

Arakel, A.V. and McConchie, D. (1982) Classification and genesis of calcrete and gypsite lithofacies in paleodrainage systems of inland Australia and their relationship to carnotite mineralization, J. sedim. Petrol., 52, 1149-1170.

Arakel, A.V., Jacobson, G., Salehi, M. and Hill, C.M. (1989) Silification of calcrete in palaeodrainage basins of the Australian arid zone, Austral. J. Earth Sci.., 36, 73-89.

Arakel, A.V., Jacobson, G. and Lyons, W.B. (1990) 'Sediments-water interaction as a control on geochemical evolution of playa lake systems in the Australian arid interior', Hydrobiologia, 197, 1-12.

Bardossy, G. (1982) 'Karst bauxites, bauxite deposits on carbonate rocks', Elsevier, Amsterdam, 441 pp.

Bardossy, G. (1989) 'Bauxites', in P. Bosak, D.C. Ford, J. Glazek and I. Horacek (eds.), Paleokarst: a systematic and regional review, Elsevier, Amsterdam, pp. 399-418.

Battiau-Queney, Y. (1984) The Pre-glacial evolution of Wales, Earth Surf. Proc. and Landforms, 9,

229-252.

Beckmann, G.G. (1983a) Development of old landscapes and soils. In: Soils: an Australian Viewpoint, Division of Soils CSJRO, pp. 51-72, Academic Press, Melbourne.

- (1983b) Development of soil landscapes. In: Soils: an Australian Viewpoint, pp. 163-169, Academic Press, Melbourne.

Bird, M.I., Chivas, A.R. and McDougall, I. (1990) An isotopic study of surficial-aluvite in Australia 2 : Potassium-argon geochronology, Isotope Geoscience.

Bird, M.I. and Chivas, A.R. (1988) Oxygen isotope dating of the Australian regolith, Nature, 331, 513-516.

Bjørkum, P.A., Mjøs, R., Walderhaug, O. and Hurst, A. (1990) The role of the late Cimmerian unconformity for the distribution of kaolinite in the Gullfaks Field, northern North Sea, Sedimentology, 37, 395-406.

Bjørlykke, K. (1984). 'Formation of secondary porosity: how important is it?' in D.A. McDonald and R.C. Surdam (eds.), Clastic Diagenesis. Amer. Assoc. Petrol. Geol., Mem., 37, 277-286.

Bjørlykke, K. and Aagaard, P. (1992). 'Clay minerals in North Sea sandstones', in D.W. Houseknecht and E.D. Pittman (eds.). Origin, Diagenesis and Petrophysics of Clay Minerals in Sandstones. Soc. Econ. Paleontol. Spec. Publ., 47, 65-80.

Bourman, R.P., Milners, A.R. and Oades, J.M. (1987) 'Investigations of ferricretes and related surficial ferruginous materials in parts of southern and eastern Australia', Zeitschrift Geomorph. N. F. Suppl.. Bd., 64, 1-24.

Carlisle, D. (1983) 'Concentration of uranium and vanadium in calcretes and gypcretes', in R.C.L. Wilson (ed.), Residual Deposits, Geol. Soc. Lond. Spec. Publ., 11, 185-195.

Chadwick, O.A. and Nettleton, W.D. (1990) 'Micromorphological evidence of adhesive and cohesive forces in soil cementation', in L.A. Douglas (ed.), Soil Morphology : a Basic and Applied Science, Elsevier, Amsterdam, pp. 207-212.

Chesworth, W. (1992) 'Weathering systems', in I.P. Martini and W. Chesworth (eds.), Elsevier, Amsterdam, pp. 19-40.

Colin, F., Brimhall, G.H., Nahon, D., Lewis, C.J., Baronnet, A. and Danti, K. (1992). 'Equatorial rainforst lateritic mantles: a geomembrance filter'. Geology, 20, 523-526.

Colman, S.M. and Dethier, D.P. (1986) (editors), Rates of Chemical Weathering of Rocks and Minerals, Academic Press, Orlando, 603 pp.

Dethier, D.P. (1986) 'Weathering rates and chemical flux from catchments in the Pacific northwest, U.S.A.', in S.M. Colman and D.P. Dethier (eds.), Rates of Chemical Weathering of Rocks and Minerals, Academic Press, Orlando, pp. 503-530.

El-Sayed, M.I., Fairchild, I.J. and Spiro, B. (1991) 'Kuwaiti dolocrete: petrology, geochemistry and groundwater origin', Sedim. Geol., 73, 59-75.

Emery, D., Myers, K.J. and Young, R.. (1990) 'Ancient subaerial exposure and freshwater leaching in sandstones', Geology, 18, 1178-1181.

Esteoule-Choux, J. (1983) 'Kaolinitic weathering profiles in Brittany: genesis and economic importance', in R.C.L. Wilson (ed.), Residual Deposits, Geol. Soc. Lond. Spec. Publ., 11, 33-38.

Fairbridge, R.W. and Finkl, C.W. Jnr. (1980) Cratonic erosional unconformities and peneplains', J. Geol.., 88, 69-86.

Fitzpatrick, R.W. (1988) 'Iron compounds as indicators of pedogenic processes: examples from the Southern Hemisphere', in J.W. Stucki, B.A. Goodman and U. Schwertmann (eds.), Iron in Soils and Clay Minerals, D. Reidel Publ. Company, Dordrecht, pp. 351-396.

Fritz, B. (1975) 'Etude thermodynamique des reáctions entre minéraux et solutions. Application à la géochimie des altérations et des eaux continentales', Sci. Géol. Mem., Strasbourg, 41, 152 pp.

Goldbery, R. and Beyth, M. (1984) 'Laterization and groundwater alteration phenomena in the Triassic Budra Formation, southwestern Sinai', Sedimentology, 31, 575-594.

Herbillon, A.J. and Nahon, D. (1988) 'Laterites and laterization processes, in J.W. Stucki, B.A. Goodman and U. Schwertmann (eds.), Iron in Soils and Clay Minerals, D. Reidel Publ. Company, Dordrecht, pp. 779-796.

Idnurm, M. and Senior, B.R. (1978) 'Palaeomagnetic ages of late Cretaceous and Tertiary weathered profiles in the Eromanga Basin, Queensland', Palaeogeogr.., Palaeoclim..., Palaeoecol., 24, 263-277.

Issac, K.P. (1983a) 'Tertiary, lateritic weathering in Devon, England, and the Palaeogene continental environment of South West England', Proc. Geol. Assoc., 94, 105-114.

Issac, K.P. (1983b) 'Silica diagenesis of Palaeogene residual deposits in Devon, England, Proc. Geol. Assoc., 94, 181-186.

Jacobson, G., Arakel, A.V. and Chen, Y. (1988) 'The central Australian groundwater discharge zone : evolution of associated calcrete and gypcrete deposits, Austral. J. Earth Sci., 35, 549-565.

Johnsson, M.J. and Meade, R.H. (1990) 'Chemical weathering of fluvial sediments during alluvial storage: the Macuapanim Island point bar, Solimóes River, Brazil', J. sedim. Petrol., 61, 827-842.

Khalaf, F.I. (1990) 'Occurrence of phreatic dolocrete within Tertiary clastic deposits of Kuwait, Arabian Gulf, Sedim. Geol., 68, 223-239.

Kronberg, B.I., Fyfe, W.S., McKinnon, B.J., Couston, J.F.., Stilianüdi Filho, B. and Nash, R.A. (1982) 'Model for bauxite formation : Paragominas (Brazil)', Chemical Geology, 35, 311-320.

Lerman, A. and Meybeck, M. (1988) Physical and Chemical Weathering in Geochemical Cycles, NATO ASI Ser. C., 251, 375 pp.

Macias, F. and Chesworth, W. (1992) 'Weathering in humid regions, with emphasis on igneous rocks and their metamorphic equivalents', in I.P. Martini and W. Chesworth (eds.), Weathering, Soils and Paleosols, Elsevier, Amsterdam, pp. 283-306.

Maizels, J.K. (1987) 'Plio-Pleistocene raised river channel systems of the western Shargiya (Wahiba), Oman', in L.E. Frostick and I. Reid (eds.), Desert Sediments : Ancient and Modern, Geol.. Soc. Lond. Spec. Publ., 35, 31-50.

McFarlane, M.J. (1976) 'Laterite and Landscape', Acad. Press, Lond., 151 pp.

McFarlane, M.J. (1983) Laterites, in A.S. Goudie and K. Pye (eds.), Chemical Sediments and Geomorphology, Acad. Press, Lond., pp. 7-58.

McFarlane, M.J. (1987) 'The key role of micro-organisms in the process of bauxitisation', Modern Geology, 11, 325-344.

Milnes, A.R. and Thiry, M. (1992) Silcretes, in I.P. Martini and W. Chesworth (eds.), Weathering, Soils and Paleosols, Elsevier, Amsterdam, pp. 349-377.

Milnes, A.R., Bourman, R.P. and Northcoate, K.H. (1985) 'Field relationships of ferricretes and weathered zones in southern South Australia: a contribution to 'laterite' studies in Australia', Austr. J. Soil Res., 23, 441-465.

Milnes, A.R., Bourman, R.P. and Fitzpatrick, R.W. (1987) 'Petrology and mineralogy of 'laterites' in southern and eastern Australia and Southern Africa', Chemical Geology, 60, 237-250.

Muller, J-P. and Bocquier, G. (1986) 'Dissolution of kaolinites and accumulation of iron oxides in lateritic-ferruginous nodules: mineralogical and microstructural transformations', Geoderma, 37, 113-136.

Muller, J-P. and Bocquier, G. (1987) 'Textural and mineralogical relationships between ferruginous nodules and surrounding clayey matrices in a laterite from Cameroon', Proceedings of the

International Clay Conference, Denver, 1985 (ed. by L.G. Schultz, H. van Olphen and F.A. Mumpton), Clay Minerals Society, Indiana, 186-194.

Mustard, P.S. and Donaldson, J.A.. (1990) Paleokarst breccias, calcretes, silcretes and fault talus breccias at the base of Upper Proterozoic "Windermere" strata, northern Canadian Cordillera, Journal of Sedimentary Petrology, 60, 525-539.

Nahon, D.B. (1976) 'Cuirasses ferrugineuses et encroûtements calcaires an Sénégal occidental et en Mauritanie', Science Géologie Mémoir, Strasbourg, 44, 232 pp.

Nahon, D.B. (1986) 'Evolution of iron crusts in tropical landscapes', in S.M. Coleman and D.P. Dethies (eds.), Rates of Chemical Weathering, Academic Press, New York, 603 pp.

Nahon, D.B. (1991) Introduction to the Petrology of Soils and Chemical Weathering, Wiley Interscience, New York, 313 pp.

Nahon, D., Janot, C., Karpoff, A.M., Paquet, H.. and Tardy, Y. (1977) 'Mineralogy, petrography and structures of iron crusts (ferricretes) developed on sandstones in the western part of Senegal', Geoderma, 19, 263-277.

Nesbitt, H.W. (1992) 'Diagenesis and metasomatism of weathering profiles, with emphasis on Precambrian paleosols', in I.P. Martini and W. Chesworth (eds.), Weathering, Soils and Paleosols, Elsevier, Amsterdam, pp. 127-152.

Nesbitt, H.W. and Young, G.M. (1984) 'Prediction of some weathering trends of plutonic and volcanic rocks based on thermodynamic and kinetic considerations', Geochim. Cosmochim., Acta, 48, 1523-1534.

Nesbitt, H.W. and Young, G.M. 1989 'Formation and diagenesis of weathering profiles', Jour. Geol., 97, 129-147.

Nicholas, J. and Bildgen, P. (1979) 'Relations between the location of karst bauxites in the North Hemisphere, the global tectonics and climatic variations during geological time', Palaeogeogr., Palaeoclimatol., Palaeoecol., 29, 205-239.

Ollier, C.D., Chan, R.A., Craig, M.A. and Gibson, D.L. (1988) 'Aspects of landscape history and regolith in the Kalgoorlie region, Western Australia', BMR J. Aust. Geol. Geophys., 10, 309-321.

Patel, P.K. (1987) 'Laterization and bentonitization of basalt in Kutch Gujarat State, India', Sedimentary Geology, 53, 327-346.

Pavich, M.J. (1986) 'Processes and rates of saprolite production and erosion on a foliated granitic rock of the Virginia Piedmont', in S.M. Colman and D.P. Dethier (eds.), Rates of Chemical Weathering of Rocks and Minerals, Academic Press, Orlando, pp. 551-590.

Purvis, K. and Wright, V.P. (1991) 'Calcretes related to phreatophytic vegetation from the Middle Triassic Otter Sandstone of South West England', Sedimentology, 38, 539-551.

Rainbird, R.H., Nesbitt, H.W. and Donaldson, J.A. (1990) 'Formation and diagenesis of a sub-Huronian saprolith : comparison with a modern weathering profile', Jour. Geol., 98, 801-822.

Ritter, D.F.. (1967) 'Rates of denudation', Geol. Education, 14, 154-159.

Ross, G.M. and Chiarenzelli, J.R. (1985) 'Paleoclimatic significance of widespread Proterozoic-silcretes in the Bear and Churchill Provinces of the Northwestern Canadian Shield', J. sedim. Petrol.., 55, 196-204.

Schellman, W. (1982) 'Eine neue Laterit-definition', Geol. Jaheb., 58, 31-47.

Senior, B.R. and Mabbut, J.A. (1979) A proposed method of defining deeply weathered rock units based on regional geological mapping in southwest Queensland', J. geol. Soc. Austr., 26, 237-254.

Shanmugam, G. and Higgins, J.B. (1988). 'Porosity enhancement from chert dissolution beneath Neocomian unconformity: Ivishak Formation, North Slope', Alaska. Amer. Assoc. Petrol. Geol.

Bull., 72, 523-535.

Sigleo, W. and Reinhardt, J. (1988) 'Paleosols from some Cretaceous environments in the southeastern United States', Geol. Soc. Amer. Spec. Publ., 216, 123-142.

Spötl, C. and Wright, V.P. (1993) 'Groundwater dolocretes from the Late Triassic of the Paris Basin, France : a case study of an arid, continental diagenetic facies', Sedimentology (in press).

Summerfield, M.A. and Goudie, A.S. (1980) 'The sarsens of southern England: their palaeoenvironmental interpretation with reference to other silcretes', in D.K.C. Jones (ed.), The Shaping of Southern England, Academic Press, London, pp. 71-100.

Tandon, S.K. and Narayan, D. (1981) 'Calcrete conglomerate, a case-hardened conglomerate and cornstone - a comparative account of pedogenic and non-pedogenic carbonates from the continental Siwalik Group, Punjab, India', Sedimentology, 28, 353-367.

Tardy, Y. (1992) 'Diversity and terminology of lateritic profiles', in I.P. Martini and W. Chesworth (eds.), Weathering, Soils and Paleosols, EWlsevier, Amsterdam, pp. 379-405.

Tardy, Y. and Nahon, D. (1985) 'Geochemistry of laterites, stability of Al-oethite, Al-hematite, and Fe^{3+}- kaolinite in bauxites and ferricretes : an approach to the mechanism of concretion formation', Amer. J. Sci., 285, 865-903.

Tardy, Y. and Roquin, C. (1992) 'Geochemistry and evolution of lateritic landscapes, in I.P. Martini and W. Chesworth (eds.), Weathering, Soils and Paleosols., Elsevier, Amsterdam, pp. 407-443.

Thiry, M. (1989) 'Geochemical evolution and palaeoenvironments of the Eocene continental deposits in the Paris Basin', Palaeogeogr., Palaeoclim., Palaeoecol., 70, 153-163.

Thiry, M. and Millot, G. (1987) 'Mineralogical forms of silica and their sequence of formation in silcretes', J. sedim. Petrol., 57, 343-352.

Thiry, M. and Milnes, A.R. (1991) 'Pedogenic and groundwater silcretes at Stuart Creek Opal field, South Australia', J. sedim. Petrol., 61, 111-127.

Valeton, I. (1983) 'Palaeoenvironment of lateritic bauxites with vertical and lateral differentiation', in R.C.L. Wilson (ed.), Residual Deposits, Geol. Soc. Lond., Spec. Publ., 11, 77-90.

Wright, V.P. and Tucker, M.E. (1991) 'Calcretes - an Introduction', Intern. Assoc. Sedimentol. Reprint Vol. 2, Blackwell Scientific, Oxford, p. 1-22.

Wright, V.P., Sloan, R.J., Garvie, L.A.J. and Rae, J. (1991) 'A polygenetic palaeosol from the Silurian (Wenlock) of Dingle, southwest Ireland', J. geol. Soc. Lond., 148, 849-859.

Wright, V.P., Sloan, R.J., Valero Garces, B. and Garvie, L.A.J. (1992) 'Groundwater ferricretes from the Silurian of Ireland and Permian of the Spanish Pyrenees', Sedim. Geol., 77, 37-49.

MICROBIAL AND ORGANIC PROCESSES

C.J. CLAYTON
Exploration and Production Division
BP Research Centre
Chertsey Road
Sunbury-on-Thames
Middlesex, TW16 7LN
UK

ABSTRACT. This paper discusses the main sedimentary bacterial reactions which produce diagenetic precipitates. Sedimentary organic matter is derived from algae, bacteria and higher land plants. Once deposited, it is rapidly reworked by bacteria and inorganic reactions into insoluble kerogen. Microbial processes in sediments are crudely stratified, with the most energy efficient processes occurring shallowest in the sediment. An aerobic bacterial assemblage dominates surface sediments, followed in sequence by suboxic bacterial groups (nitrate, iron and manganese reduction), sulphate reduction, and eventually pure fermentation. Fermentation occurs at all depths providing the starting biochemicals for the above respiratory processes. The tangible by-products of the microbial reactions (*e.g.* carbonates) are rarely the product of a single bacterial reaction: they reflect the combined effect of bacterial consortia, frequently cycling inorganic materials between bacterial zones. The final products also depend on both lithology and sedimentary environment.

1. Introduction

1.1. WHY STUDY ORGANIC COMPOUNDS?

Sediment diagenesis is primarily concerned with inorganic mineral phases. Most studies have concentrated on the inorganic precursors to these and, of course, to the products, but few have paid more than passing attention to associated bacterial reactions. But bacterial reactions are the driving force behind many of the reactions occurring during early diagenesis, and organic matter is generally the fuel for such reactions.

All organic matter is ultimately the product of photosynthesis, the storage of solar energy into complex organic molecules. All bacterial reactions rely on this source of energy. They progressively decompose the complex molecules into simpler and simpler compounds and use the resulting energy released for motility and growth. In doing so they frequently convert some inorganic compounds into different energy states, offering other bacteria the opportunity to gain energy by inorganic reactions converting these back to their original state. In this sense, the bacterial community in sediments acts as a vast, complex energy generator. Many of the reactions which are observed would also occur abiotically, since they always result in a *net* release of energy and hence are thermodynamically favourable. But usually such reactions are

A. Parker and B. W. Sellwood (eds.),
Quantitative Diagenesis: Recent Developments and Applications to Reservoir Geology, 125–160.

either inhibited because one key step in the pathway requires an input of energy, or they are too slow, even on geological timescales. We can therefore consider the bacteria as catalysts to the key reactions of early diagenesis. They by-pass energy barriers in the general flow from complex precursors to simple products and speed up other reactions. If we are to explain the array of reactions which occur during diagenesis, we must both acknowledge and understand the vital role played by the bacteria.

1.2. WHAT WILL YOU FIND IN THIS CHAPTER?

In order to understand bacterial diagenesis we must first know something of their food. We will start with a description of organic matter. This itself is not a simple task. Organic matter found in sediments is not just a mixture of the compounds found in the organisms living in the water column above but is altered into a range of new compounds, some simpler than the starting material, some more complex. Ultimately, the main organic material remaining is *kerogen*, an apparently insoluble residue, whose mechanism of formation is still incompletely understood. Nevertheless, we will touch on the subject of kerogen formation as far as it influences our understanding of inorganic reactions.

Next we will move on to consider the bacteria themselves. This also is a vast subject but a superficial look will be sufficient to show how the different groups of organisms function before we move on to the more difficult subject of how they interact. As we shall see, *all the by-products of diagenesis we observe as mineral phases are the result of bacteria working as consortia, not as individuals*. This is illustrated by two simple examples, although we must bear in mind that in reality the situation is much more complex than even these can show.

2. Organic matter in sediments: what are we made of?

2.1. ORGANIC MATTER SUPPLY TO SEDIMENTS

Anything which dies can become the focus for bacterial reworking in the sediment. However, quantitatively we need only concern ourselves with a relatively small range of organisms which totally dominate sedimentary organic matter (Fig. 1).

The basis of all living organisms is photosynthesis, which uses solar energy to fix carbon from atmospheric carbon dioxide. All primary productivity is therefore from photosynthetic organisms: algae, higher land plants, and some specialised photosynthetic bacteria (such as the so called "blue-green algae" and the purple bacteria). These are the main contributors to sedimentary organic matter. Secondary production consists of organisms which feed on the primary producers (*e.g.* us!). Copepods (e.g. krill) are the most important of these and contribute a little to sediments although not very much, for two reasons. Firstly, the food chain is very inefficient, with only about 10% of the carbon being passed on (the rest is converted to energy, which is, after all, why we eat). By the time a whale eats the krill which fed on the algae, only 1% of the carbon is left. Secondly, many of the chemical components are passed on directly without alteration, so the chemistry of the secondary producers is not very different to that of the primary

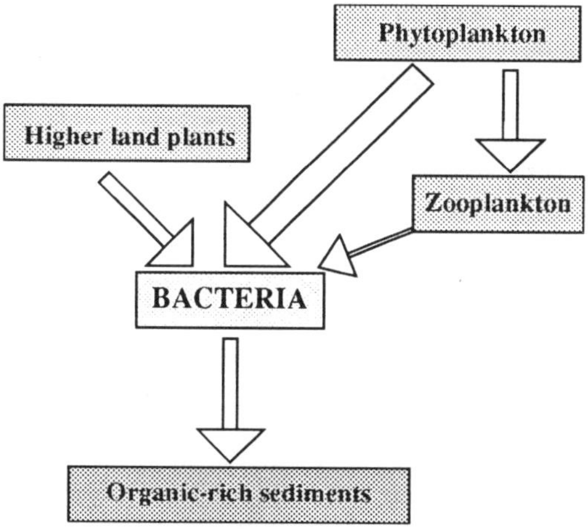

Figure 1. Main organic contribution to sediments

producers, with the exception of a few very specific trace compounds. Finally, much of the organic matter in sediments consists of bacteria themselves which, while oxidising primary organic matter, convert some of it to bacterial biomass. Therefore, we are concerned almost entirely with algae, bacteria and higher land plants.

2.2. BIOCHEMICAL CONSTITUENTS

Organic matter in sediments is composed principally of carbon, hydrogen and oxygen, with lesser amounts of nitrogen, sulphur and, in recent sediments, phosphorus. Numerous other trace elements occur in living organisms (see Table 1) although most are rapidly recycled into pore waters soon after deposition. Quantitatively these can be considered insignificant to organic diagenesis, although they will obviously contribute the elemental concentrations of pore water and may end up in diagenetic phases.

TABLE 1. The elemental composition of food (after Robins, 1980)

	Element	Occurrence	Association with other elements
Major:	Carbon Hydrogen Oxygen	Carbohydrates, lipids, essential oils	C,H and O always occur together in food, never singly
	Nitrogen Phosphorus Sulphur	Amino acids, nucleic acids, flavour and odour components	Always with C,H, O, either singly or in any combination
Minor:	Iron	Haemoglobin	C,H,O,N
	Calcium	Bone	P,O
	Sodium Potassium Chlorine	Sodium & potassium chloride in body fluids	
	Fluorine	Tea	Na
	Iodine	Thyroxin, fish	C,H,O
Trace:	Cobalt	Vitamin B_{12}	C,H,O,N
	Copper Manganese Zinc	traces occur in most foodstuffs	

In considering bacterial reactions it is more convenient to deal with organic matter in terms of the major compound groups involved, although the elemental composition *is* used as a basis of geochemical classification (see below). The main compound groups of interest are the *carbohydrates, proteins, lipids,* and in the case of higher land plants, *lignins* and related tannins. The contents of these in the main biological contibutors to sedimentary organic matter are shown in tables 2 & 3.

TABLE 2. Average dehydrated composition of main biological groups.

	Lipid	Lignin	Carboh'te	Protein	Other*
Wood & leaves	4	31	55	3	7
Spores	50	-	8	8	34
Phytoplankton	10	-	60	23	7
Zooplankton	15	-	5	53	27

*Mainly cell wall components such as sporopollenin, cutin etc.

TABLE 3. The main constituents of marine plankton and selected bacteria as percent dry weight. The remainder are inorganic components (Krey, 1970; Schlegel, 1969).

	Lipids	Proteins	Carbohydrates
Diatoms	2-10	24-48	0-31
Dinoflagellates	2-6	41-48	6-36
Copepods	5-19	71-77	0-4
Bacteria:			
Micrococcus sp.	28-37	50	15-20
Purple bacteria	40-50	50	5-30

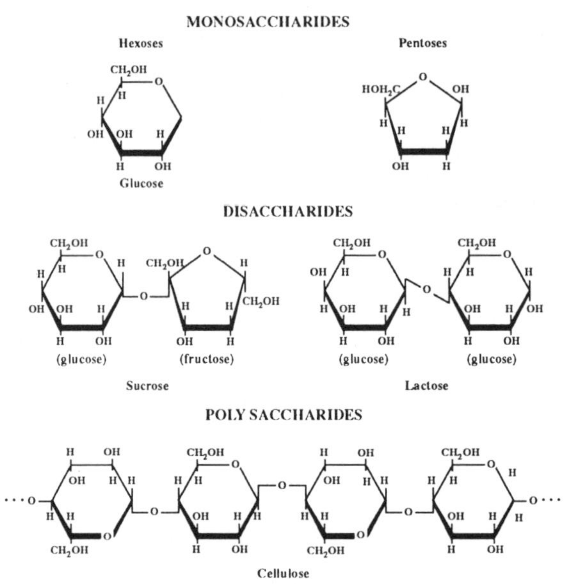

Figure 2. Structures of typical carbohydrates

Carbohydrates. Also known as the saccharides, or simply sugars, they are the main sources of energy for organisms and also form part of the supporting structure of most algae. They include compounds such as sucrose (table sugar), lactose (in milk), starch and cellulose.

The basic carbohydrate unit is an oxygen-containing ring structure with either five members (pentoses) or six (hexoses). Most commonly occurring carbohydrates are polymers of the basic *monosaccharide* units, either with two rings joined (*e.g.* sucrose is a glucose unit joined to a fructose unit) or more (Fig. 2). It is the manner in which the individual sugars are linked that provides the basis of classification. In addition, some carbohydrates may contain nitrogen (*e.g.* chitin) or other organic acid groups (*e.g.* pectin).

Carbohydrates are generally chemically neutral so can be dissociated by fermentation into an alcohol and CO_2, a factor exploited by many groups of bacteria and yeasts.

Proteins. Proteins are highly ordered polymers comprising a small number of individual amino acids (Fig. 3). These individual components are linked into chains via peptide bonds to give a structure which is generally fibrous. Protein is best known as muscle ("meat"), although silk and sponge are also proteins, as are most enzymes (biological washing powders contain proteins as well as washing soda). There are about 20 main amino acids and the full variety of proteins are composed of specific sequences of these, each protein frequently containing only a few of the amino acids. Many bacteria are able to break down proteins to their individual amino acids which may then be assimilated by other bacteria into their own cell mass.

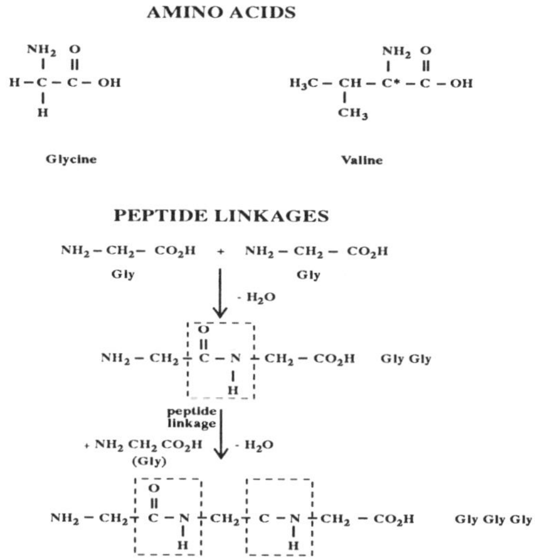

Figure 3. Structure of amino acids and proteins.

Lipids. Geochemically these are the most important compounds because of their relative stability and variability, although they often represent only a few percent of the organism. They are better known as fats, waxes and oils and show a very great diversity (Fig. 4), grouped together only because they are practically insoluble in water but soluble in one or more of the common organic solvents. They are present in algae as stiffeners in the cell membrane and in some higher land plants as waxy leaf coatings which prevent water loss.

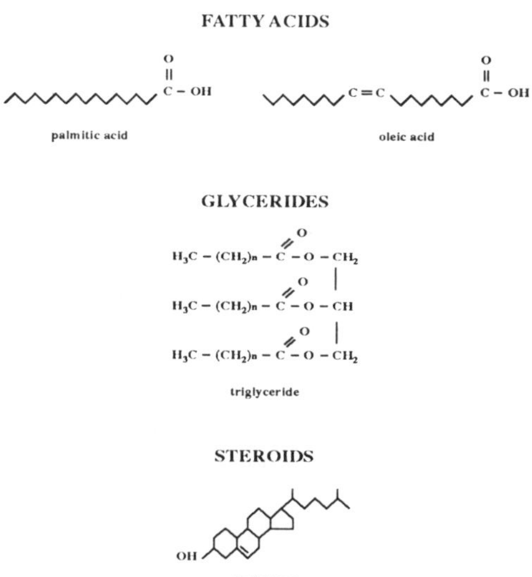

Figure 4. Structure of some typical lipids.

Lipids consist of long chains or rings of carbon atoms, composed of sub-units of the isoprene unit. Normal alkanes are lipids, as are all fatty acids (including cholesterol for example) and lipids are an important part of many natural pigments such as chlorophyll, haemoglobin, and carotene. Two groups of lipids of particular interest to geochemists are the hopanes (or triterpanes) and the steranes, which are derived ultimately from bacteria and algae respectively.

Lipids are considered to be the main precursors of oil when organic-rich sediments reach thermal maturity.

Lignins and tannins. These are highly complex compounds consisting prodominantly of aromatic ring complexes linked by short alkyl chains (Fig. 5). The detailed structure is not known with any certainty and a number of models have been proposed, although the details are of little significance to understanding most organic reactions in sediments. Lignin is particularly resistant to bacterial degradation (although it is degraded by some aerobic fungi), and so is an important constituent of most sediments after bacterial reworking and removal of the more labile algal/bacterial components. Lignin is the most important constituent of coals, and is the reason that coals generally yield gas rather than oil on reaching thermal maturity for petroleum generation.

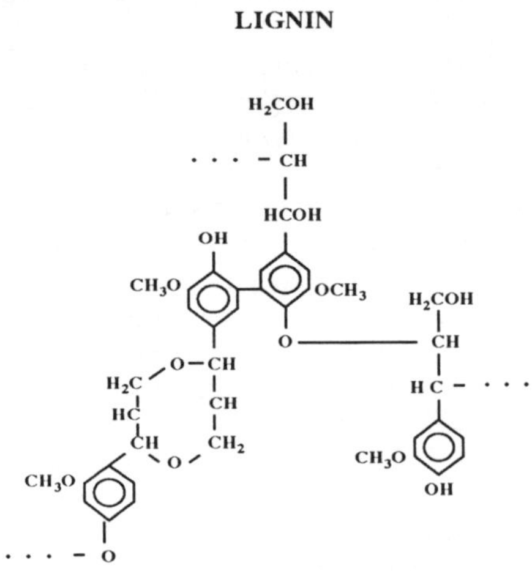

Figure 5. Structure of lignin.

2.3. FORMATION OF KEROGEN

Kerogen is defined as the organic material in sedimentary rocks which is insoluble in normal organic solvents such as hexane of dichloromethane Durand (1980). Nearly all organic matter in sediments, below the upper few tens of metres, is present as kerogen, along with a small amount of soluble organic matter (the "total soluble extract").

Kerogen is a highly complex, high molecular weight "geopolymer" whose structure is still poorly understood, and its mechanism of formation even less so. Part of the problem lies in the fact that there is a race within the sediment between bacteria trying to recycle the biological components (particularly carbohydrates and proteins) and the formation of more resistant kerogen (Fig. 6). We know for example that proteins are rapidly lost, as demonstrated by a decrease in the N/C ratio of the bulk organic matter (Fig. 7) and if the concentrations of the main biochemical groups are measured we see a comparable decrease in the carbohydrates and, to a lesser extent,the lipids (Fig. 8). However, just because a certain precursor disappears from a sediment during diagenesis is not a reason for assuming that it has passed into the kerogen structure. We do know, however, that kerogen is comparatively enriched in lipids and lignin compared with organisms. Current work is even suggesting the possibility that kerogen is really just a residue which is left after bacterial reworking and is little changed from some of the original biochemical constituents. Other work has attibuted kerogen formation to the "melanoidin" reaction in which sugars (from carbohydrates) and amino acids (from proteins) condense to form a more complex mixed polymer (similar to the manufacture of chocolate). Quite how lipids and lignins fit into the picture is less clear however.

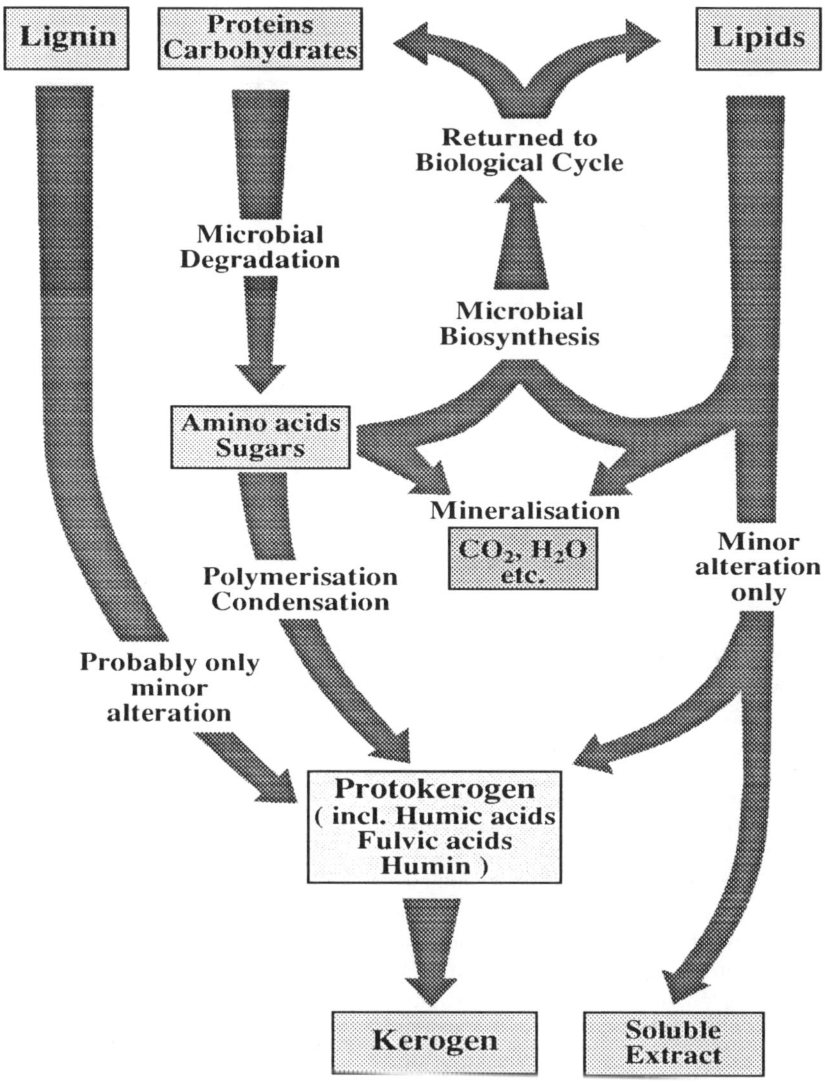

Figure 6. Early diagenetic process occurring during kerogen formation.

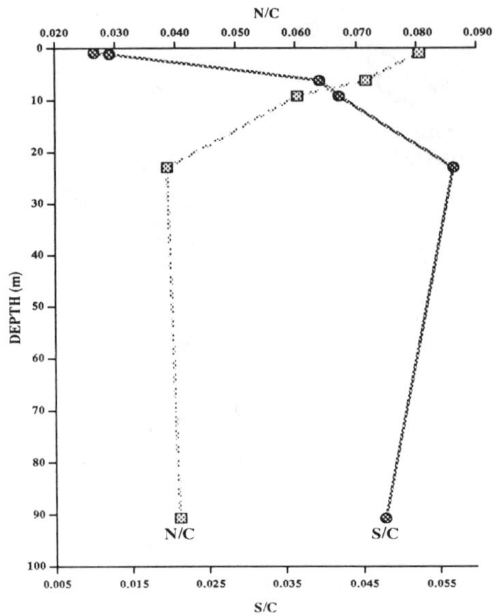

Figure 7. Variation in N/C and S/C ratio in organic matter from the Peru margin. (After Patience *et al.*, 1990)

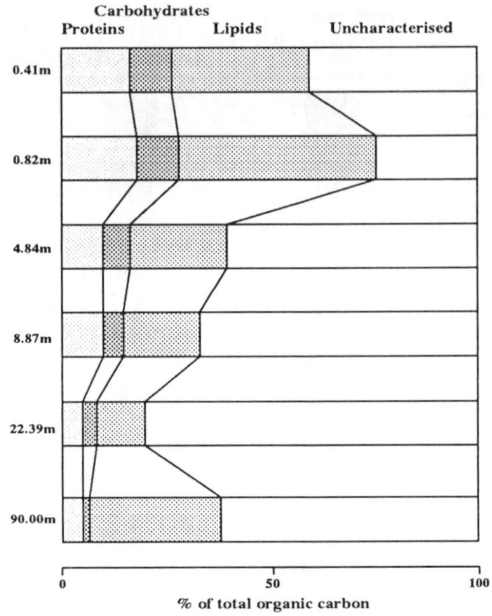

Figure 8. Variation of recognisable biochemical compounds in Peru Margin sediments (Patience *et al.*, 1990).

Soil scientists, who were the first to study sedimentary organic matter, have an alternative approach (Fig. 9). They use chemical separation in the laboratory to define *fulvic acids* (soluble in acids), *humic acids* (insoluble in acids but soluble in NaOH) and *humin* (insoluble in both). In principle, organic monomers condense first to fulvic acids, then to humic acids, humin and eventually kerogen with increasing molecular weight and complexity. Variations on this theme exist with, for example, some authors suggesting that fulvic and humic acids form concurrently and are both present in humin. An immense amount of work has gone into resolving the relationships between these components to little avail, probably because the classification is based primarily on a convenient laboratory preparation scheme rather than on real compound classes.

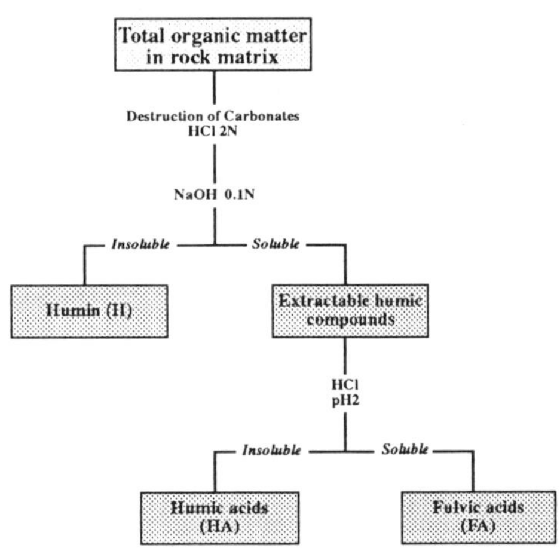

Figure 9. Classification of organic matter commonly used in soil science.

2.4. CLASSIFICATION OF KEROGEN

The composition of kerogen is widely variable. For practical purposes, however, it is convenient to break down this variability into a limited number of groups, based on one or more of its physical or chemical properties. There are three such schemes in common use today:

a) microscopic analysis of organic remains (organic petrography)

b) chemical (elemental) analysis

c) examination of the thermal breakdown products, either naturally or from laboratory pyrolysis.

2.4.1. *Organic petrography.* This is based on the identification macerals, equivalent to the description of rocks in terms of individual minerals. Table 4 shows, as an example, how coal petrologists classify organic remains. Analagous schemes are also used by palynologists and to some extent by soil scientists. There is a basic three-fold classification into exinite (oil-prone, lipid-rich), vitrinite (gas prone, lignin-rich) and inertinite (inert) and each of these is then further subdivided into individual macerals, many of which can be related back to a distinct biological precursor.

TABLE 4. Visual kerogen classification employed by organic petrologists.

Maceral Group	Maceral	Botanical Origin
	alginite	algae
	cutinite	cuticle
EXINITE	sporonite	spores/pollen
(oil prone)	resinite	resin
	subertinite	bark/cork
	liptodetrinite	any/all of above
VITRINITE	telinite	plant tissues
(gas prone)	colinite	humic gels
	fusinite	charred plant tissue
	semi fusinite	"
INERTINITE	pyrofusinite	"
(inert)	sclerotinite	fungi
	micrinite	amorphous - no
	macrinite	clear precursor

2.4.2. *Elemental analysis.* This depends on the relative contents of carbon, hydrogen and oxygen in a kerogen. In the widely adopted scheme of Tissot and Welte (1978) the atomic ratios of H/C and O/C are cross plotted in a Van Krevelen diagram (Fig. 10). Although a continuous spectrum of composition exists, it is possible to define three end members: Types I, II and III.

Type I kerogens are relatively rare. They frequently correspond to lacustrine oil shales, where much of the organic matter is derived from green algae (*e.g. Botryococcus* and related genera) and have undergone little degradation in the bacterial respiration zones. This results in the preservation of the hydrogen-rich lipid fraction and such kerogens contain a high abundance of long alkyl chains. Typical examples are the algal coals from the Central Valley of Scotland (torbanites) and parts of the Eocene Green River Formation of the Uinta Basin, Utah.

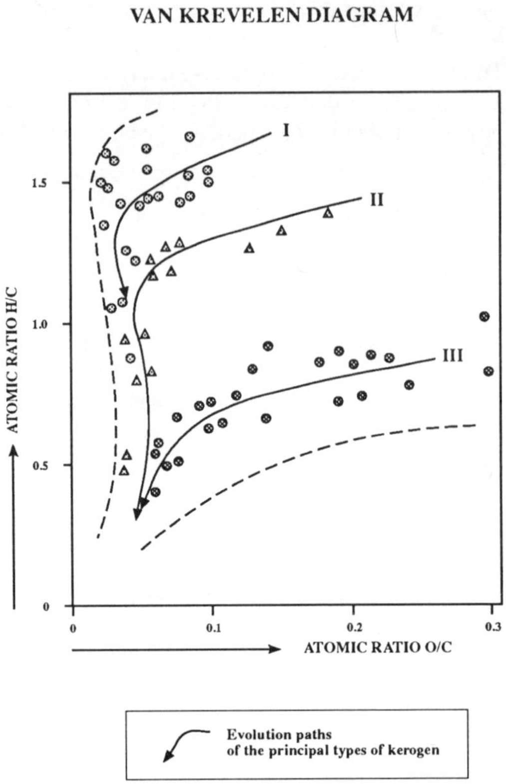

Figure 10. Classification of kerogen according to elemental ratios.

Type II kerogens are mixed algal/bacterial-sourced, generally from a marine environment (*e.g.* The Kimmeridge Clay of the North Sea and the Lower Toarcian shales of the Paris Basin). Other sources of this kerogen include the more hydrogen-rich components of higher land plants such as cuticle, resins, spores and pollen. Sometimes type II kerogens are subdivided into a low-sulphur and a high-sulphur (IIS) category.

Type III kerogens are derived largely from the lignin components of higher land plants, with only minor amounts of cuticle, resin and spores. The kerogens are mostly aromatic in character and characterised by low hydrogen contents. Most of the world's coals are dominantly type III kerogen.

Sometimes a type IV kerogen is recognised corresponding to the inertinite maceral group. It comprises plant tissue that has been oxidised, either by fungal attack or during palaeo-forest fires.

2.4.3. *Breakdown products.* Here we will restrict ourselves primarily to classification based on laboratory pyrolysis, since the other methods are of little consequence to diagenetic studies because they only reflect kerogen properties at higher temperatures.

The technique depends on using high temperatures over short times, in the laboratory, to simulate natural petroleum generation at lower temperatures over millions of years. The principal instrument for these measurements is a French pyrolyser known as the Rock-Eval (Fig. 11). A small sample of powdered rock is heated in an inert atmosphere, initially at 250°C for three minutes, and then the temperature is steadily raised to 550°C. The first stage volatilises any bitumen in the rock, while the second stage thermally decomposes the kerogen, simulating petroleum generation. Products from both stages are swept to a hydrocarbon-sensitive FID

PRINCIPLE OF THE ROCK EVAL

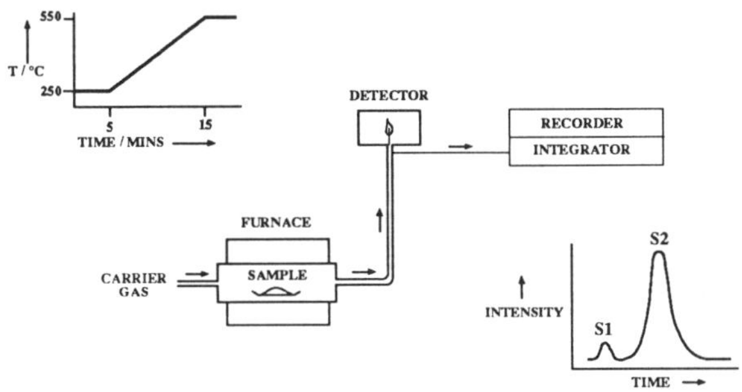

Figure 11. Schematic diagram of the Rock Eval pyrolysis instrument

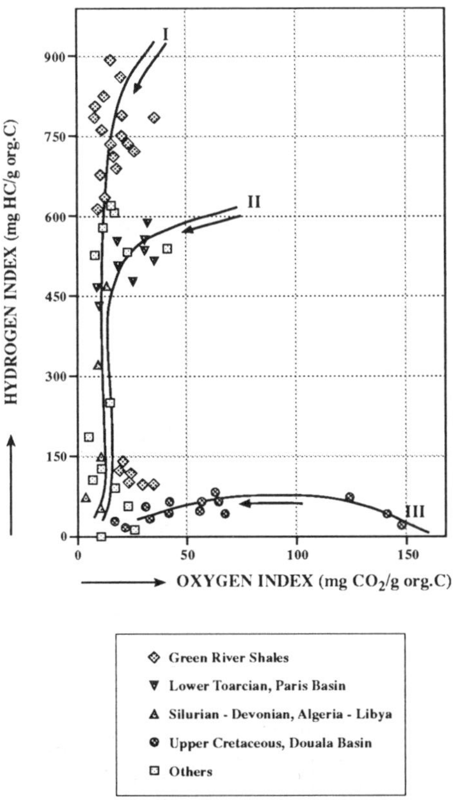

Figure 12. Modified Van Krevelen diagram based on pyrolysis HI and OI parameters (Espitalie *et al.* 1977).

detector and two peaks (S1 and S2) are recorded which correspond to the bitumen and the source potential respectively. These peaks are usually calibrated directly into kg yield per tonne of source rock (*i.e.* mg/g). Meanwhile, any carbon dioxide released by the kerogen early during pyrolysis (thermal decarboxylation) is temporarily trapped and then also passed to a detector to produce a third peak (S3) which reflects the oxygen content of the kerogen.

To use these measurements for classification, both the S2 peak and the S3 peak are normalised to the total organic carbon content (TOC or C_{org}) to yield the hydrogen index (HI) and the oxygen index (OI):

$$HI = S2/TOC.100$$

$$OI = S3/TOC.100$$

These parameters can then be cross plotted in a manner similar to the Van Krevelen diagram to classify the kerogen into types I, II and III (Fig. 12).

2.5. CARBON ISOTOPE RATIOS IN ORGANIC MATTER

It is relevant now to discuss the isotope ratio of the organic matter since this will have an influence on the isotope ratios of any diagenetic carbonates formed as by-products of bacterial reworking.

All organic matter depends ultimately on photosynthesis which discriminates strongly against ^{13}C. Hence all kerogen is isotopically negative, typically between -20 and -30 $^{o}/oo$. The exact value depends on the environmental conditions at the time of photosynthesis, the species concerned (*i.e.* the metabolic pathways) and the extent of bacterial reworking. The specific details need not concern us here except to say that the isotopic ratio of organic matter in any environment is not yet predictable.

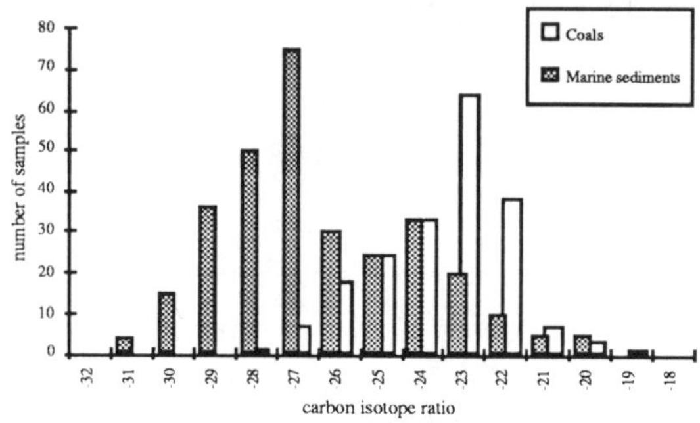

Figure13. Carbon isotope ratio of ancient kerogens.

Figure 13 shows the range of isotopic ratios of ancient kerogens. Note that, on average, ancient terrestrial organic carbon is isotopically heavier than that from marine sources. This is in contrast to the present day where land plants typically have $\delta^{13}C$ of about -27 $^{o}/oo$, and marine algae have $\delta^{13}C$ of about -20 $^{o}/oo$. Clearly things have not been constant over geological time: the major change in fact took place in the Miocene.

In addition to the bulk $\delta^{13}C$ ratio it is important to know a little of the differences in $\delta^{13}C$ between the main biochemical groups, since this will also influence $\delta^{13}C$ of CO_2 released during bacterial reworking. Different biochemical constituents are synthesised at different stages in the metabolic chain and hence the magnitude of isotopic fractionation during biosynthesis varies between components. Within a single organism, the lipids are typically

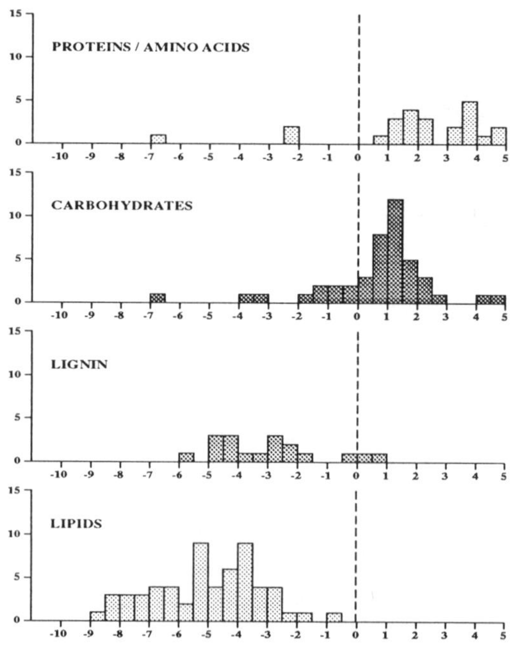

Figure 14. Carbon isotope ratios of main biochemical groups relative to $\delta^{13}C$ of bulk organism (Deines, 1980; Galimov, 1985; Benner *et al.*, 1987)

several units ‰ lighter than the bulk of the organism, with the carbohydrates and proteins progressively heavier (Fig. 14). The result is that the $\delta^{13}C$ of CO_2 released during bacterial reactions does not usually match directly that of the bulk organic matter. The first formed CO_2, for example, will be isotopically a few parts‰ heavier, since it will be derived dominantly from the more easily degraded carbohydrates and proteins. The residual organic matter will thus become isotopically lighter. Our knowledge of exactly what happens in any given situation is rudimentary, but certainly we cannot expect a simple release of CO_2 with $\delta^{13}C$ equal to that of the original organic matter.

3. Bacterial assemblages: who eats what (and whom)?

3.1. BACTERIAL CLASSIFICATION

Bacteria are single-celled organisms, usually classified as a separate kingdom, distinct from animals and plants. They are ubiquitous, extremely diverse in their metabolism and most are invisible without a microscope. Morphologically, they are also variable but they are more normally classified according to their metabolism (Fig. 15).

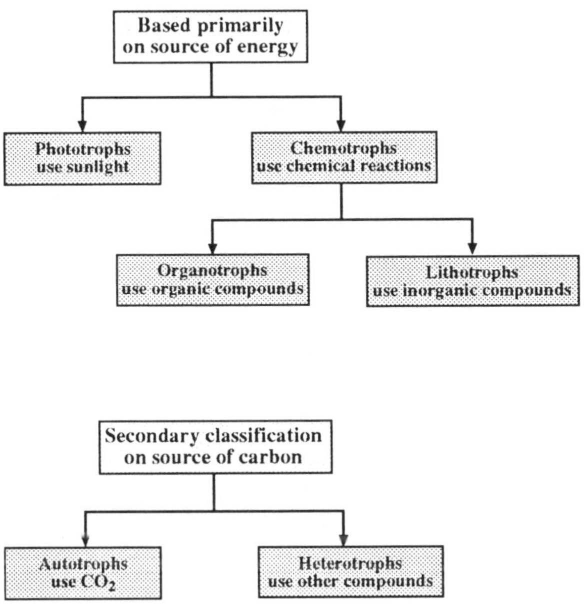

Figure 15. Simplified classification scheme for bacteria based on metabolism.

All living things must both gain energy to live and assimilate carbon in order to grow. When considering bacteria, it is usual to differentiate these two tasks. Bacteria can gain energy either through chemical reactions (fermentation or respiration) in which case they are termed *chemotrophs*, or by photosynthesis in the case of the *phototrophs*. By far the majority of bacteria are chemotrophs, and these are the organisms with which we are concerned here. Phototrophs include the "blue green algae" (which are now classified with the bacteria), the so called purple bacteria, and two groups of sulphur bacteria (green and purple), and all can be thought of as primary organic producers, rather than organisms which rework sediments.

The chemotrophs are subdivided according to which types of chemical reactions are involved in order for them to gain energy. *Chemoorganotrophs* obtain their energy from the metabolism of organic compounds, and *chemolithotrophs* derive energy from oxidation of inorganic compounds or elements. They are then further classified as either *autotrophs* if their sole source of carbon for growth is CO_2, or *heterotrophs* if they can assimilate carbon from other compounds. Hence we can have, for example, heterotrophic chemolithotrophs, or autotrophic chemolithotrophs.

Finally, and probably most important, bacteria rarely work in isolation. Bacterial reactions in sediments are the result of many different types of bacteria working together, each feeding off of the by-products of others. Such bacterial assemblages are frequently classified according to the bulk chemical environment in which they operate such as the *aerobic* organisms which function in oxic environments or the *anaerobic* organisms which live in anoxic conditions.

3.2. METABOLISM IN THE CHEMOTROPHIC BACTERIA

It is helpful, before we go futher, to understand a little more of the biochemical reactions performed by bacteria.

Metabolic pathways can be thought of as a series of chemical reactions orchestrated by a range of protein catalysts or enzymes. Many pathways are common to many bacterial groups (and indeed eukaryotic organisms) while others are more specific. The end result of most of the pathways is a *net* gain in energy, although many individual reactions require an input of energy to get them started. Clearly then, the "energy currency" within the bacterial cell is fundamental to understanding what happens.

ADENOSINE 5' - TRIPHOSPHATE (ATP)

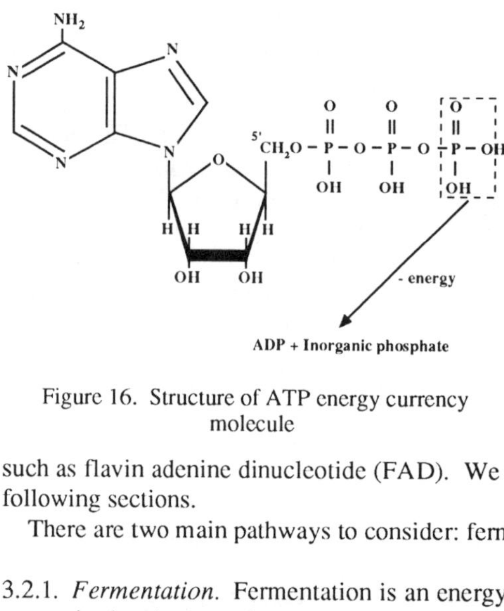

Figure 16. Structure of ATP energy currency molecule

The most important of the energy-carrying molecules is adenosine 5'-triphosphate, or ATP (Fig. 16). ATP is able to release energy by losing its terminal phosphate group to form adenosine 5'-diphosphate (ADP), or a cell can store energy by converting ADP into ATP (a process known as substrate-level phosphorylation).

Another type of "currency molecule" stores energy in the form of "reducing power". Such molecules are able to accept electrons (and hence energy) from intermediate reactions in the metabolic pathway, and later release them to another substrate or use them in the generation of ATP. Nicotinamide adenine dinucleotide (NAD) is one of the most common of these, along with NAD phosphate (NADP), and flavin containing compounds such as flavin adenine dinucleotide (FAD). We will see the importance of these chemicals in the following sections.

There are two main pathways to consider: fermentation and respiration.

3.2.1. *Fermentation.* Fermentation is an energy-yielding pathway which typically, but not only, occurs in the absence of oxygen, and without the need of an external electron acceptor. In other words, fermentation is a simple dissociation reaction of a complex organic molecule into two simpler molecules, such as the formation of CO_2 and ethanol from sugars during brewing of beers and wines. Because there is no external electron "sink" (*i.e.* an oxidising agent) fermentation can never involve the net oxidation of the substrate (in contrast to respiration), which limits the potential substrates to a number of compounds which are chemically neutral and easily dissociated. These include most carbohydrates and a number of proteins, but excludes most lipids which can only be attacked as part of a net oxidation reaction (respiration).

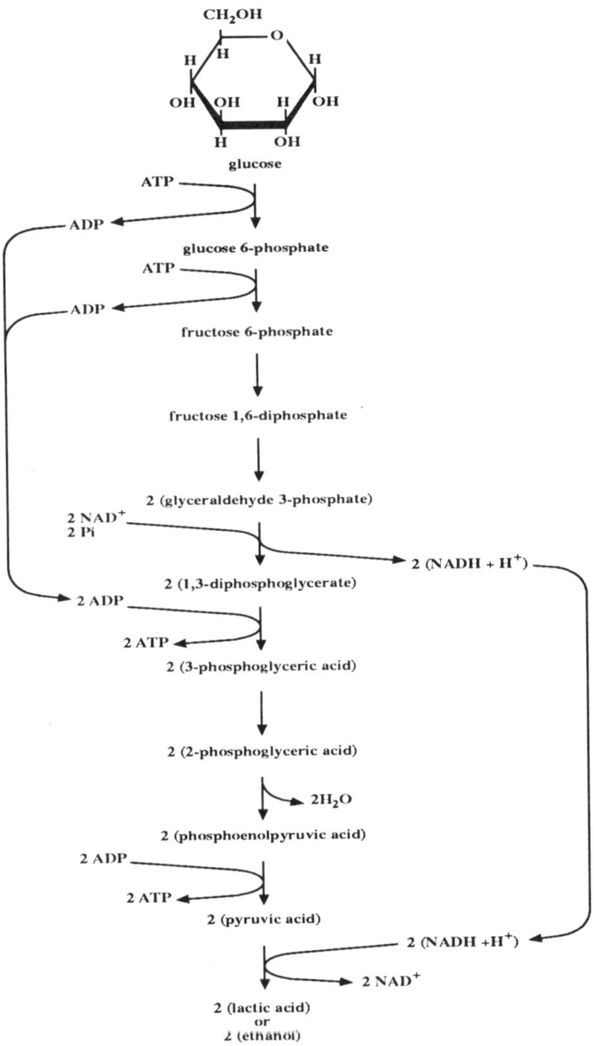

Figure 17. The Embden-Meyerhoff pathway of fermentation. Pi indicates inorganic phosphorus.

We can understand this principle by looking at a typical (and in this case the most important) fermentation pathway: the Embden-Meyerhoff (E-M) pathway commonly called glycolysis (Fig. 17). The intermediate compounds need not concern us here, but note that the reaction consists of the conversion of a single glucose molecule into two molecules of the chemically simpler lactic acid or ethanol with the *net* gain of two molecules of ATP. Such fermentation reactions are central to the breakdown of large organo-polymers in the complex bacterial ecosystems encountered in sediments.

The most important fermentation reaction we shall come across is methanogenesis, the dissociation of acetate to methane and CO_2:

$$CH_3COOH \rightarrow CH_4 + CO_2$$

3.2.2. *Respiration.* In contrast to fermentation, respiration is an energy yielding process in which the organic, or in some cases inorganic, substrate is oxidised by means of an external oxidising agent obtained from the environment. Oxidation occurs in a number of stages in which electrons are typically transferred to NAD, and then ultimately passed on to the external oxidising agent (or terminal electron acceptor). The external agent is usually oxygen, but may also be sulphate, nitrate, or even ferrous iron or managese (the latter two zones are often referred to as *suboxic diagenesis*). Respiration is much more energy efficient than fermentation and hence dominates microbial reactions when a suitable oxidant is available. This is also the pathway which gives rise to most of the tangible by-products which we find as mineral precipitates in sediments.

The early stages of the respiration cycle are typically the same as fermentation with conversion of, for example, glucose to pyruvic acid. Beyond this, however, the process is fundamentally different. Pyruvic acid is converted to acetyl-coenzyme A which is fed into the tricarboxylic acid cycle (also known as the Kreb's cycle or citric acid cycle). During this cycle the organic substrate is effectively completely converted to carbon dioxide (Fig. 18). The cell finishes up with 4 NADH and 1 $FADH_2$ molecules which must be oxidised back to NAD and FAD for futher metabolism to take place. This is where the electron acceptors in the environment come into play. In aerobic oxidation, oxygen plays this role and is converted to water, for example. Under progressively more reducing conditions, then, nitrate, manganese, iron and, eventually, sulphate take over. This sequence is not arbitrary. It is dictated by the free energy yield obtained in the process (Fig. 19). The dominant reaction will be the most efficient of the possible pathways available. In addition, as the ultimate electron acceptor changes, then so do the bacterial species. For example *Pseudomonas aeruginosa* converts nitrate to nitrite (while for other bacteria the result may be nitrogen or ammonia) but in the absence of nitrate the dominant bacteria will be *Desulfovibrio* and *Desulfotomaculum* which use sulphate and convert it to sulphide or H_2S.

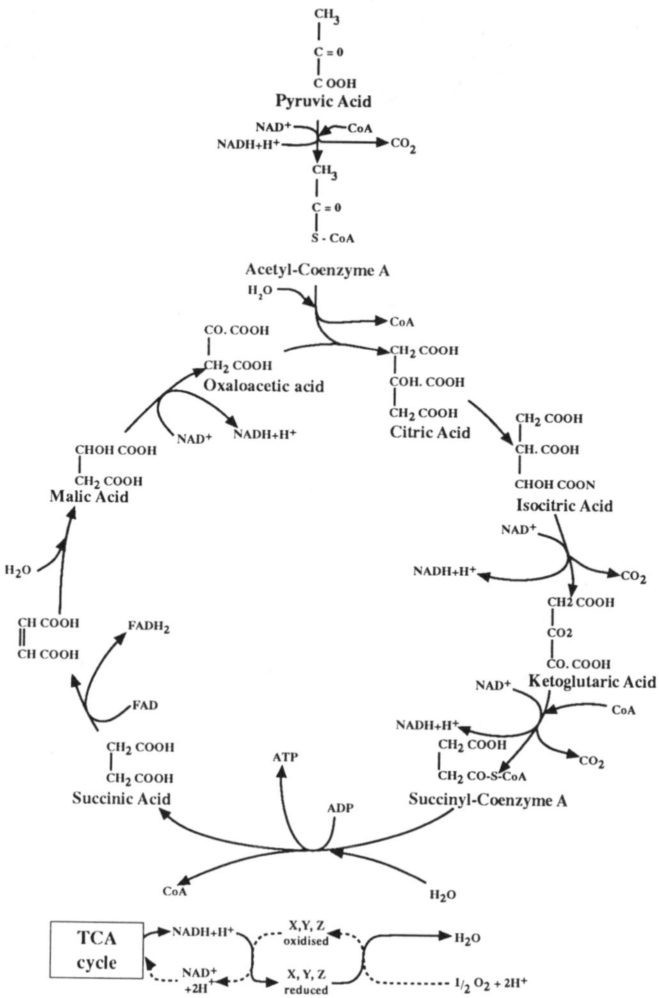

Figure 18. The tricarboxylic acid (or Krebs) cycle of bacterial respiration.

AEROBIC RESPIRATION :

$$(CH_2O)_{106}(NH_3)_{16}(H_3PO_4) + 138\,O_2 \rightarrow 106\,CO_2 + 16\,HNO_3 + H_3PO_4 + 122\,H_2O$$

$\Delta G^{o\prime}$ = -3190 kJ/mole of glucose

MANGANESE REDUCTION :

$$(CH_2O)_{106}(NH_3)_{16}(H_3PO_4) + 236\,MnO_2 + 472\,H^+ \rightarrow 236\,Mn^{2+} + 106\,CO_2 + 8N_2 + H_3PO_4 + 366\,H_2O$$

$\Delta G^{o\prime}$ = -3090 kJ/mole (birnessite)
-3050 kJ /mole (nsutite)
-2920 kJ/mole (pyrolusite)

NITRATE REDUCTION :

$$(CH_2O)_{106}(NH_3)_{16}(H_3PO_4) + 94.4\,HNO_3 \rightarrow 106\,CO_2 + 55.2N_2 + H_3PO_4 + 177.2\,H_2O$$

$\Delta G^{o\prime}$ = -3030 kJ/mole of glucose

$$(CH_2O)_{106}(NH_3)_{16}(H_3PO_4) + 84.8\,HNO_3 \rightarrow 106\,CO_2 + 42.4N_2 + 16\,NH_3 + H_3PO_4 + 148.4\,H_2O$$

$\Delta G^{o\prime}$ = -2750 kJ/mole of glucose

IRON REDUCTION :

$$(CH_2O)_{106}(NH_3)_{16}(H_3PO_4) + 212\,Fe_2O_3 + 848\,H^+ \rightarrow 424\,Fe^{2+} + 106\,CO_2 + 16\,NH_3 + H_3PO_4 + 530\,H_2O$$

$\Delta G^{o\prime}$ = -1410 kJ/mole (nematite)
-1330 kJ/mole (goethite)

SULPHATE REDUCTION :

$$(CH_2O)_{106}(NH_3)_{16}(H_3PO_4) + 53\,SO_4^{2-} \rightarrow 106\,CO_2 + 16\,NH_3 + 53S^{2-} + H_3PO_4 + 106\,H_2O$$

$\Delta G^{o\prime}$ = -380 kJ/mole

FERMENTATION :

$$(CH_2O)_{106}(NH_3)_{16}(H_3PO_4) \rightarrow 53\,CO_2 + 53\,CH_4 + 16\,NH_3 + H_3PO_4$$

$\Delta G^{o\prime}$ = -350 kJ/mole

Figure 19. Organic matter oxidation reactions during early diagenesis. Stoicheiometry is based on the Redfield (1958) formula for average organic matter and free energy yield is based on 1 mole of glucose ($C_6H_{12}O_6$). After Froelich *et al.* (1979).

3.2.3. *Metabolism in Chemolithotrophs.* These bacteria obtain energy by the oxidation of inorganic substrates, usually in the presence of oxygen. The organisms often play a central role in the cycling of various compounds. For example, nitrogen released by nitrate reducing organisms can be re-oxidised by the nitrifying bacteria (such as *Nitrosomonas* and *Nitrosococcus*) or ammonia may be oxidised to nitrite or nitrate by *Nitrobacter* and *Nitrococcas*. *Thiobacillus* derive energy from oxidation of sulphur compounds back to sulphate, and *Thiobacillus denitrificans* can use nitrate as the ultimate electron acceptor when oxidising sulphide. Other organisms can also oxidise hydrogen, sometimes to water or, in the case of an important group of methanogenic bacteria, to methane, using CO_2 as the oxidant:

$$4H_2 + CO_2 \rightarrow CH_4 + 2H_2O$$

Other bacteria are then able to oxidise the methane back to CO_2 using oxygen, or in some cases sulphate.

3.3. BACTERIAL ASSEMBLAGES IN SEDIMENTS

The bacterial assemblage found in any given environment is dictated by a few very simple rules:

1. It will be a large number of bacterial groups working together.
2. For any reaction which is energetically favourable, there will be a bacterium to do it.
3. The dominant bacterial assemblage will be that which is most favourable energetically.

Inorganic nutrients in sediments are not unlimited and so must be replenished, usually from the water column. There is a stratification in sediments as each nutrient, supplied from above, is progressively used up and the bacterial assemblage changes to use the next most efficient electron acceptor (Fig. 20). Usually oxygen will be available near the surface and an aerobic bacterial assemblage develops. The oxygen is only able to diffuse into the sediment to a certain distance before it is consumed and below that depth other compounds are used in order of preference: nitrate, manganese, iron and eventually sulphate. Below this, no further electron acceptors are available and only bacterial fermentation and methanogenesis from CO_2 and H_2 are possible.

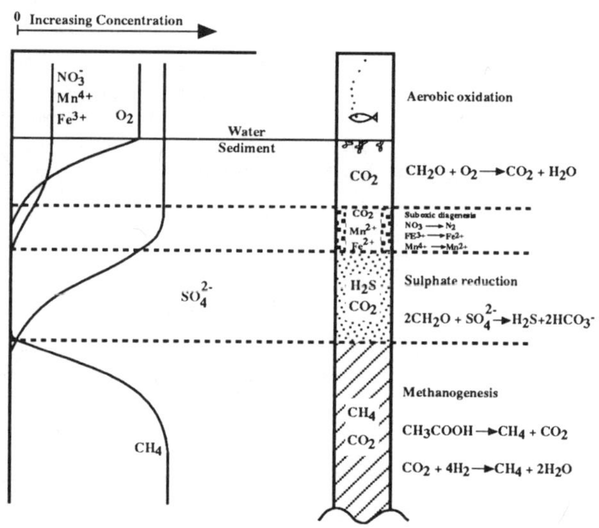

Figure 20. Schematic depth zonation of bacterial oxidation zones in sediments. In freshwater sediments the sulphate reduction zone is suppressed.

This depth-related sequence of processes is often reflected in diagenetic precipitates as a temperature series revealed by, for example, oxygen isotopes. The deeper in the sediment the reactions occur, then the higher will be the temperature (see below).

However, this simple picture hides a multitude of other reactions. For example, fermentation will occur all through the sediment but will only be apparent where it is not masked by respiration reactions. In addition, the boundaries between the zones are blurred as different organisms compete for the available substrates. Many local microenvironments will develop such as localised sulphate reduction in a generally aerobic environment. Finally, most of the by-products of the individual reactions will be recycled into other zones (*e.g.* methane from the fermentation zone may be re-oxidised at the base of the sulphate reduction zone). It is these anomalous situations which leave most of the evidence behind in the form of diagenetic products. We shall see examples of this in the following sections.

3.4. ISOTOPIC EFFECTS

The majority of the above reactions produce CO_2 in one form or another, which can result in diagenetic carbonates if other factors are conducive. It is possible to determine in which zone the carbonate formed from the carbon isotope value of the carbonate although there is considerable variability in the values found.

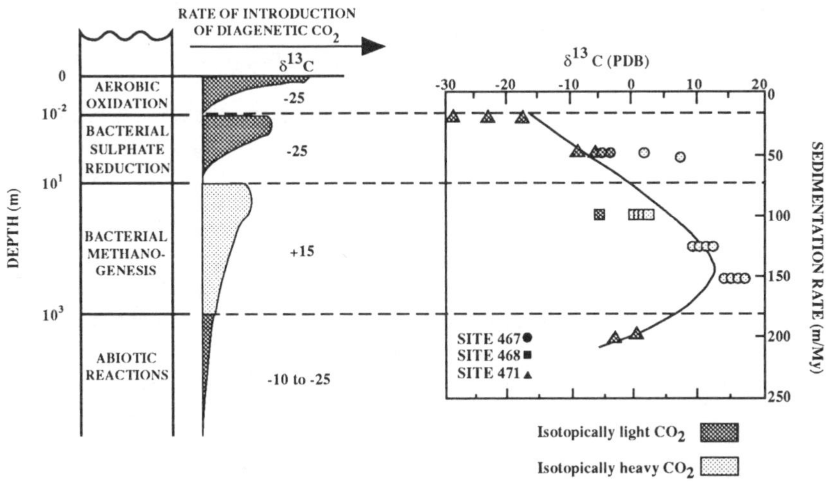

Figure 21. Carbon isotopic ratio of CO_2 released in each bacterial zone and the influence of sedimentation rate on the zone where diagenetic carbonates are preferentially precipitated (from Clayton, 1992).

Carbon dioxide released by the various respiration processes should have a carbon isotope ratio similar to that of the organic matter being attacked (Fig. 21), approximately -20 to -30 $^{o}/_{oo}$. In practice, there is little evidence of carbonate precipitation in the aerobic and nitrate reduction zones (probably because carbon is released as CO_2 which will dissolve carbonate rather than precipitate it in most cases) but carbonates from the other zones are indeed depleted in ^{13}C. However, typically we only see values of about -15 $^{o}/_{oo}$ (see for example Fig. 22). This is usually interpreted as indicative of some dilution by marine carbonate ($\delta^{13}C \sim 0$ $^{o}/_{oo}$), although this is not

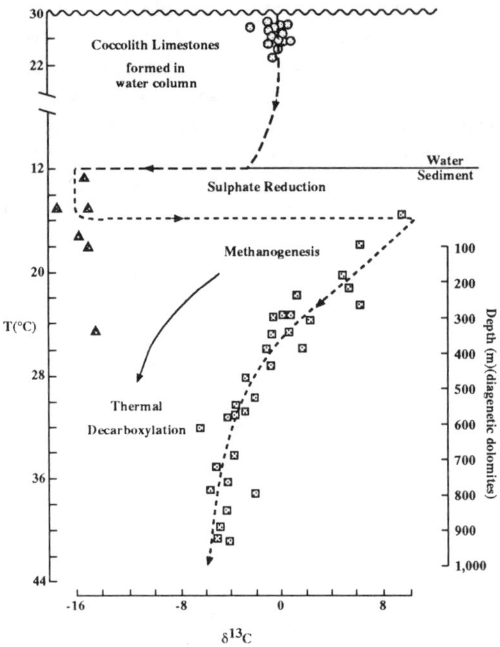

Figure 22. Carbon isotope ratios versus depth of formation (based on oxygen isotope temperature) for carbonates in the Kimmeridge Clay of southern England (Irwin *et al.*, 1977).

necessarily the case. Bear in mind that the carbohydrates and proteins are the most readily metabolised of the biochemical constituents and that these are isotopically enriched relative to bulk organic carbon. We should not therefore be too surprised if the carbonate is anomalously enriched also. This conclusion is supported by laboratory bacterial cultures (Fig. 23) which also show that the CO_2 released during anaerobic organic matter oxidation is significantly enriched in ^{13}C relative to the starting organic matter. Carbonates precipitated in this zone are usually found in sediments deposited at low to moderate sedimentation rates (Fig. 21). This is because at higher sedimentation rates the organic matter passes rapidly through this bacterial zone and most of the diagenetic alteration occurs in the deeper methanogenic zone.

Carbonate phases precipitated in the suboxic zone are typically iron and manganese carbonates (see for example Coleman *et al.*, 1982; Jenkyns & Clayton, 1986). In the sulphate reduction zone we usually see precipitation of non-ferroan calcite since and all the available iron is removed as sulphides by the H_2S which is formed, yet sufficient sulphate is still present to inhibit dolomite formation.

During methanogenesis $\delta^{13}C$ of the CO_2 depends on the specific microbial pathway involved. If fermentation is the dominant pathway then the CO_2 is derived from acetate and inherits the isotopic signature of the precursor carboxyl groups, typically 5 to 10 *o/oo* heavier than the bulk carbon (see Galimov, 1985). If we are dealing with CO_2 reduction then the isotopic ratio of the CO_2 depends on a kinetic isotope fractionation which leaves residual CO_2 progressively enriched in ^{13}C, typically up to +15 *o/oo*. However, in the early stages this CO_2 is actually depleted in ^{13}C as it is derived directly from organic matter. Thus CO_2 present in the methanogenic zone may have $\delta^{13}C$ anywhere between -25 and +15 *o/oo*. The fact that we typically observe carbonates from this zone with positive isotope values implies that we are looking at unusual circumstances, not typical methanogenic activity. This aspect is considered in more detail in Section 4.

Below this, when temperatures become too high for biological activity, we often see abiotic reactions contributing to the carbonate pool. This CO_2 is derived from residual carboxyl groups in the organic matter by *thermal decarboxylation.* Carboxyl groups in organic matter are typically enriched in ^{13}C by 5 to 10 *o/oo* relative to the bulk organic matter (Games *et al.*, 1978) and so the CO_2 released by this mechanism is isotopically heavier. The process is accompanied by a kinetic isotope effect which results in the first formed CO_2 having $\delta^{13}C$ of about -25 *o/oo*, and increasing with progressive generation. Diagenetic products from this zone are found mostly in sediments characterised by very high sedimentation rates (fig. 21), and usually define a trend from isotopically heavy ferroan dolomites associated with methanogenesis to decarboxylation cements formed at even higher temperatures (Fig. 22).

Carbon is not the only isotope fractionated during bacterial activity. Sulphur is also affected. Typically we observe up to 40 to 50 $^{o}/_{oo}$ depletion in ^{34}S during sulphate reduction. In contrast, there is negligible fractionation during sulphide oxidation and hence sulphate which is continually being recycled between the aerobic and sulphate reduction zones will become progressively more depleted in ^{34}S. We will see an example of this curiosity in Section 5.

Finally, we should also bear in mind that bacterial oxidation releases hydrogen from the organic matter. Organic-bound hydrogen is highly depleted in deuterium (D), by up to 300 $^{o}/_{oo}$, and this light hydrogen will ultimately finish up as pore water, modifying δD values away from seawater values (δD ≈ 0 $^{o}/_{oo}$).

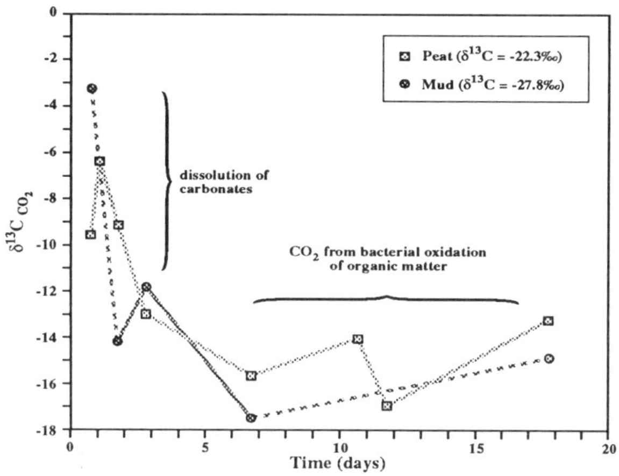

Figure 23. Carbon isotopic ratio of CO_2 released during anaerobic bacterial oxidation of sediments in the laboratory (Clayton, unpublished data). Cultures were incubated in sea water using a natural marine bacterial assemblage. The first formed CO_2 is the result of carbonate dissolution associated with iron sulphide oxidation. Later CO_2 is derived from organic matter.

4. The Methane Cycle

The concept of bacterial recycling is well illustrated by the process of generation and subsequent destruction of biogenic methane. In this cycle, methanogenic bacteria produce methane in the fermentation zone which can then migrate to more oxidised sediments where it may be consumed by respiration processes in the aerobic or sulphate reduction zones.

4.1. BIOGENIC METHANE GENERATION

Biogenic (bacterial) gas is formed in organic-rich environments by a specialised group of bacteria, the methanogens. These bacteria require anoxic, sulphate-free conditions at temperatures below about 75°C. (If sulphate is present they are outcompeted by the sulphate reducing bacteria for available substrates.) These conditions are met in paddy fields, the intestinal tracts of higher organisms and in marine sediments below the zone of sulphate reduction.

Two distinct metabolic processes contribute to methane generation (Fig. 24): fermentation and CO_2 reduction using hydrogen (chemolithotrophy). In the first, simple organic substrates such as acetate are fermented to CH_4 and CO_2:

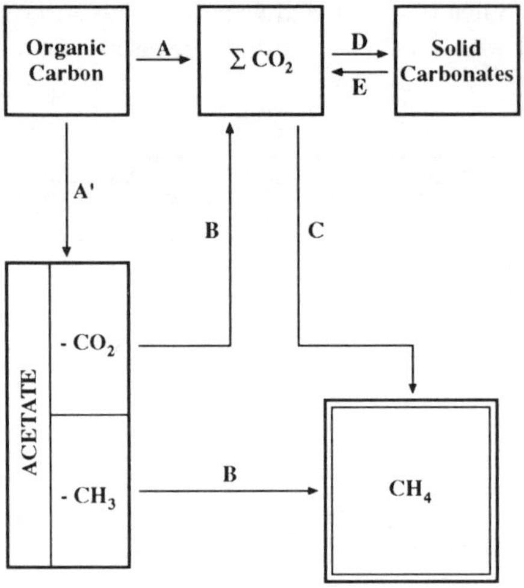

Reaction A : Fermentation of organic matter by non-methanogens. Acetate (A') is one by-product.
B : Fermentation of acetate
C : Reduction of CO_2 using hydrogen
D/E : Equilibrium with solid carbonates.

Figure 24. Processes occurring in the methanogenic zone (after Lazerte, 1981).

$$CH_3COOH \rightarrow CH_4 + CO_2$$

In the second, CO_2 from other bacterial reactions (including that above) is reduced by hydrogen from the environment (itself a product of other bacterial reactions):

$$CO_2 + 4H_2 \rightarrow CH_4 + 2H_2O$$

The former reaction tends to be the dominant pathway in freshwater environments (swamps, marshs *etc.*) while the latter is dominant in the marine environment. The difference in pathway is apparent when considering the CO_2 content of the resulting gas (Fig. 25).

The two processes can also be differentiated by the hydrogen isotope ratios of the methane (Fig. 26). This is because all of the hydrogen in the CO_2 pathway comes from the surrounding water and hence is in isotopic equilibrium with it, whereas for fermentation three quarters of the hydrogen comes from the organic matter. For CO_2 reduction this gives methane with δD in the region of -200 $^o/_{oo}$ for unmodified seawater. For fermentation only one of the hydrogen atoms is in equilibrium with water (the terminal, loosely bonded one in the acetate) and the rest comes from organic matter which is isotopically very light. The result is methane with δD typically in the region of -320 $^o/_{oo}$ with an absolute value dependent on δD of the water. This difference is very apparent in the histogram of hydrogen isotope ratios in biogenic methane accumulations (Fig. 27). The separation of the two processes in terms of environment is not perfect, however, as both pathways contribute to some extent, in all cases depending on local conditions such as organic matter type, temperature (*e.g.* seasonal variations in shallow water) and the hydrogen isotope ratio of the pore waters (Jenden & Kaplan, 1986).

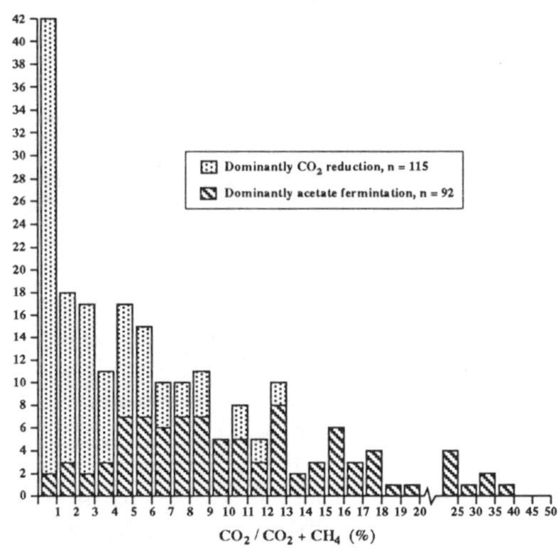

Figure 25. CO_2 content of biogenic gases worldwide.

4.2. CARBON ISOTOPE RATIOS

4.2.1. *Isotopic Fractionation.* Carbon isotope effects during methanogenesis are of particular interest since much of the CO_2 may form carbonate. There are two fractionation mechanisms which control $\delta^{13}C$ of methane and carbon dioxide during methanogenesis. *Kinetic fractionation* depends on the relative reaction rates of molecules containing only ^{12}C and those which contains a ^{13}C atom. During CO_2 reduction,for example, the first formed methane is enriched in ^{12}C and the residual CO_2 becomes enriched in ^{13}C. Methane formed later contains more ^{13}C, but is still lighter than the CO_2

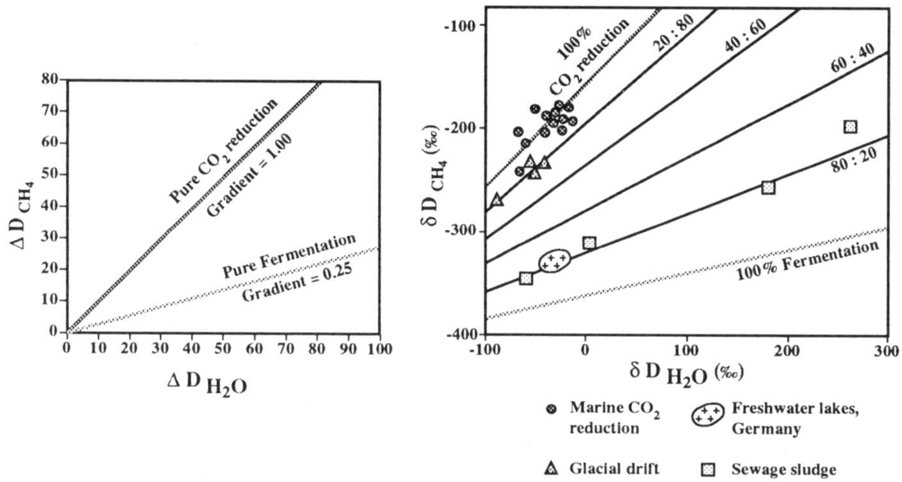

Figure 26. Hydrogen isotope fractionation during bacterial methanogenesis (after Woltemate *et al.*, 1984)

from which it came. The magnitude of this effect can be measured in the laboratory (Table 5) and gives CH_4 which is between 25 and 60 $^{o}/_{oo}$ lighter than the CO_2 depending on the environmental conditions and the microbial species concerned. In natural environments the original CO_2 is derived from microbial decomposition of organic matter and so is typically in the order of -20 to -30 $^{o}/_{oo}$. The first formed methane therefore has $\delta^{13}C$ between -50 and -90 $^{o}/_{oo}$. As methanogenesis proceeds, both CO_2 and CH_4 become progressively enriched in ^{13}C until the consumption of CO_2 is balanced by addition from other sources and a steady state is reached. The net result is natural methane isotope ratios typically between -55 and -80 $^{o}/_{oo}$.

For acetate fermentation the "equilibrium state" is rather different. Kinetic isotope

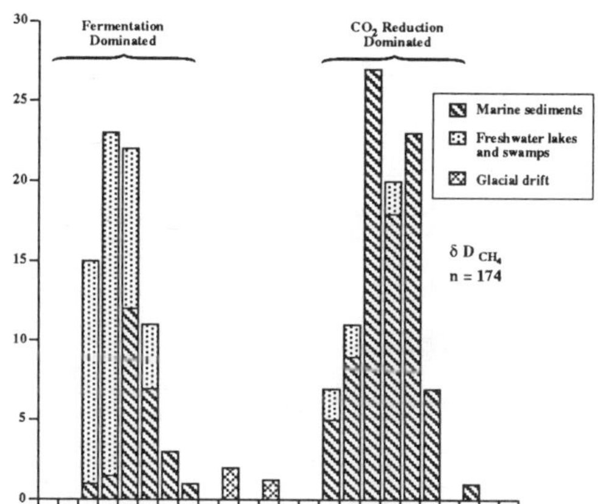

Figure 27. Hydrogen isotope ratios of worldwide biogenic methanes according to depositional environment.

effects first lead to depletion in ^{13}C in both CO_2 and CH_4, with the methane depleted by up to 80 $^{o}/oo$. But the supply of acetate is usually limited and is eventually consumed entirely. At this point the isotopic ratio of the CO_2 and CH_4 depend on the *intramolecular isotope distribution* of the acetate: CO_2 inherits $\delta^{13}C$ of the precursor carboxyl groups and CH_4 inherits $\delta^{13}C$ of the methyl groups. For biochemical reasons, the methyl groups in acetate are lighter than the carboxyl (Games *et al.*, 1978), and methane from this source often has an isotopic ratio of -55 to -60 $^{o}/oo$, a value commonly found in freshwater sediments where acetate fermentation is dominant (see below).

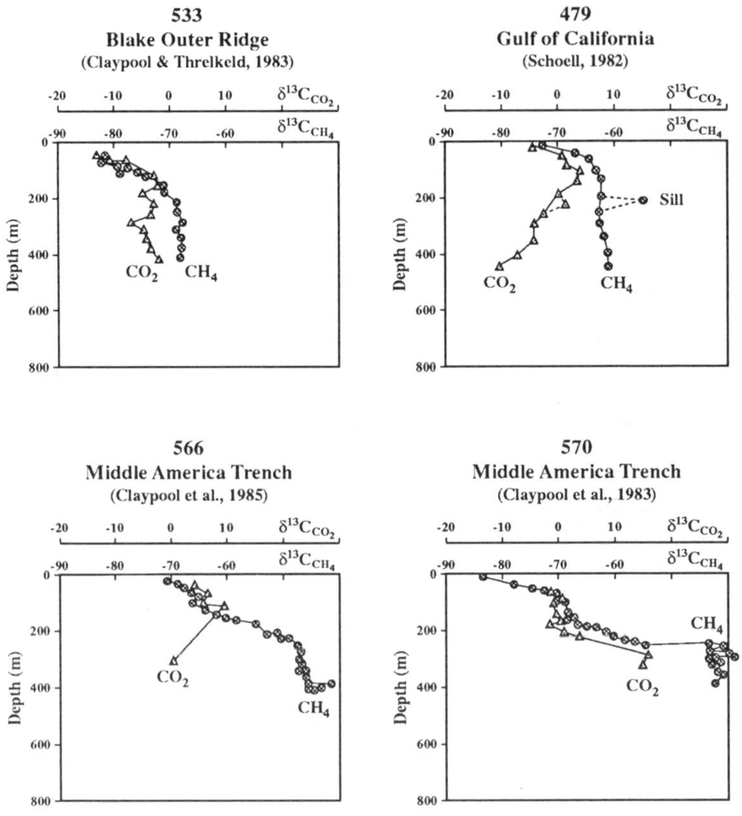

Figure 28. Carbon and hydrogen isotope ratios of gases in methanogenic sediments from DSDP wells.

4.2.2. *Depth profiles*. The interplay between CO_2 reduction and fermentation is well illustrated by data from Deep Sea Drilling Project boreholes (Fig. 28). Shallow samples can be assumed to represent the early stages of methanogenesis; deeper samples reflect the more advanced stages.

Regardless of the main methane-producing process, the earliest formed methane should be isotopically the lightest. This is indeed the case.

TABLE 5. Isotopic fractionation in laboratory methanogenic bacterial cultures

Bacterial culture	Medium	Temp (°C)	$\alpha_{medium\text{-}CH_4}$*	Reference
Methanobacterium sp.	CO_2/H_2	37	1.035-1.036	Belyaev *et al.* (1977)
Methanobacterium sp.	CO_2/H_2	46	1.034-1.036	Belyaev *et al.* (1977)
Methanobacterium sp.	CO_2/H_2	40	1.061 ± .002	Games *et al.* (1978)
Methanosarcina barkeri	CO_2/H_2	40	1.045 ± .002	Games *et al.* (1978)
Methanosarcina barkeri	CO_2/H_2	37	1.046 ± .003	Krzycki *et al.* (1987)
Methanobacterium thermoautotrophicum	CO_2/H_2	65	1.025 ± .002	Games *et al.* (1978)
Methanobacterium formicum	CO_2/H_2	34	1.048	Balabane *et al.*, (1987)
N. Pacific Ocean culture	0.1M methanol	23	1.094	Rosenfeld & Silverman (1959)
N. Pacific Ocean culture	0.1M methanol	30	1.081	Rosenfeld & Silverman (1959)
N. Pacific Ocean culture	0.25M methanol	30	1.083	Rosenfeld & Silverman (1959)
Not reported	0.1M methanol	30	1.080	Nakai (1962)
Not reported	0.1M acetic acid	30	1.080	Nakai (1962)
M. barkeri	0.2M methanol	37	1.074 ± .008	Krzycki *et al.* (1987)
M. barkeri	0.2M acetate	37	1.025+	Krzycki *et al.* (1987)

* $\alpha_{A\text{-}B}$ is the isotope ratio of A with respect to B

i.e. $\alpha_{A\text{-}B} = \frac{(^{13}C/^{12}C)_A}{(^{13}C/^{12}C)_B} = \frac{1000 + \delta_A}{1000 + \delta_B}$

or to a first approximation:

$\delta_A\text{-}\delta_B \cong 1000.(\alpha\text{-}1)$

\+ relative to total substrate (1.021 relative to methyl precursor)

Considering first the two examples from the Blake Outer Ridge and the Gulf of California, which are quite typical of observed profiles, $\delta^{13}C$ increases with depth before reaching an equilibrium value somewhere between -60 and -70 $^o/oo$. The CO_2 data closely parallel $\delta^{13}C_{methane}$ in the higher parts of each well, with the methane depleted by about 70 $^o/oo$ relative to the CO_2. This is in reasonable agreement with the laboratory measurements, especially since the latter were measured at higher temperature where fractionation effects are smaller. In the deeper parts of the wells, the $\delta^{13}C$ trend of the CO_2 either levels off in advance of the methane (533), or reverses (479). This suggests an additional input of CO_2 with $\delta^{13}C$ somewhere below -5 $^o/oo$ indicating an increased contribution from acetate fermentation at depth. An end point of -10 $^o/oo$ would be consistent with measured $\delta^{13}C$ values of carboxyl groups in acetate. The corresponding methyl group (and hence methane produced) should have an isotopic ratio of -40 to -50 ‰, but in field measurements CH_4 from this source is apparently swamped by that from CO_2 reduction.

The two wells from the Middle America Trench show a rather different, and less common, trend. Initial isotope ratios for CO_2 and CH_4 are similar to the above examples but with increasing depth the methane value continue to increase, ultimately reaching values close to -40 ‰. Generally the CO_2 values keep track with this, suggesting that the cause is a runaway CO_2 reduction process. Note also, that it is only in these unusual situations that the very heavy CO_2 isotope values are observed which would lead to isotopically heavy carbonates.

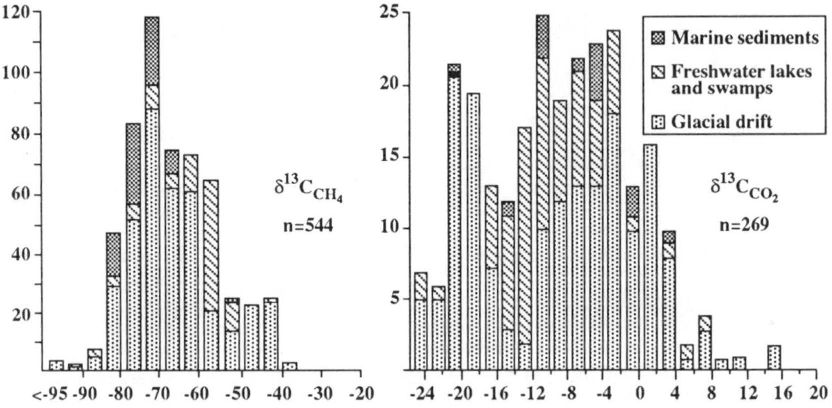

Figure 29. Carbon isotope ratio of methane and CO_2 from methanic sediments and biogenic gas accumulations worldwide according to depositional environment.

4.2.3. *Variations with depositional environment.* Figure 29 shows a collation of methane isotopic data from biogenic gases worldwide, arranged according to depositional environment. Marine sediments show the greatest spread in $\delta^{13}C$, from below -95 ‰ to about -40 ‰. This reflects the range of mixtures of CO_2-derived and acetate-derived CH_4. Most values (80%) fall in the range of simple mixtures between the two. The particularly heavy samples result from a very few localities which exhibit extreme CO_2 reduction as in the example of the Middle America Trench. The very light values come from particularly "young" gases. Samples from freshwater lakes and swamps tend to be a little heavier (-50 to -70 ‰) than the marine gases, as would be expected in environments dominated by acetate fermentation. The glacial drift samples also contain a significant contribution of fermentation-derived CO_2 but because they are so young (<30,000 radiocarbon years) they have not yet attained their heavier equilibrium signature.

4.3. CARBONATE PRECIPITATION

The sites of intense methanogenic activity are frequently marked in ancient sediments by authigenic ferroan dolomites and ankerites. These are usually attributed to "fermentation" although the picture is not quite this simple. Methane-associated carbon dioxide isotopic ratios show a dramatic spread in $\delta^{13}C$ (Fig. 29). Probably the most significant point to note however is that the +15 ‰ end member value assumed in many diagenetic studies (*e.g* Irwin *et al.*, 1977) is achieved only in exceptional circumstances (less than 1% of the samples). Values between +2 and -22 $^o/_{oo}$ are dominant. The only cases where the highly positive isotope ratios are observed are in rare cases of runaway CO_2 reduction. It is thus the anomalous situations which give rise to the extensive carbonate cementation as observed, for example, in the Kimmeridge Clay of Dorset.

This makes sense if we consider the carbonate equilibrium system. During fermentation, CO_2 is released, which in the absence of any other buffering, leads to slightly acidic conditions which would inhibit carbonate precipitation. This effect however is usually buffered to some extent by other bacterial reactions which keep the sediments at more or less neutral pH. If however CO_2 reduction takes over, after a time the CO_2 will be partially removed, lowering acidity and eventually promoting precipitation of what is left of the dissolved carbonate. If this *is* the case, we should expect to see a break in carbonate precipitation between the sulphate reduction and methanogenic zones, which does appear to be the case in the Kimmeridge examples studied by Irwin *et al.* (1977).

4.4. METHANE DESTRUCTION SHALLOWER IN THE SEDIMENT

4.4.1. *Bacterial Reactions.* Once formed, the methane does not last indefinitely. It will diffuse along concentration gradients, or float by buoyancy if a free gas phase forms, to shallower levels in the sediment. Here, other bacterial groups are available to take advantage of the energy available from oxidation of the methane.

The most common of the bacterial methane oxidation reactions is aerobic oxidation:

$$CH_4 + 2O_2 \rightarrow CO_2 + 2H_2O$$

More recently, evidence has grown for a process of anaerobic methane oxidation, using sulphate as the oxidant:

$$CH_4 + SO_4^{2-} \rightarrow CO_3^{2-} + H_2S + H_2O$$

This mechanism is still unproven since the bacteria responsible have not yet been isolated, but there is strong circumstantial evidence. For example, the boundary between the methanogenic and sulphate reduction zones is usually very abrupt and it is common to find anomalous sulphate reduction *rates* at this level.

This process can apparently occur at a range of scales from a local, few cm thick zone in typical sediments, to large areas associated with the escape of methane seeping from accumulations at depth. In the latter case it is also common to find authigenic carbonate nodules, sheets, or in one case large columns up to four metres high (Jensen *et al.*, 1992). Such carbonate, frequently dolomitic, typically has $\delta^{13}C$ ratios as negative as -60 $^o/_{oo}$ (more typically -20 to -40 $^o/_{oo}$), indicating a source of carbon from methane. Where such carbonates are exposed at the surface, they are frequently corroded and it seems likely that anaerobic methane oxidation is responsible for the formation of the carbonate, but CO_2 from aerobic methane oxidation when the carbonates are exposed leads to their dissolution (CO_2 forms a weak acid in water).

Associated with carbonate formation in deep water environments there is often a characteristic faunal assemblage which is fed from the energy released from methane and sulphide oxidation (*e.g.* see Kennicutt *et al.*, 1985; Dando *et al.*, 1991). These include *Beggiatoa* and related species of filamentous sulphur oxidising bacterial mat, large mussels containing methanotropic bacteria in their gills and giant tube worms containing various sulphur bacteria.

4.4.2. *Carbon and Hydrogen Isotope Effects*. Methane oxidation is also associated with substantial carbon and hydrogen kinetic isotope fractionation. The magnitude of the effect depends on the microbial species involved and the environmental conditions (probably mediated via a rate control). This process has been studied in the laboratory (Alperin *et al.*, 1988; Barker & Fritz, 1981; Coleman, *et al.*, 1981: Fig. 30) and in the field by several authors (Whiticar &Faber, 1985: Fig. 31). The CO_2 released is depleted in ^{13}C leaving the residual methane progressively enriched, frequently by a substantial amount. The actual value found is highly sensitive to environmental conditions since the fractionation factor depends on the methane oxidation rate (A. Zyakun, manuscript in preparation) and at present cannot be predicted for any given environment. This variability is clear when one considers the isotopic ratios of carbonates precipitated during methane oxidation (Fig. 32) which show values of anywhere between -10 and -60 $^o/oo$, and no clear correlation with mineralogy.

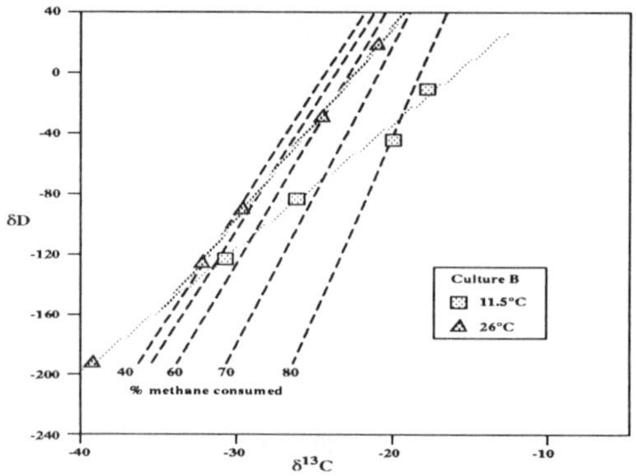

Figure 30. Carbon and hydrogen isotope fractionation during bacterial methane oxidation in the laboratory (data from Coleman *et al.*, 1981).

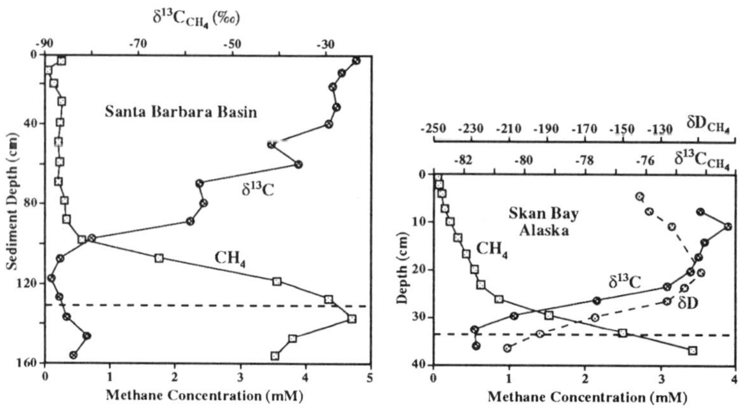

Figure 31. Carbon and hydrogen isotope variations in natural methane undergoing bacterial oxidation (Whiticar & Faber, 1985).

The effect on the hydrogen isotope values is even greater (Fig. 30), with the light hydrogen passing into both H_2S and H_2O. These latter phases have not yet been investigated, but in theory at least, this provides a mechanism for creating pore waters which are anomalously depleted in deuterium.

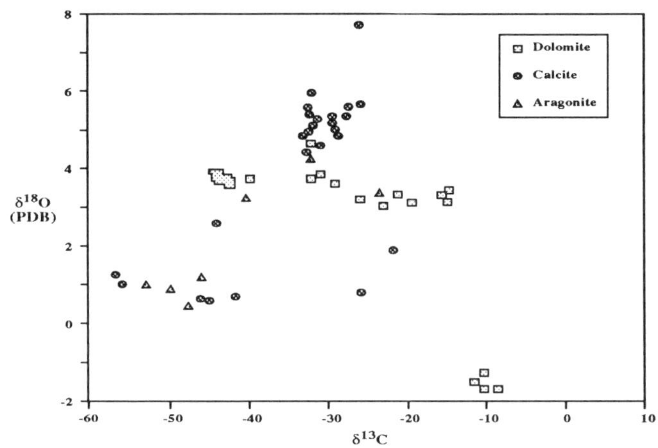

Figure 32. Carbon versus oxygen isotope ratio for carbonates associated with bacterial methane oxidation. Data from Jorgensen (1989), Roberts *et al.* (1989), Matsumoto (1990) and unpublished data for the North Sea (UKCS block 15/25) and the Kattegat, Denmark.

The effect of the methane cycle on carbonate precipitation is therefore twofold. Carbonates can precipitate from the by-products of methane generation (usually as dolomites) and during reoxidation of the methane nearer the surface. In the case of the former it appears to be the unusual cases associated with "runaway" CO_2 reduction which are the most important; in the case of the latter precipitation occurs at localised sites associated with extensive methane seepage.

5. The Sulphur Cycle in Carbonate Sediments

In clastic sediments, most H_2S released during bacterial sulphate reduction is fixed by iron into sulphides. The iron is introduced into the sediment largely as iron oxides adsorbed on the surface of clay minerals. In carbonate sediments, where clay minerals are present in only minor amounts, there is less iron available so it is normal for a large excess of H_2S to be formed which results in a different suite of bacterial reactions. These reactions have been implicated in nodular chert formation (Clayton, 1986). The evidence for such reactions comes from geochemical analysis of "paramoudras".

5.1 FORMATION OF PARAMOUDRAS

Paramoudras are round, barrel or columnar shaped flints, typically a metre or so across, which have formed around a central organic rich burrow in otherwise organically lean chalks (Bromley *et al.*, 1975). The central burrow is associated with phosphatisation, glauconitisation, pyritisation, and calcite cementation of a chalk core within the flint annulus. Around the central burrow all the main diagenetic products observed in chalks are closely associated, and all are obviously related in some way to the burrow itself. Dilution of clay minerals in the cemented chalk core relative to the host sediment, suggests that the paramoudra formed in a sediment of about 75-80% porosity, in the uppermost ten metres or so, soon after deposition. Clearly, the mineral assemblage is related to bacterial reactions during early diagenesis.

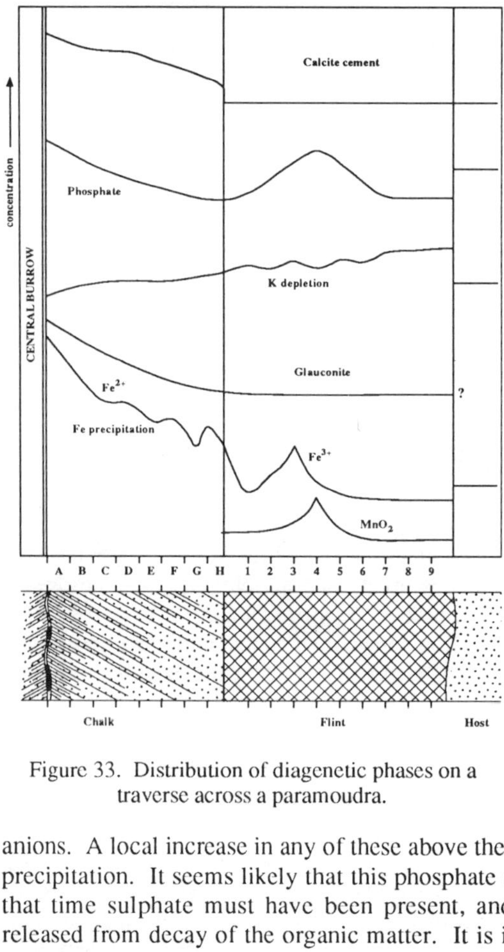

Figure 33. Distribution of diagenetic phases on a traverse across a paramoudra.

Figure 33 shows schematically the distribution of trace elements and diagenetic phases along a traverse from the central burrow, through the cemented chalk core, the flint, and into the host sediment (from Clayton, 1986). The presence of pyrite (shown as Fe^{2+}) in association with calcite cement suggests that bacterial sulphate reduction of organic matter, in the central burrow, is a key process in the formation of the paramoudra. The carbon isotope ratios of the carbonate (Fig. 34) also point to an input of isotopically light carbon from organic matter, although the effects are very subtle, reflecting dilution of the organic-derived carbon with inorganic carbonate. (The anomalously light carbon and oxygen isotope ratios in the host chalk are caused by a later diagentic overprint from meteoric waters, commonly observed in the chalk.)

The glauconite contains both Fe^{2+} and Fe^{3+}, suggesting that it formed before the pyrite, in the iron reduction suboxic zone, or possibly during alternating oxidising/reducing conditions.

For the phosphate mineral to precipitate phosphate, carbonate and sulphate must be present since this mineral (carbonate fluorapatite) contains all three of these anions. A local increase in any of these above the solubility product of fluorapatite would lead to precipitation. It seems likely that this phosphate also formed during sulphate reduction, since at that time sulphate must have been present, and carbonate and phosphate would have been released from decay of the organic matter. It is interesting to compare the concentration of the phosphate within the paramoudra to that in the host. If phosphates precipitated only around the central burrow the concentration should be higher than in the host sediment. This is not the case however. There is more phosphate in the host than in the outer part of the cemented chalk core of the paramoudra or in the flint. This suggests growth of phosphate in the host sediment after the paramoudra formed. In other words, the reducing conditions which led to phosphate formation around the burrow developed later in the surrounding sediment. The burrow represented a locally reducing environment in an otherwise more oxidised sediment, which did not become reducing itself until later following further deposition.

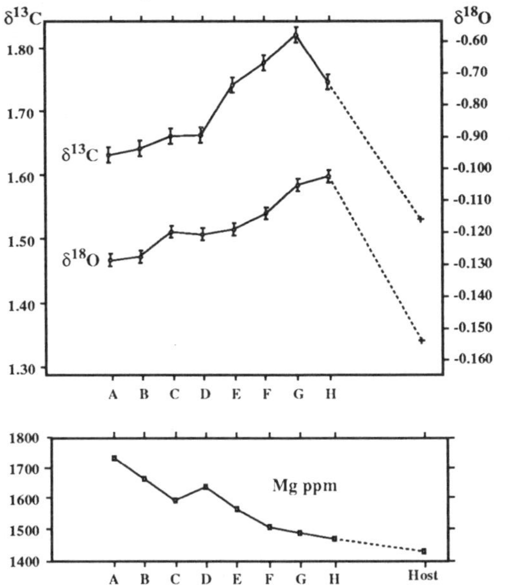

Figure 34. Carbon and oxygen isotope ratios within the cemented chalk core of a paramoudra and cement content as revealed by elevated Mg concentration.

There is further support for the burrow causing localised reducing conditions in the distribution of other trace elements. Consider potassium for example. This shows a systematic depletion towards the burrow in both the flint and the chalk core, which is even more pronounced in the host sediment. This depletion is thought to represent leaching of potassium from clay minerals (illite and probably the glauconite) during intense sulphate reduction, implying that sulphate reduction in the host sediment did not occur until after the signature had been frozen into the paramoudra. Iron and manganese also both show evidence of a change in redox conditions between the central burrow and the host sediment. Both of these elements are highly soluble in their reduced state (Fe^{2+} and Mn^{2+}) but insoluble in their oxidised state ($Fe^{3+}$and Mn^{4+}) Bacterial reduction of these elements in the vicinity of the burrow would lead to dissolution and migration away from that site to precipitate at the boundary with oxidising conditions. A concentration of these elements is found within the flint annulus around the burrow, suggesting that the flint formed around the oxic-anoxic boundary.

Therefore paramoudras appear to have formed around a central burrow where bacterial oxidation of organic matter led to locally reducing conditions within a generally oxidised environment.

5.2 FLINT FORMATION

In order to understand this process, it is important to consider the mass balance during sulphate reduction in the burrow. From dilution of clay minerals, it can be shown that the amount of cement in the chalk core varies from about 66% near the burrow to 54% close to the flint. Volumetrically, there is about 62 cm^3 of cement per cm of burrow length, equivalent to 1.67 moles if it is pure calcite. Assuming an input of organic-derived carbon of -25 $^o/_{oo}$, the carbon isotope ratios suggest that about 6% of the cement is derived from organic matter, which equates to a minimum of 0.10 moles of CO_2 released per cm of burrow. From the stoicheiometry of the sulphate reduction process, this implies that about 0.05 moles of sulphide were also released:

$$2CH_2O + SO_4^{2-} \rightarrow S^{2-} + 2CO_2 + 2H_2O$$

The pyrite content of the chalk core can be estimated from the iron concentration: 0.43 g/cm of burrow, or 0.01 moles of sulphide. At least 0.04 moles of sulphide has been lost from the system for each cm of burrow and this must have diffused away to the surrounding environment.

The H_2S would have arrived at the boundary with oxygenated conditions surrounding the cementing chalk core. Under these conditions, another specialised group of bacteria, the chemolithotrophic sulphur oxidising bacteria, will take advantage of the situation. The bacteria concerned are principally the *Thiobacilli* (Roy and Trudinger, 1970) and these oxidise sulphide, via a range of intermediates, to sulphate:

$$H_2S + 2O_2 \rightarrow SO_4^{2-} + 2H^+$$

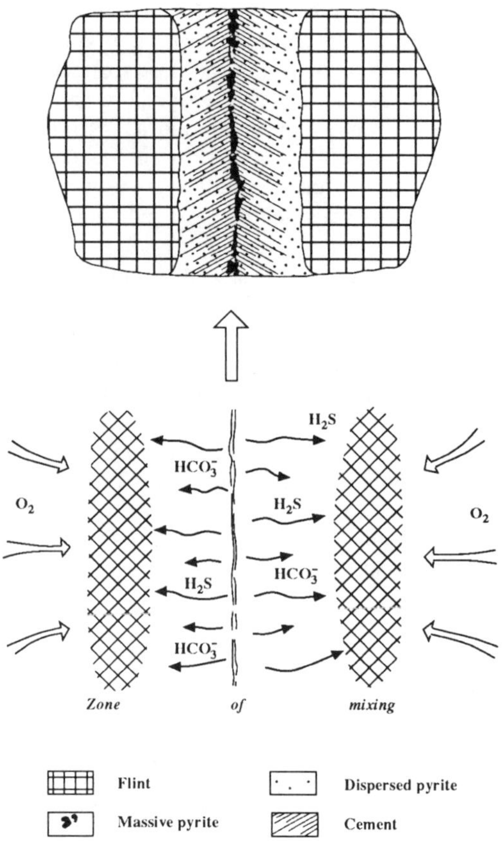

Figure 35. Model for silicification in paramoudras.

Such a release of acidity would be buffered by chalk dissolution occuring in an annulus around the cemented chalk core. It is the consequent release of carbonate into solution which probably caused seeding of silica from supersaturated solution at this site to form the flint, since silica would have been abundant from dissolution of biogenic opal. The high concentrations of dissolved carbonate and regenerated sulphate apparently also caused localised phosphate precipitation in the centre of the zone of silicification (Fig. 33).

There is further evidence of this process from sulphur isotope ratios. Sulphate reduction discriminates against ^{34}S, resulting in H_2S which is up to about 40 $^{o}/oo$ lighter than the sulphate, but sulphide oxidation results in negligible fractionation (Nakai and Jensen, 1964). If the newly regenerated sulphate from the site of flint formation returns to the burrow to be reduced once again (*i.e.* sulphide-sulphate reflux) anomalously light sulphur would be preserved in pyrite around the burrow. This is indeed the case. The $\delta^{34}S$ of burrow-associated pyrite varies between -22.5 and -41.4 $^{o}/oo$, corresponding to 38 to 57 $^{o}/oo$ lighter than contemporary seawater sulphate.

The complete sequence of microbial processes associated with the burrow is therefore as follows. The burrow represented a locally reducing environment within an otherwise more oxidising environment. As the burrow passed into the suboxic zone there was a period of glauconitisation while the (limited) iron was being reduced. Iron reduction was not necessarily mediated by bacteria but this seems likely; bacteria such as *Micrococcus lactilyticus* are known to reduce ferric hydroxides (Woolfolk and Whiteley, 1962). With continued sedimentation the burrow passed into sulphate reducing conditions while the surrounding sediment remained oxygenated (Fig. 35). Sulphate reducing bacteria led to calcite cementation and phosphate precipitation around the burrow and leaching of potassium from clay minerals and the glauconite. This produced an excess of H_2S which diffused outwards to be oxidised back to sulphate by other bacteria at the boundary with aerobic conditions. This sulphate was then available to diffuse back to the burrow to continue in the sulphate reduction process. The reduction in pH associated with sulphide oxidation caused large scale calcite dissolution and consequent seeding of silica from solution to form the flint. Meanwhile, reduced iron and manganese also diffused outwards and precipitated during oxidation at the oxic-anoxic boundary.

More widespread sulphate reduction, with upward diffusion of H_2S and oxidation at the base of the aerobic zone, has been suggested as the cause of bedded chert nodules in fine grained carbonate sediments (Clayton, 1986). This does not happen in clastic-rich sediments because there is sufficient iron available to fix the excess H_2S as sulphides.

The important points to note from this example are that it is *local* reducing conditions which lead to the anomalous conditions which are reflected as diagenetic products, and it is the *recycling* of sulphate in the sulphide-sulphate reflux process which controls the magnitude of the effects.

6. Summary

There are a number of general points we can draw from the above discussion.

Firstly, bacterial alteration of organic matter during early diagenesis is a very important process in controlling many inorganic precipitation reactions. Carbonates and pyrite are the most common of these, but glauconite, phosphates and silica may also be controlled in this way.

Secondly, it is rare that a single bacterial species is involved in the process. It is usually the result of bacterial groups working together. Furthermore, it tends to be the anomalous situations which have most effect, such as extreme CO_2 reduction during methanogenesis, or repeated recycling of sulphide or methane.

Finally, we cannot take a simple estimate of the carbon isotope ratio of bulk organic matter as a guide to $\delta^{13}C$ of CO_2 released during bacterial reactions. The variation in $\delta^{13}C$ between different compound groups, along with kinetic and intramolecular isotope effects are significant as is dilution with inorganic carbonate.

References

Alperin, M.J., Reeburgh, W.S. & Whiticar, M.J. (1988) Carbon and hydrogen isotope fractionation resulting from anaeorobic methane oxidation. *Global Biogeochemical Cycles*, 2, 279-88.

Balabane, M., Galimov, E.M., Hermann, M. & Letolle, R. (1987) Hydrogen and carbon isotope fractionation during experimental production of bacterial methane. *Org. Geochem.*, 11, 115-19.

Barker, J.F. & Fritz, P. (1981) Carbon isotope fractionation during microbial methane oxidation. *Nature*, 293, 289-291.

Belyaev, S.S., Laurinavichus, K.S. & Gayton, V.J. (1977) Modern microbial formation of methane in Quaternary and Pliocene rocks of the Caspian Depression. *Geochem. Int.*, 14, 172-6.

Benner, R., Fogel, M.L., Sprague, E.Kent & Hodson, R.E. (1987) Depletion of ^{13}C in lignin and its implications for stable carbon isotope studies. *Nature*, 329, 708-710.

Bromley, R.G., Schultz, M.-G. & Peake, N.B. (1975) Paramoudras: giant flints, long burrows and the early diagenesis of chalks. *K. Dansk. Vidensk. Selsk. Geol. Skr.*, 20, pt. 10.

Coleman, D.D., Risatti, J.B. & Schoell, M. (1981) Fractionation of carbon and hydrogen isotopes by methane-oxidising bacteria. *Geochim. Cosmochim. Acta*, 45, 1033-1037.

Coleman, M.L., Fleet, A. & Donson, P. (1982) Preliminary studies of manganese-rich carbonate nodules from leg 68, site 503, eastern equatorial Pacific. In Prell, W.L., Gardner, J.V., *et al., Init. Repts., DSDP*, 68: Washington (US Govt. Printing Office), 481-489.

Claypool, G.E. & Threlkeld, C.N. (1983) Anoxic diagenesis and methane generation in sediments of the Blake Outer Ridge, DSDP site 533, Leg 19. In Sheridan, R.E., Gradstein, F.M. *et al., Init. Repts. DSDP*, 76, Washington (US Govt. Printing Office), 391-402.

Claypool, G.E., Threlkeld, C.N., Mankeiwicz, P.N., Arthur, M.A. & Anderson, T.F. (1985) Isotopic composition of interstitial fluids and origin of methane in slope sediments of the Middle America Trench, DSDP Leg 84. In von Heune, R., Auboine, J., *et al., Init. Repts. DSDP,* 84, Washington (US Govt. Printing Office), 681-691.

Clayton, C.J. (1986) The chemical environment of flint formation in Upper Cretaceous Chalks. *In* Sieveking, G. de G. & Hart, M.B., *The Scientific Study of Flint and Chert,* Cambridge University Press, Cambridge, 43-54.

Clayton, C. J. (1992) Source volumetrics of biogenic gas generation. In Vially, R. (ed.), *Bacterial Gas*. Editions Technip, Paris.

Dando, P.R., Austen, M.C., Burke, R.A.Jr., Kendall, M.A., Kennicutt, M.C., Judd, A.G., Moore, D.C., O'Hara, S.C.M., Schmaljohann, R. & Southward, A.J. (1991) Ecology of a North Sea Pockmark with an active methane seep. *Mar. Ecol. Prog. Ser.*, 70, 49-63.

Deines, P. (1980) The isotopic composition of reduced organic carbon. In Fritz, P. & Fontes, C.H. (eds.) *Handbook of Environmental Isotope Geochemistry v.1.* Elsevier, New York. 329-406.

Durand, B. (1980) *Kerogen. Insoluble Organic Matter in Sedimentary Rocks.* Edition Technip, Paris, 519pp.

Espitalie, J., Laporte, J.L., Madec, M., Marquis, F., Leplat, P., Paulet, J. & Boutefeu, A. (1977) Methode rapide de characterisation des roches meres, de leur potential petroleur et de leur degre d'evolution. *Rev. Inst. Fr. Pet.*, 32, 23-42.

Froelich, P.N., Klinkhammer, G.P., Bender, M.L., Luedtke, N.A., Heath, G.R., Cullen, D., Dauphin, P., Hammond, D. Hartman, B. & Maynard, V. (1979) Early diagenesis of organic matter in pelagic sediments of the eastern equatorial Atlantic: suboxic diagenesis. *Geochim. Cosmochim. Acta*, 43, 1075-90.

Galimov, E.M. (1985) T*he Biological Fractionation of Isotopes.* Academic Press, London, 261pp.

Games, L.M., Hayes, J.M. & Gunsalis, R.P. (1978) Methane producing bacteria: natural fractionation of the stable carbon isotopes. *Geochim. Cosmochim. Acta,* 42, 1295-7.

Irwin, H., Curtis, C. & Coleman, M.L. (1977) Isotopic evidence for the source of diagenetic carbonates formed during burial of organic-rich sediments. *Nature*, 269, 209-13.

Jenden, P.D. & Kaplan, I.R. (1986) Comparison of microbial gases from the Middle America Trench and Scripps Submarine Canyon: implications for the origin of natural gas. *Appl. Geochem.* 1, 631-646.

Jenkyns, H.C. & Clayton, C.J. (1986) Black shales and carbon isotopes in pelagic sediments from the Tethyan Lower Jurassic. *Sedimentology*, 33, 87-106.

Jensen, P., Aagaard, I., Burke, R.A.Jr., Dando, P.R., Jorgensen, N.O., Kuijpers, A., Laier, T., O'Hara, S.C.M. & Schmaljohann, R. (1992) "Bubbling reefs" in the Kattegat: submarine landscape of carbonate cemented rocks support a diverse ecosystem at methane seeps. *Mar. Ecol. Prog. Ser.*, 83, 103-112.

Jorgensen, N.O. (1989) Holocene methane-derived dolomite cemented pillars from the Kattegat, Denmark. *Mar. Geol.*, 88, 71-81.

Kennicutt, M.C., Brooks, J.M., Bidigare, R.R., Fay, R.R., Wade, T.L. & McDonald, S.J. (1985) Vent-type taxa in a hydrocarbon seep region on the Lousiana slope. *Nature*, 317, 351-352.

Krey, J. (1970) Die Urproduction des Meeres. In Dietrich, G. (ed.), *Erforschung des Meeres.* Umschau, Frankfurt, 183-195.

Krzycki, J.A., Kenearly, W.R., De Niro, M.J. & Zeikus, J.G. (1987) Stable isotope fractionation by *Methanosarcina barkeri* during methanogenesis from acetate, methanol, or carbon dioxide-hydrogen. *Appl. Environ. Microbiol.*, 53, 2597-9.

Lazerte, B.D. (1981) The relationship between total dissolved carbon dioxide and its stable carbon isotope ratio in aquatic sediments. *Geochim. Cosmochim. Acta*, 45, 647-656.

Lee, J.L. (1983) *Microbiology*. Barnes & Noble, New York, 282pp.

Matsumoto, R. (1990) Vuggy carbonate formed by hydrocarbon seepage on the continental shelf of Baffin Island, northeast Canada. *Geochem. J.*, 24, 143-158.

Metzner, H. (1973) *Biochemie der Pflanzen*. Enke, Stuttgart.

Nakai, N. (1962) Geochemical studies on the formation of natural gases. *Jour. Earth Sci. Nagoya Univ.*, 10, 71-111.

Nakai, N. & Jensen, M.L. (1964) The kinetic isotope effect in the reduction and oxidation of sulfur. *Geochim. Cosmochim. Acta*, 28, 1893-1912.

Patience, R.L., Clayton, C.J., Kearsley, A.T., Bishop, A.N., Rees, A.W.G., Bibby, K.G. & Hopper, A.C. (1990) An integrated biochemical, geochemical and sedimentological study of organic diagenesis in sediments from leg 112. In Suess, E. & von Huene, R. *et al.*, *Proc. ODP*, *Sci. Results*, 112: College Station, TX (Ocean Drilling Program), 135-153

Redfield, A.C. (1958) The biological control of chemical factors in the environment. *Am. Sci.* 46, 206-226.

Roberts, H.H., Sassen, R., Carney, R. & Aharon, P. (1989) ^{13}C-depleted authigenic carbonate buildups from hydrocarbon seeps, Lousiana continental slope. *GCAGS Trans.*, 39, 523-530.

Robins, G.V. (1980) *Food Science in Catering*. Heinemann Ltd., London, 234pp.

Rosenfeld, W.D. & Silverman, S.R. (1959) Carbon isotope fractionation in bacterial production of methane. *Science*, 130, 1658-9.

Roy, A.B. & Trudinger, P.A. (1970) *The Biochemistry of the Inorganic Compounds of Sulphur*. Camb. Univ. Press, Cambridge, 400pp.

Schlegel, H.G. (1969) *Allgemeine Mikrobiologie*. Thieme, Stuttgart.

Schoell, M. (1982) Stable isotopic analyses of interstitial gas in Quaternary sediments from the Gulf of California. In Curray, J.R., Moore, D.G. *et al., Init. Repts. DSDP*, 64(2), Washington (US Govt. Printing Office), 815-7.

Singleton, P. & Sainsbury, D. (1981) *Introduction to Bacteria.* Wiley & Sons, Chichester, 167pp.

Tissot, B.P. & Welte, D.H. (1978) *Petroleum Formation and Occurrence.* Springer Verlag, Berlin, 538pp.

Ward, D.M. & Winfrey, M.R. (1985) Interactions between methanogenic and sulphate reducing bacteria in sediments. In *Advances in Aquatic Microbiology v.3*, Academic Press, London, 141-179.

Whiticar, M.J. & Faber, E. (1985) Methane oxidation in sediment and water column environments - isotope evidence. *Org. Geochem.*, 10, 759-768.

Woltemate, I., Whiticar, M.J. & Schoell, M. (1984) Carbon and hydrogen isotope compositions of bacterial methane in a shallow freshwater lake. *Limnol. Oceanogr.*, 29, 985-992.

Woolfolk, C.A. & Whiteley, H.R. (1962) Reduction of inorganic compounds with molecular hydrogen by *Micrococcus lactilyticus. J. Bacteriol.*, 82, 647-658.

CLAY MINERAL DIAGENESIS

H. CHAMLEY
Dynamique sédimentaire et structurale, SN 5, URA 719 CNRS
Lille 1 University
59655 Villeneuve d'Ascq
France

ABSTRACT. A condensed review of the main processes of clay diagenesis in sedimentary series is presented by using specific examples issuing from the literature. The Mexico Gulf Coast drill sites illustrate clay diagenesis with depth of burial, involving mainly the illite-to-smectite transition and becoming active when overburden exceeds 2 km thickness. The Galapagos spreading center documents the case of glauconite-like minerals developed from illite-smectite in hydrothermal, chemically-concentrated environment. North Sea wells allow to study the formation of kaolinite, illite and/or chlorite in sandstones submitted to fluid migrations. The significance of Atlantic Cretaceous to Paleogene series rich in Al-Fe smectite is discussed in terms of transformation of volcanic materials, early diagenesis of common mud, and detrital supply from dominantly pedologic exposed formations. A special attention is paid to the relationships between clay diagenesis and petroleum generation, and to the transition from the paleoenvironmental to the diagenetic expression of clay suites.

1. Introduction

The purpose of this paper is to provide information on the different aspects of the clay mineral diagenesis, by focusing on some recent and quantitative approaches emphasizing the progress reached during the last 15 years. Priority is given to practical examples concerning sedimentary series.

The diagenetic history of a sediment is conventionally considered as starting after the deposit is buried and permanently protected from the direct influence of the open aqueous environment. Diagenetic, i.e. post-sedimentary effects therefore differ form syngenetic, i.e. syn-sedimentary or hydrogenous effects, the latter being characterized by free exchanges between liquid and solid environments. Of course the boundary of syngenetic and diagenetic processes is transitional. In the same way the boundary between non-diagenetic and diagenetic processes is transitional, and we will briefly explore this transition.

The basic clay diagenetic processes wich we successively consider include the following types : diagenesis with depth of burial, involving an overburden more than 2 km thick, except in the case of abnormally high geothermal gradient ; diagenesis in hydrothermal, chemically concentrated environment ; diagenesis duc to fluid migration allowed by high interstitial porosity and active ion exchanges ; diagenesis of volcaniclastic material favored by high specific surface and/or heat transfers. The last topic of the paper is devoted to the distinction between actual clay diagenetic effects, non-diagenetic effects and intermediate situations.

Most information in the present paper is extracted from Clay Sedimentology (Chamley 1989) pulished by Springer-Verlag who is gratefully acknowledged, as well as from textbooks by Kisch (1983), Singer and Müller (1983), Velde (1985), Odin (1988) and Weaver (1989). Notice that the basic data on clay mineralogy are supposed to be known (see Brindley and Brown 1980, Velde 1985, Chamley 1989, Weaver 1989).

A. Parker and B. W. Sellwood (eds.),
Quantitative Diagenesis: Recent Developments and Applications to Reservoir Geology, 161–188.

2. Diagenesis with Depth of Burial

2.1. THERMODYNAMIC RANGE AND GENERAL TREND

The most prevalent and best-known diagenetic changes affecting the clay minerals with increasing burial depth involve the conversion of smectite to illite, and the improvement of illite crystallinity. The smectite-to-illite conversion actually concerns the diagenetic range, whereas the illite crystallinity improvement mainly applies to the early stages of metamorphism, i.e. very low (anchizone) and low (epizone) grades. This is the reason why Weaver et al. (1984) proposed to divide the clay diagenetic evolution into three stages based on the illite/smectite (I/S) ratio in the less than two micrometers fraction :

Early Diagenesis : Ranges from below the water-mud contact to the stage where a regular mixed-layer I/S is, or should be, present (approximately 60 % illite layers ; 90 to 140° C), i.e., at least 3000 to 5000 m below the sediment surface.

Middle Diagenesis : Ranges from the first development of a regular mixed-layer phase to the disappearance of a discrete glycolated peak for the mixed-layer phase (≈ 90 % illite layers ; ≈ 200° C). This corresponds to a burial depth of 6 to 7 km.

Late Diagenesis : Stage where the glycolated mixed-layer phase appears as an integral part of the illite peak (< 10 % smectite layers ≈250° to 280° C).

Many investigations have been performed on the subject the last ten years, and the time is still too early for an extensive coverage. Let us consider the average case of clay-rich series where the burial diagenesis primarily consists of the progressive illitization of smectite minerals. Subregular illite-smectite mixed-layer and allewardite-like minerals usually develop above 100° C in series buried more than 3 kilometers, but higher temperature may allow a clay ordering in much less deeply buried sediments. The last stages of clay mineral evolution under increasing thermodynamical constraints are characterized by the superlattice ordering defined by Reynolds and Hower (1970 ; Type IISI, three illite and one smectite layers), and by the illite/chlorite facies. The latter facies, that appears to depend on high temperatures more frequently than on high pressures, correlates with an increase of mineral grain size and crystallinity.

2.2. AN EXAMPLE : GULF OF MEXICO COAST SEDIMENTS

The sediments deposited form Eocene to Pleistocene off Mexico to Florida coasts comprise shale-rich series extending over several kilometers thickness and extensively drilled by oil companies. The sediments chiefly belong to Anahuac and expecially Frio formations. As early as 1959, Burst noted a progressive modification in X-ray diffraction pattern of smectite and the subsequent disappearance of the mineral with increasing depth of burial. First quantitative data on the mineral and chemical evolution in Gulf Coast sediments were reported by Perry and Hower (1972). The smectite-rich layers dominate the clay mineralogy in shallow-buried deposits and are progressively replaced by illite-rich layers, down to 5500 meter depth. The decrease of expandability of illite-smectite minerals ranges from about 80 % to 20 % smectite layers with increasing depth. The clay fraction shows a downward increase of potassium attributed to an interlayer fixation of the element in illite-smectite as the proportion of illite layers increases. The diagenetic reaction depends on the depth of burial mainly through temperature effects. Reynolds and Hower (1970) reported from various examples the existence of three types of mixed-layering : random, ordered (allevardite-like organization), and superlattice units (three illite and one smectite layers, labeled as type IISI). Illite-smectite minerals with 40 to 100 % smectite layers belong to random mixed-layers, and mostly characterize sediments buried at less than 3.5 kilometers.

Allewardite-like and superlattice varieties correspond to highly evolved sediments poor in expandable layers (about 10 %) and often buried deeper than about 6 kilometers.

A detailed examination of burial diagenetic reactions in Oligocene-Miocene Gulf Coast shales is performed by Hower et al. (1976), Aronson and Hower (1976), and Yeh and Savin (1977). These authors consider the mineralogy, the inorganic chemistry and the radiogenic and stable isotope chemistry of bulk materials and different grain-size fractions from the 1250 to 5500 m depth-interval of the well CWRU-GC6. The major mineralogic changes, mainly marked by and increase in illite layers within smectite-illite, occur over the depth interval 2000 to 3700 meters, after which no significant modifications are detectable. The reduction of illite-smectite expandability is more important for very small grain-sized fractions than for larger fractions (Fig. 1A). Between 2000 and 3700 meters, calcite abundance decreases from about 20 % of the rock to almost zero, disappearing from progressively coarser fractions with increasing depth. Potassium feldspar decreases to zero, while chlorite appears to slightly increase in amount (Fig. 1B). The distribution of chemical elements shows an opposite behavior of potassium in coarse and fine fractions (Fig. 1C), suggesting a transfer from sandy minerals to newly formed illitic clay-layers. The strong increase of potassium in the smallest size-fraction (< 0.1 µm) is associated with an increase in aluminum and a decrease in silicon, while quartz abundance increases in coarser fractions. The potassium and aluminum appear to be derived from the diagenetic decomposition of potassium-feldspar and of mica, and the excess of silicon probably allows the formation of quartz.

Hower et al. (1976) consider that the shale acts as a closed system for all components excepts H_2O, CaO, Na_2O and CO_2. The reality of the diagenetic process is confirmed by isotopic dating (Aronson and Hower 1976). The whole-rock apparent K-Ar ages decrease with increasing depth, pointing to the existence of dominantly detrital minerals, derived from old formations in the upper part of the well, and of diagenetically rejuvenated minerals in the lower part. Yeh and Savin (1977) investigate the oxygen-isotope composition in three Gulf Coast wells. Their results indicate that the Cenozoic sediments do not constitute isotopically equilibrated systems, even those buried at 5600 meters where temperatures are as high as 170° C. The oxygen-isotope disequilibrium among the clay fractions decreases as burial temperature increases. The isotope exchange between clay and pore water becomes more important in the temperature range 55 to 90° C, in which the most extensive conversion of smectite layers to illite layers occurs within mixed-layer phases.

Notice that the estimated temperatures (from geothermal gradient maps) at which the first illite layers develop at the expense of smectite vary from 60 to 100° C (2 to > 3 km depth), and those at which ordered I/S mixed-layers appear vary from 120 to 160° C (4 to > 5 km). Most wells are not deep enough for allewardite-like minerals or superlattice ordered groups to be observed. Notice also that the residence time of a given series at a sufficiently high temperature may control in a significant way the clay diagenetic reactions. For instance Ramseyer and Boles (1986) show in Miocene rocks from the San Joaquin basin, California, that series buried less than 900 meters display clay diagenetic changes much more important than contemporary series buried more than 2500 meters but submitted at critical temperatures during a shorter residence time (Fig. 2). For example, areas containing I/S with > 95 % smectite had a residence time between 100° and 120° C of 2.0 to 2.1 m.y., whereas areas containing I/S with 30 % smectite had a residence time at these temperatures of 2.5 m.y. The near horizontal distribution of various I/S ratios leads the authors to conclude that the change from highly expandable I/S to a slightly expandable I/S occurred over a narrow temperature interval (10° to 20°).

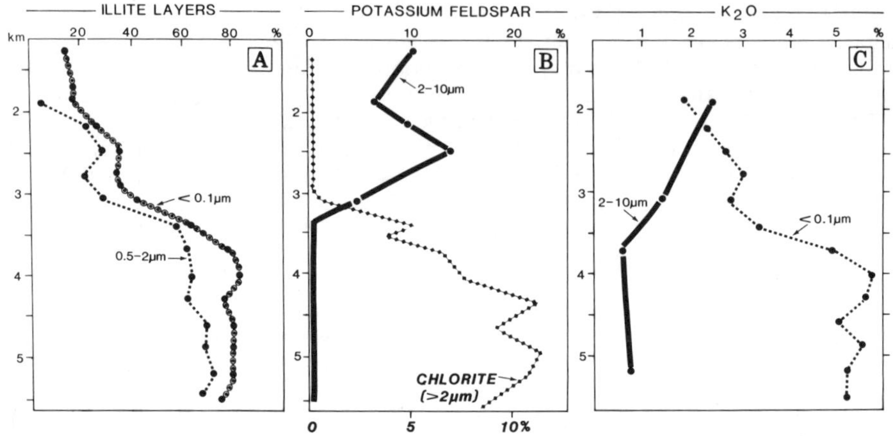

Figure 1. CWRU Gulf Coast well 6, late Oligocene-Miocene. Vertical distribution of : A illite layers in illite-smectite ; B chlorite and K-feldspar percent ; CK_2O in different ignited, CaO free fractions (% weight) (Afer Hower et al. 1976).

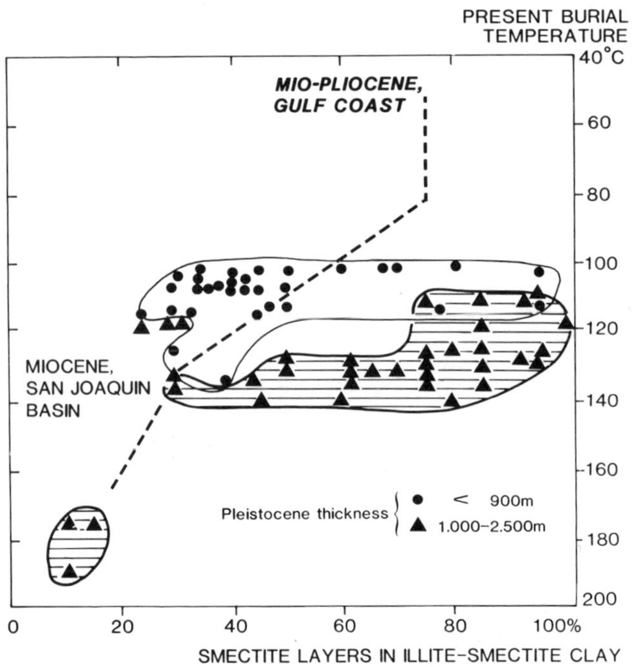

Figure 2. Present-day burial temperature vs percentage of smectite layers in < 2 µm illite-smectite from Miocene rocks of the San Joaquin basin, California (After Ramseyer and Boles 1986). Gulf Coast reference curve from Perry and Hower (1972).

2.3. MECHANISM

2.3.1. *Open Versus Closed Systems*

Most modern investigations conclude to short-distance transportation processes of chemical elements needed for the smectite-to-illite conversion in shaly sediments. This is confirmed by high-resolution transmission electronmicroscopy, that shows the growth of small packets of illite layers within subparallel layers of smectite matrix (Ahn and Peacor 1986). Even at 5500 meters non-equilibrated assemblages of relatively reactive clay minerals exist. The coexistence, at a distance of a few tens of angströms, of smectite and illite layers and of structural imperfections implies that ionic homogenization is very limited. As water is the principal medium for allowing ion transport, this means that diagenetic sediments have not been affected by important water migration, at least not since their initial compaction. Ahn and Peacor (1986) conclude that the argillaceous Gulf Coast deposits behaved essentially as a nearly closed system, water being present in very small amounts compared to solid materials, and acting as a catalyst for short-distance ion transport and for local clay reconstruction at the reaction interfaces.

By contrast to Gulf Coast shales, open systems may occur in more permeable or more active geothermal environments. This is the case of the shales from the Salton Sea geothermal field (Yau et al. 1987). Individual, euhedral to subhedral illite crystals develop in open pore space during burial diagenesis, and probably result from a direct crystallization from solutions. The components derive partly by dissolution of detrital smectite, whose abundance decreases with increasing depth. As smectite is not observed by transmission electronmicroscopy in the immediate proximity of neoformed illite, the ions necessary for illite growth issue at least partly from relative distant sites. This supposes a high relative water/rock ratio, high proportions of pore space, and active flow of solutions. In the shallower sediments of Salton Sea geothermal field detrital smectite is also observed to be in part directly replaced by illitic layers, similarly to what happens in nearly closed microenvironments of Gulf Coast sediments (Yau et al. 1987). This points to the existence of intermediate situations between closed and open systems.

Other evidence for the mainly local character of chemical exchanges in deeply buried sediments arises from the existence of local microenvironments where specific mineral formations take place. Pye et al. (1986) perform backscattered electronmicroscopic observations and energy-dispersive X-ray analyses on Gulf Coast CWRU no.6-well sediments already studied by Hower et al. (1976). Above 2750 meter depth, calcite foraminifera tests are partly filled with authigenic pyrite, kaolinite and chlorite, whith some unoccupied pore space. Such new minerals do not occur in other shale constituents adjacent to the tests. Between 2750 and 3400 meters the remaining voids are filled with authigenic chlorite or illite, while the calcareous tests themselves are dissolved. In the same depth interval iron-rich chlorite replaces zoned carbonate rhombs formerly formed in sulfate reduction and fermentation zones.

2.3.2. *Transformation Versus Solution-Precipitation*

a) **Transformation through mixed-layering**

Several mechanisms have been put forward to explain the progressive passage from smectite-rich to illite-rich minerals at increased depths of burial. Hower et al. (1976) first proposed a precise mechanism of mineral transformation for explaining the increase in the proportion and ordering of illite layers at depth. The destruction of detrital potassium-feldspars and perhaps of mica was suggested to provide the potassium and aluminum required for the progressive transformation of smectite to illite. The substitution of octahedra-derived Al^{3+} for Si^{4+} in the tetrahedral layers of smectite was envisaged to be responsible for the fixation of K^+ in the interlayer positions by increasing the layer negative charge. The correlative release of Si^{4+}, Fe^{2+} and Mg^{2+} was supposed to participate in forming other new minerals like quartz and illite.

b) **Transformation with mixed-layering and cristallization**

Pye et al. (1986) observe that main depth-controlled mineralogical changes occur in the zone of organic matter decarboxylation. The morphological, mineralogical and chemical variations recorded by punctual microscopic and geochemical analyses lead these authors to propose a five-step illitization process, starting in the decarboxylation zone : (1) reduction of pH due to decarboxylation, and simultaneous iron reduction ; (2) carbonate dissolution ; (3) dissolution of detrital K-feldspars and micas ; (4) appearance of negative octahedral layer charge in mixed-layer clays, due to reduction of Fe^{3+} to Fe^{2+} ; (5) fixation of interlayered K^+ and substitution of Al^{3+} for tetrahedral Si^{4+}, the supplementary K and Al being derived from non-clay detrital silicates. In addition, a direct precipitation of illite and chlorite is reported in pore space, as well as a replacement of kaolinite by chlorite. K. Pye and colleagues therefore envisage the coexistence of transformation and dissolution-precipitation processes, the former being dominant and responsible for the downward decrease of expandability in illite-smectite layers.

c) An alternative mechanism of illitization, intermediate between transformation and dissolution-precipitation, is referred to as "smectite cannibalization" and documented from field studies and laboratory experiments by several authors. For instance Pollastro (1985) investigates the late Cretaceous-early Tertiary interbedded sandstones and shales in the Green River basin, Wyoming, as well as the late Cretaceous chalks to chalky shales of the Niobrara Formation, Colorado. Scanning electronmicroscopic observations on deeply buried rocks reveal the presence of illite-rich, ordered illite-smectite minerals growing on a substrate with a typically smectitic morphology. These textural characteristics suggest the **combined inheritance and solution-recrystallization of smectitic minerals**, with only slight chemical modification. A part of illite-smectite clay is supposed to have been destroyed under high temperature conditions, through a selective cannibalization of smectite layers. This partial, structural destruction would provide some or all of the chemical elements needed to build illite-smectite with a higher illite content. As a result, the number of clay layers in a given particle would tend to decrease in the course of time and with increasing depth, by progressive destruction of smectite layers and correlative feeding of illite layers.

d) **Dissolution-precipitation without mixed-layering**

McHardy et al. (1982) observed that a mineral displaying on X-ray diffraction diagrams an illite-smectite pattern looked on electronmicrographs like dioctahedral illite made up of very thin particles (about 3 nanometers thick). They supposed that smectite interlayers of the "illite-smectite" minerals were in fact the spaces existing between individual platy or lathed crystals of very thin illite. Within the areas where such small illite crystals overlap, ethylene-glycol or water could be adsorbed and swelling could take place as within true smectite interlayers. These ideas were tested on < 0.1 µm-sized fractions by Nadeau et al. (1984), who experimentally produced a wide range of materials that yielded X-ray diffraction patterns typical of mixed-layer minerals simply by making mixtures of elementary particles of different minerals in various combinations and proportions. Nadeau et al. (1984) deduced that many common mixed-layered clay minerals really consist of the superimposition of a few very thin particles of a single mineral species, that behave on X-ray diffraction diagrams like mixtures of this mineral and of smectite. These small and thin illite particles are therefore believed to result from the dissolution of detrital smectite followed by a recrystallization. When smectite particles are entirely dissolved, the remaining population would essentially consist of two to five elementary illite particles, that appear through X-ray diffraction to be regularly interstratified illite-smectite with at least 50 % illite particles. With increasing depth of burial and diagenesis, the thickness of fundamental illite particles increases within the population. When the particles become thick enough to allow dominant intraparticle diffraction effects (more than five stacked elementary layers), the clay material is identified by X-ray diffraction as illite. Nadeau et al. (1984) concluded that the diagenetic conversion of smectite to illite constitutes a recrystallization mechanism, where smectite dissolution and illite precipitation occur within a population of extremely small crystals without intermediate stage.

e) **Transformation without mixed-layering**

The micromorphological study of thin sections of Gulf Coast buried clays whose plane is normal to the clay layering provides spectacular new information (Ahn and Peacor, 1986). Smectite presents a texture of wavy 10- to 13-Å layers with a high density of edge dislocations, while illite is made up of relatively defect-free straight 10-Å layers. The microstructures of smectite and illite are not continuously parallel to (001) planes at smectite-illite interfaces. The relations between both mineral species are therefore characterized by the widespread occurrence of chemical, textural and structural discontinuities, and by the presence of separate discrete domains of either smectite or illite. These data suggest the absence of true mixed-layered clay minerals, the random or regular alternation of illite and smectite layers being replaced by small packets of either illite or smectite texturally discontinuous. Instead of a dissolution-precipitation process, Ahn and Peacor (1986) envisage a transformation process acting in the following way (Fig. 3) :
1) Initial smectite contains anastomosing layers with variable thickness and abundant dislocations, forming continuous structural units referred to as megacrystals. By contrast, illite consists of well-defined, straight and relatively defect-free layers arranged in packets within the subparallel imperfect smectite matrix. 2) Layers of illite grow discontinuously with layers of smectite in the along-layer direction. This growth is determined by the diffusion of K and Al trough the smectite matrix to the transition boundary, and by the diffusion of Na, Si, Fe and Mg away this boundary. The abundant defects in smectite are though to serve as pathways for ion transport. The transformation probably involves the partial breakdown of both octahedral and tetrahedral sheets, facilitated by the presence of interstitial and interlayered water. As a consequence, chemical arrangements occur in both

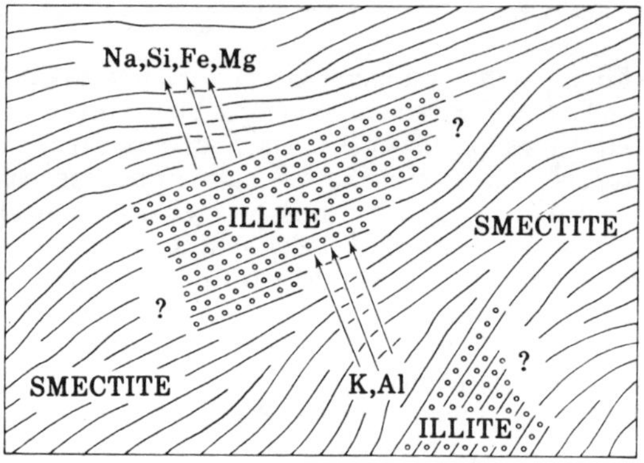

Figure 3. Interpretation of the direct transformation of smectite to illite in Gulf Coast argillaceous sediments, from high resolution transmission electron microscopic observations (After Ahn and Peacor 1986).

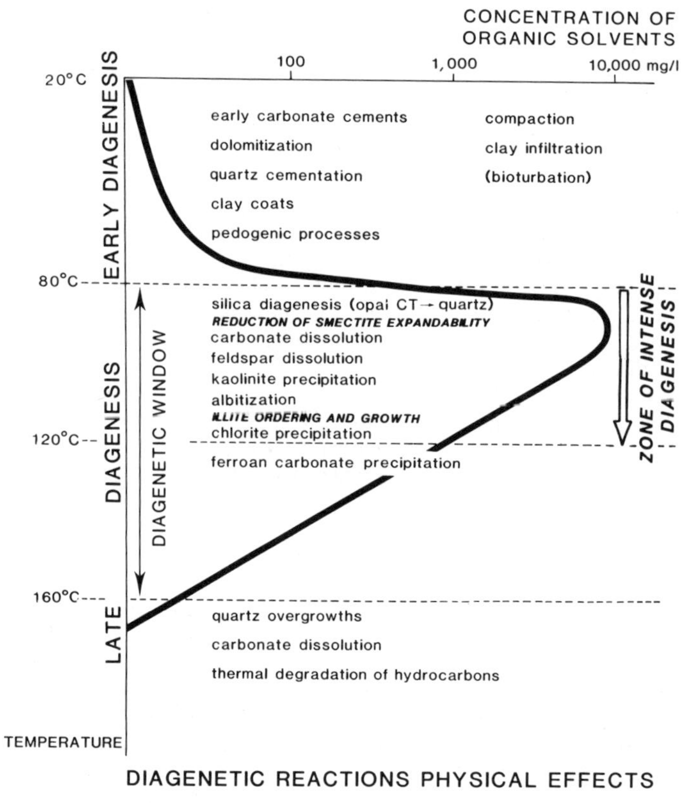

Figure 4. Correlations between diagenetic reactions and organic acids versus temperature, in oil field waters (After Surdam and Crossey 1987).

octahedral and tetrahedral sheets and lead to relatively defect-free illite-layers. 3) As the reaction progresses, the boundaries of illite packets advance until they coalesce, as indicated on high-resolution electron micrographs. Because illite inherits the layer orientation of the smectite where it growths, and because initial smectite layers are divergent, layers of illite packets become subparallel where they intersect, resulting in small-angle grain boundary-like features. The final texture at 5500 meters depth is dominated by thick illite packets with subparallel orientations.

Notice that similar high-resolution electronmicroscopical observations arise from the recent literature about the formation of noticeable amounts of chlorite, or even kaolinite, through burial diagenetic processes.

2.4. CLAY MINERAL CHANGES AND PETROLEUM GENERATION

The physical, mineralogical and chemical modifications experienced by clay minerals during burial diagenesis determine fundamental changes in the water and organic matter content, and in the potential of fluid migration. The nature and abundance of clay minerals in a given rock may significantly affect its reservoir or cap-rock properties. Clay minerals are therefore considered since a long time as useful tools in search for oil. New techniques develop through both borehole and laboratory investigations in order to better understand the control of burial constraints on organic-inorganic relations. Clay research concerns either argillaceous sediments possibly acting as parent rocks and transitory reservoirs, or sandy rocks serving as final reservoirs where clay coatings or infillings can take place. In addition to information on the petroleum history, clay mineral studies may supply data on the paleogeographic evolution of oil-related basins, which may possibly facilitate further exploration.

Argillaceous materials submitted to increasing depth of burial fundamentally experience a change in fabric due to compaction. This progressively tends to destroy the porosity and permeability down to a certain depth (generally about 3 km). Beyond this depth the chemical environment often becomes corrosive ; some minerals dissolve with consequently the development of a secondary porosity. Such an opposite evolution with increasing depth mainly characterizes sandy rocks subject to acidic groundwater migration, but may also affect more argillaceous series marked by relatively high pore pressures and water flows. Surdam and Crossey (1987) indicate that main diagenetic reactions involving inorganic constituents in sediments correlate with the maximum concentrations of organic solvents (Fig. 4). Such a correlation emphasizes the interdependance existing between mineral diagenesis and petroleum history.

The late diagenetic evolution of clay minerals consists essentially of the conversion of smectite to illite. This evolution is accompanied by a water explusion from smectite interlayers. The release of water during illitization is important in flushing out hydrocarbons from source rocks and transporting dissolved ions for the dissolution-precipitation of cements in adjacent reservoir sandstones. In addition, the water expulsion may determine overpressuring and hydro-fracturing in the rocks submitted to clay diagenesis. The clay dehydration stages apparently develop at temperatures higher than 80-90° C and characterize burial thicknesses of more than 2 to 3 kilometers. The importance of water release from clay for oil formation is proved by the frequent coincidence observed in the decrease of smectite expandable layers and the hydrocarbon appearance.

The presence of abundant smectite in sedimentary rocks submitted to burial effects appears to noticeably favor the production and migration of hydrocarbons (in Chamley 1989). First smectite interlayers may incorporate huge amounts of organic products that constitute potential precursors for hydrocarbons. Second smectitic clays act as important water reservoirs, that can provide through diagenesis the carrier necessary for hydrocarbon migration. In addition, the water adsorbed in smectite interlayers may help in resisting burial compaction and permeability decrease of clays and other sediments ; this could explain why smectite-bearing calcareous rocks commonly exhibit higher porosity than kaolinite-containing rocks. Finally, smectite constitutes a possible catalyst for decarboxylation, a very important step in petroleum generation.

The crystallinity of illite generally increases in the lower part of the late diagenetic zone only, but sometimes displays an earlier amelioration. Significant and irreversible increase of illite crystallinity statistically correlates with an increase of the hydrocarbon potential at depth (in Weaver 1989). Guthrie et al. (1986) demonstrate that illite crystallinity is significantly related to vitrinite reflectance, a very useful index in the search for oil within deeply buried series. They suggest that in the absence of vitrinite in a given buried series, illite crystallinity can be used quantitatively in order to estimate the levels of thermal maturity, and even to approximate hydrocarbon generation-preservation stages.

3. Diagenesis in the hydrothermal environment of the Galapagos Spreading Center

3.1. INTRODUCTION

Warm and ion-enriched fluids migrating through magmatic and sedimentary formations tend to influence, and often to accelerate the diagenetic reactions. For instance the conversion of smectite to illite, that usually occurs at depths of burial exceeding 2.5 kilometers, is almost completed at much shallower depth in Jurassic sediments of Scotland due to local convective hot-water circulation cells associated with Tertiary igneous intrusions (Andrews 1987 ; Yau et al. 1987). The clay minerals forming at the basalt/sediment interface or in sediments overlying hydrothermal exudations may be of different types : Mg-smectites (stevensite, saponite), fibrous clays (palygorskite, even sepiolite), celadonite, iron-smectite (nontronite), glauconite. The clay minerals are often minor sedimentary constituents relative to sulphides and oxides (Cu, Fe, Mn, Zn, Co ...), and sometimes to silica. Let us consider the Fe-smectite - glauconite transition in hydrothermal green clays of the mounds developped in the Galapagos Spreading Center, Eastern Central Pacific Ocean (Buatier et al. 1989).

3.2. GEOLOGICAL SETTING, SAMPLING AND METHODS

Hydrothermal mounds of the Galapagos Spreading Center are made up of several layers of green clays interbedded with siliceous, foraminiferal and nannofossil oozes, which are overlain by Fe-Mn oxyhydroxide crusts. The mounds are located in the Eastern Equatorial Pacific in areas of high heat flow (8 to 10 heat flow units) above small, vertical basement faults. Hole 509B, drilled during DSDP leg 70, coincides with IPOD-DSDP leg 54, site 424, and is located near 1°N latitude and 86°W longitude on the southern flank of the spreading center about 22 km from the spreading axis. Two main sedimentary units occur : an upper unit of green clay interbedded with pelagic oozes (3 to 16 m beneath the sediment-water interface at the top of the mound) and a lower unit of pelagic oozes (16 to 33.4 m). Mn oxide fragments occur in the uppermost 4.3 m of the section. Calcite is the predominant mineral in the pelagic oozes, whereas Mn oxide crusts, which cap the sequence, consist chiefly of todorokite.

Analytical methods performed on both bulk sediments and clay fraction (< 2, 2-0.5, 0.5-0.2, < 0.2 µm) include X-ray diffraction (XRD), Major element analysis (spectrometry), transmission electron miroscopy (TEM : bright field images and electron diffraction patterns), high resolution TEM (HRTEM), and energy-dispersive X-ray (EDX) allowing elemental analysis of single clay particles. Analyses are restricted to green clay layers (10 samples) from various depths in the 30 m-thick sedimentary column.

3.3. RESULTS

Chemical analyses of the bulk clay samples show that Si and Fe are the predominant elements. K contents tend to increase with depth in the core. X-ray diffraction indicates that, beside Fe and Mn oxyhydroxide, and calcite, phyllosilicates are dominantly of a dioctahedral nature. They consist of random illite-smectite mixed-layers, the illite layer percentage of which increases toward the bottom of the drill core, and more rapidly in the smallest grain-size fractions (< 0.2 µm) than in the largest fractions (< 2 µm).

Transmission electron microscopy shows the presence of two major types of particles, associated with particles of intermediate shape (Fig. 5) : 1) filmy or veil-like clays, 1-2 µm in diameter, with curled edges, mostly located in the upper part of the core ; 2) laths as long as 0.5 µm, thinner than the veil-like particles, rare in the shallowest samples, and more abundant and crystalline (strong electron diffraction pattern) in the deepest part of the 30 m-long core.

Elemental analyses (EDX) confirm the dioctahedral character of clay particles. Two groups of composition are identified : 1) upper part of the core : lower K content and larger charge, with much Fe, indicating a Fe-smectite type close to nontronite composition ; 2) lower part of the core : same Fe-smectite, associated with a Fe-mica/illite enriched in K in the interlayers and having a low tetrahedral charge pointing to a composition of glauconite.

HRTEM. Sequences of particle stacking were studied by high resolution TEM of ≈ 500 Å thick, ultramicrotom thin sections. Glauconite-like layers (10 Å) are rigid, straight, and dislocation-free, whereas Fe-smectite-like layers are generally curled and commonly display linear defects such as linear dislocations. Electron diffraction patterns of glauconite indicate a well-organized structure, those of Fe-smectite a poorer crystallinity (weaker and more diffuse 001 reflections). In the section examined glauconite crystallites show a 10 Å regular stacking sequence of 6-7 layers, whereas Sm-smectite crystallites show 10-13 Å irregular stacking sequences of more than 10 layers.

3.4. INTERPRETATION

The analyses performed suggest that the Fe-smectite particles were replaced by glauconite. This replacement corresponds to a transition from veil-like to lath-like texture, which also appears to have been accompanied by a chemichal change as shown by the increased amount of interlayer K in the glauconite. The occurrence of discontinuities between Fe-smectite having highly imperfect, discontinuously anastomosing layers, and glauconite crystallites having straight and relatively defect-free 10-Å layers, suggests a dissolution-precipitation mechanism (Buatier et al., 1989). The downhole increase of potassium is attributed to the reaction of smectite and upwelling solutions or pore fluids. The mineralogical change represents a trend towards more perfect, homogeneous, dislocation-free crystallites.

4. Clay diagenesis in highly porous rocks

4.1. MAIN CLAY FACIES AND MINERALS

The ability of clay minerals to form diagenetically in the voids of permeable rocks is suspected since accordeon-like, vermicular or blocky kaolinite has been described from

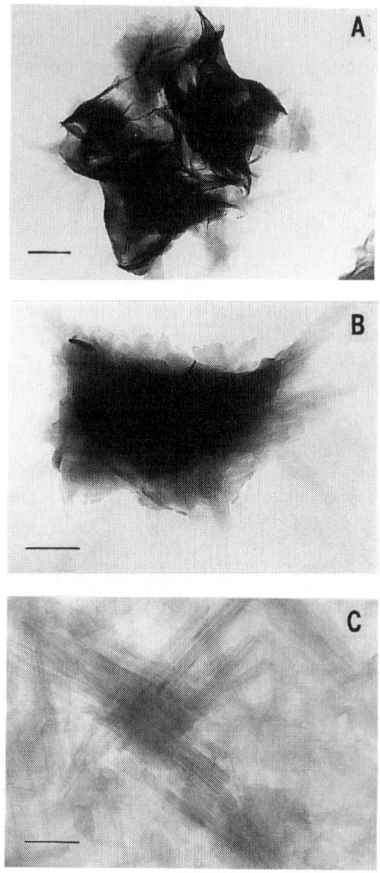

Figure 5. Transmission electron micrographs showing clay morphologies in hydrothermal sediments of the Galapagos Spreading Center (courtesy M. Buatier) : A filmy, veil-like particles with curled edges ; B intermediate particles ; C lath-like particles. (Bar = 0.2 μm).

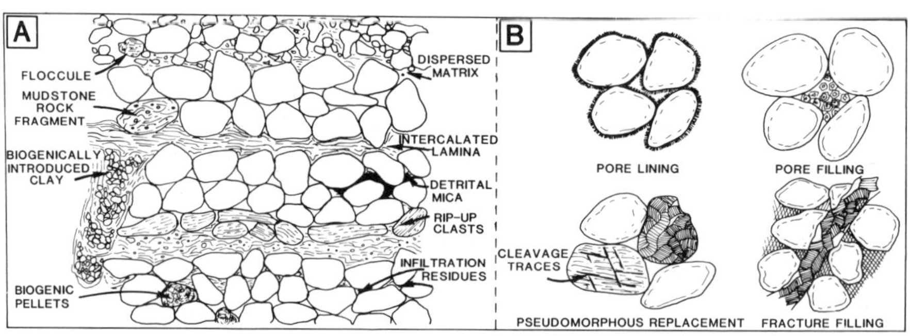

Figure 6 Modes of occurrence of allochthonous (A) and autochthonous (B) clay in sandstones (After Wilson and Pittman 1977). The size of individual flakes and aggregates is exaggerated.

thin-section studies in various sandstones. These delicate particles can hardly result from a genetic process similar to that responsible for the supply of the typically-detrital minerals among which they occur (quartz, mica, feldspar, etc). Because their high porosity and permeability favor the post-sedimentary migration of fluids, the sandstones represent particularly suitable environments for the diagenetic formation of secondary minerals like carbonates, silica and clays.

The clay mineral suite in sandstones may have either a detrital or an autochthonous origin. Detailed optical studies help greatly to distinguish primary clay minerals from secondary species. One of the first relevant studies was presented by Wilson and Pittman (1977), from data on thin sections and scanning electronmicroscopic preparations (Fig. 6). Autochthonous clays are reported as pore linings, pore fillings, pseudomorphic replacements and fracture fillings. Arguments for an authigenic origin consist mainly in the delicacy of clay morphology, the absence of pore-lining clay only at grain contacts, and a mineral composition different from that of typically autochthonous minerals. By contrast, allochthonous clays mainly occur as dispersed matrix, sand-sized floccules, sand- to cobble-sized mud, shale clasts, specific laminae, or as a result of post-depositional infiltration or bioturbation.

Kaolinite

The most frequent authigenic clay minerals in sandstones comprise kaolinite, illite and chlorite. Some mixed-layered clays are also attributed to a post-sedimentary formation. Kaolinite is widely distributed in sandstones, and may derive from either allochthonous sources of autochthonous genesis. For instance, the Carboniferous series of Scotland contain variable amounts and types of kaolinite according to the age and lithology (Wilson et al. 1972). Late Carboniferous shales and limestones contain abundant detrital kaolinite, that is small-sized, moderately ordered and mainly derives from the erosion of a tropically-weathered lateritic cover. Early Carboniferous sediments are rich in dolomites and sandstones which contain highly crystalline, morphologically-perfect kaolinite that formed during diagenesis. In most cases, the chemical elements needed for kaolinite formation, namely silicon and aluminum, are thought to derive from the leaching of some minerals pre-existing in the sandstones. Potassium feldspars and micas constitute the more probable Si and Al sources.

Illite

Hairy illite represents the most convincing diagenetic facies for illite, since the fragile and long mineral needles developing at the periphery of quartz or other substrates may hardly have experienced significant transportation. Güven et al (1980) report the frequent occurrence of authigenic **lathed illite** in the pores of many sandstone reservoirs from the United States, such as those existing in the Eocene of Texas (Wilcox formation), the Paleocene of Wyoming (Fort Union formation) and the Jurassic of Mississippi (Norphlet formation). The laths develop perfect morphologies, and are up to 30 μm long, 0.1 to 0.3 μm wide and up to 200 Å thick. The presence of hairy illite in sandstone pores increases the microporosity and pore tortuosity, and therefore decreases the permeability, which may significantly modify the reservoir properties.

Morad and AlDahan (1987a) identify hair-like, filamentous illites in sandstones as old as those of the late Proterozoic Visingsö group, Southern Sweden, which indicates the stable nature of this facies. Visingsö sandstones also contain flaky diagenetic illites. Both facies appear to derive from a dissolution- crystallization process, the source minerals probably consisting of potassium feldspars. The reason for the preferential development of laths or sheets appears difficult to assess. The cause of the preferential formation of illite instead of kaolinite seems to be related to the chemical composition and pH of migrating groundwaters, illite apparently developing under less acidic conditions.

Chlorite

Diagenetic chlorite appears to consist chiefly of **coatings** around detrital grains. For instance Curtis et al. (1984) study chlorite minerals in Cretaceous sandstones from the Tuscaloosa formation in Louisiana, USA (burial depths 2390 and 5500 m), and from the Fahler formation in Alberta, Canada (1876 m). Scanning transmission electronmicroscopic analyses indicate the dominant presence of true chlorites, the chemical composition of which varies relatively slightly and corresponds to a magnesian chamosite (= Mg-bearing Fe chlorite). In addition, some coatings include swelling chlorite-like minerals that contain a vermiculitic component. Chloritic coatings can be so continuous within the sandstone that they may prevent further overgrowth of framework grains from saturated pore waters, thereby preserving the initial pososity. **Replacement chlorite** also exists in some sandstones, the clay mineral being substituted for detrital microcline or albite. The iron and magneisum needed for the chlorite growth may be derived from minerals such as biotite and from metal oxide grains or pigments (Morad and AlDahan 1987b). The fairly homogeneous chemical composition of chloritic rims in sandstones suggests noticeable migrations of the fluids responsible for diagenetic exchanges and reactions. The formation of chlorite appears to be favored by reducing and late diagenetic conditions.

4.2. DIAGENETIC PROCESSES AND HISTORY

Detailed mineralogical and electronoptical studies show that clay diagenesis in sandstones may be much more complex than previously described. For instance, Huggett (1984) investigates the fluvial sandstones and associated siltstones, shales, coals, and occasional marine sandstones from the Westphalian Coal Measures in East Midlands, UK. The sandstone bodies contain 10 to 15 % clay minerals, that consist of kaolinite, chlorite and illite (Fig. 7).

Kaolinite may be blocky or platy, some of the grains showing cleaving figures that apparently result from compaction and pressure due to other crystals growth. Kaolinite, considered as entirely authigenic, is overgrown by quartz, feldspar and illite, which suggests an early diagenetic formation. Locally, however, almost all the kaolinite is blocky and occurs in open pores on top of chlorite rims, indicating a late formation. The preferential formation of kaolinite in Carboniferous sandstones is clearly related to leaching by freshwater and to the lack of brackish- or marine-water influence.

Diagenetic illite displays plate-, curved lath- and straight blade habits, that clearly differ from tangentially arranged, grain-lining, ragged plates of detrital illite. Illite occurs as pore filling and pore riming clay, and as inclusions in mineral overgrowths. The diagenetic replacement of feldspar by illite is ubiquitous, that of muscovite is favored in clean, washed sands, and that of kaolinite is neither abundant nor facies-related. The alteration of feldspar to illite instead of kaolinite appears to be favored by ion-enriched, less acidic pore waters. The radial illite grain-rim cements have no immediately adjacent precursor phase and represent true precipitation minerals. Illite and quartz authigenesis occurred concurrently, the development of the former species probably being favored by the presence of K-feldspar. Unlike illite blades, illite laths are observed overgrown by quartz, which suggests the have formed before the blades. From a general point of view physical and chemical

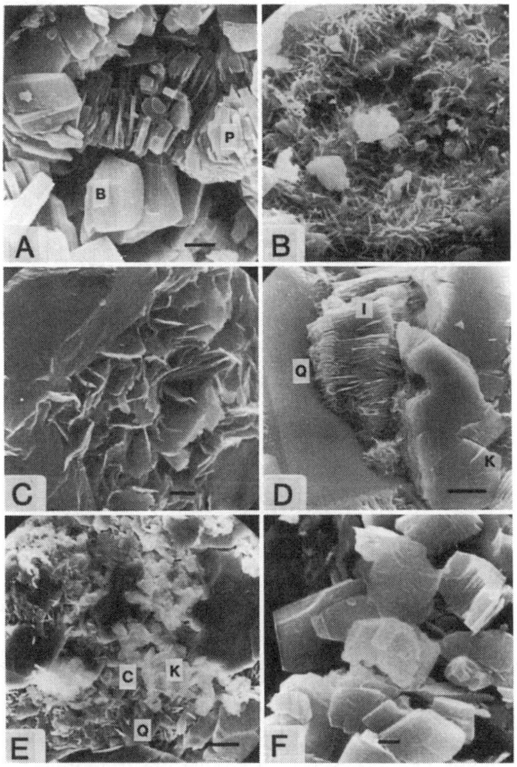

Figure 7. Scanning electronmicrographs of diagenetic clay minerals in sandstones (courtesy J.M. Huggett. See Huggett 1984). **A** Mixture of blocky (*B*) and platy (*P*) kaolinite. Bar = 5 µm. **B** Radial illite laths. Bar = 5 µm. **C** Illite plates. Bar = 1 µm. **D** Illite (*I*) replacing kaolinite (*K*) and overgrown by quartz (*Q*). Bar = 5 µm. **E** Chlorite coatings (*C*) overgrown by quartz (*Q*) and blocky kaolinite (*K*). Bar = 0.5 µm. **F** Blocky pseudohexagonal chlorite. Bar = 1 µm.

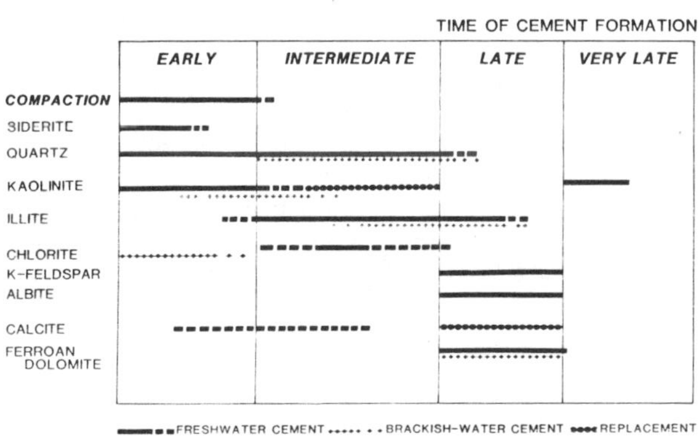

Figure 8. Tentative timing of authigenic cement formation in Coal Measures sandstones, Westphalian of East Midlands, UK (After Huggett 1984).

constraints are much more important than depth constraints. Nevertheless illite tends to form predominantly under later diagenetic conditions than kaolinite.

Chlorite is both detrital and authigenic, the latter type being only locally present in noticeable amount. Diagenetic chlorite consists of either small grain rims or blocky-pseudohexagonal crystals. The first type is the most frequently encountered. The presence of kaolinite-overgrowing chlorite suggests a mostly early-diagenetic formation. The local abundance of chlorite could be related to the local presence of ferromagnesian minerals (e.g., biotite) or of marine to brackish environment.

Huggett (1984) proposes from his investigations a tentative timing for the mineral cement formation, the authigenic clays being situated among carbonate and other silicate precipitates (Fig 8).

4.3. CLAY DIAGENESIS AND RESERVOIR PROPERTIES

Many sandstones from productive oil well contain diagenetic kaolinite only, while unproductive wells filled by saline water show the presence of both kaolinite and illite. As kaolinite is mostly considered as an early-diagenetic species that tends to be destroyed through illitization, its preservation in a given sandstone reservoir is attributed to oil irruption. The filling up of a rock reservoir by oil probably prevents late-diagenetic illitization. Similar results arise from gas emplacement. By contrast, the filling up of the reservoir by saline water favors illitization of detrital or early-diagenetic minerals. The study of the distribution of different diagenetic clays in a given sandstone body may therefore help to understand the chronological evolution of oil and water migration. For instance Thomas (1986) calculates from potassium/argon dating in Jurassic sandstones of the Central Viking graben, North Sea, that the stopping of illitization by the hydrocarbon charge occurred between late Eocene (44 my-40 my BP) and Oligocene (28 my).

The development of secondary clays during sandstone diagenesis tends to decrease the permeability through pore space occupancy, and therefore to impede the hydrocarbon migration and diminish the reservoir volume. The more evolved the diagenetic clay cementations and precipitations, the less favorable the reservoir properties. For instance Bjorlykke and Brendsdal (1986) show that the leaching of feldspar and the formation of kaolinite in Jurassic Brent sandstone, North Sea, have been caused by the flow of meteoric water at moderate burial depth during early diagenesis, while more isochemical changes occurred during late diagenesis. Most of the diagenetic kaolinite results from the alteration of micas rather than of feldspar. Muscovite that altered to kaolinite expanded and flared into the pores, causing dramatic reduction in the sandstone permeability. As detrital mica preferentially settled in low-energy depositional environments, the corresponding lithofacies display the maximum reduction of permeability after the diagenetic evolution. The recognition of the different types of depositional facies therefore greatly helps to predict the reservoir quality of sandy bodies that experienced clay diagenetic modifications. Recent three-dimensional modeling of reservoir performance by using geological, lithological and mineralogical data, combined with wireline logs, greatly helps the commercial exploitation of a given hydrocarbon-bearing field.

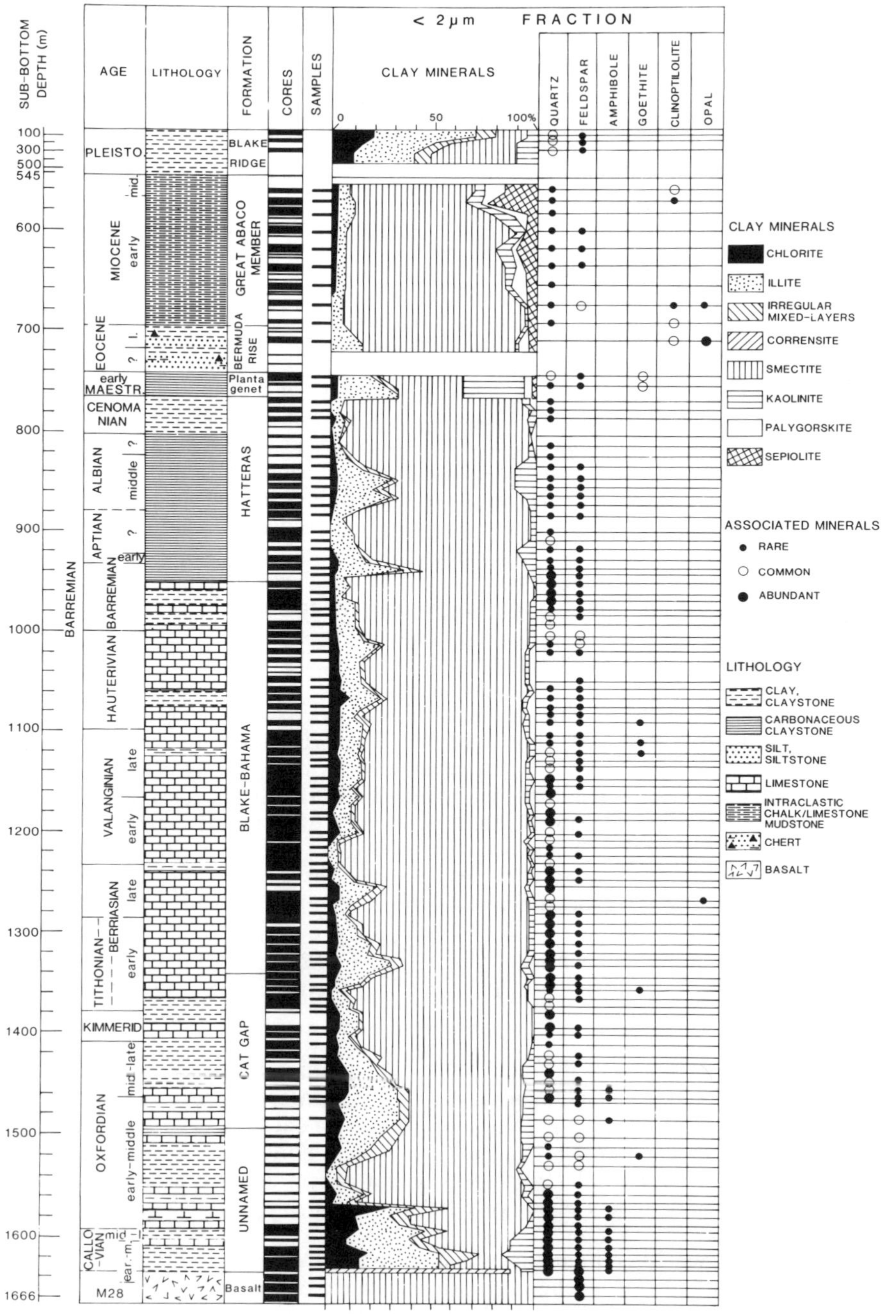

Figure 9. Blake-Bahama basin, Northwestern Atlantic ocean. Lithology and mineralogy of the < 2 μm sedimentary fraction of DSDP Hole 534A (After Chamley et al. 1983), with reference to site 391 for the Pleistocene.

5. Smectite, volcanism and diagenesis in ocean sediments

5.1. INTRODUCTION - VOLCANOGENIC SMECTITE

Among the clay minerals buried in common marine sediments, smectites constitute the group the most frequently quoted as possibly subject to post-sedimentary modifications. While illite, chlorite, kaolinite, vermiculite, random mixed-layers are mostly considered as stable in sediments buried at distances exceeding one kilometer below the sea floor, smectites are often attributed to early diagenetic effects. Such interpretations mainly result 1) from the unexpected abundance of smectite in many Mesozoic and Cenozoic marine deposits, 2) from both the diversity and continuum of chemical composition of smectites implying possible in situ modifications, 3) from the sensitivity of smectitic minerals to cation-rich environments, and 4) from the well-known dependence of smectite on volcanic or hydrothermal activity.

An early diagenetic formation of smectite occurs in various marine sediments characterized by effectively abundant volcanic glass of explosive origin. Such sediments, particularly investigated in South and West Pacific Ocean and in Eastern Mediterranean Sea, display a clear correlation between the development of smectite and the occurrence of volcanic glass. Synsedimentary smectites appear to hardly form from massive glassy debris, and to be favored by the combination of small-sized, pumiceous ash, and of easily available biogenic silica. Volcano-derived submarine smectite may present various chemical compositions and shapes depending on the nature of initial volcanic materials, but appears to differ from most common soil-derived terrigenous clay minerals.

5.2. ORIGIN OF ATLANTIC SMECTITES

Ordinary Atlantic deposits of late Mesozoic-early Cenozoic age constitue favorable materials to test the hypotheses of either volcanogenic, early-diagenetic or detrital origin of smectite, for the following reasons : 1) The Atlantic was a fairly young ocean marked by an important volcanic activity, the expression of which should be easily recognized in contemporaneous marine sediments ; 2) these sediments have usually endured a moderate depth of burial (less than 2 km) and were submitted to moderate geothermal gradients (around 30° C/km) ; 3) they mostly comprise biogenic to clayey components allowing to readily recognize secondary, in situ-formed constituents ; 4) they were deposited at noticeable sedimentation rates (usually > 1 cm.10^3yr), which precluded appreciable hydrogenous genesis at the sediment-sea water interface ; (5) their richness in smectite (e.g., Fig; 9) allows a precise identification of its mineralogy, geochemistry, and morphology (see Chamley 1989).

Mineralogy. Thousands X-ray diffraction analyses performed on Cretaceous-Paleogene sediments from the Atlantic domain indicate that most smectites contain few illite-like layers, and constitute a dioctahedral mineral. X-ray diffraction characters vary little from one area or geologic period to another. On transmission electronmicrographs, smectites mostly display flaky outlines very similar to those encountered in continental soils. Some laths occur sometimes at the periphery of smectite flakes.

Chemical Composition. Microprobe investigations on isolated particles show that smectites from common Cretaceous-Paleogene clays to limestones consist of beidellites marked by the presence of aluminum in both tetrahedra and octahedra, and by a partial substitution of ferric iron and magnesium for octahedral aluminum (Debrabant et al. 1985). The relative amounts of Al, Fe, and Mg may vary in a moderate manner, which characterizes a rather

Samples	Tetrahedra Si	Al	Octahedra Al	Fe^{3+}	Mg	Ti	Interlayers Na	K	Mg	Ca
Senegal Basin, Kafoutine										
N° 620, Maastrichtian	3.12	0.88	1	0.50	0.30	0.02	0.62	0.42	–	0.33
N° 1190, Santonian	3.22	0.78	1.31	0.44	0.23	0.02	0.32	0.30	–	0.20
Cape Verde Basin, DSDP 367										
19-1-100, late Cretaceous	3.74	0.26	1.23	0.43	0.44	0.02	0.03	0.17	–	0.06
22-6-91, late Albian	3.31	0.69	1.07	0.69	0.32	0.02	0.05	0.40	0.05	0.07
23-2-46, Albian	3.57	0.43	1.47	0.38	0.18	0.02	0.04	0.16	0.08	0.04
Mormoiron Basin, SE France										
Al–Fe smectite (wyoming)	3.92	0.08	1.21	0.40	0.30	0.03	?	0.21	–	0.16
West African Soils (parent rock)										
M-5D (gneiss w/amphibole)	3.67	0.33	1.07	0.58	0.47	0.03		?		
GB 92 (granite)	3.38	0.62	1.43	0.49	0.21	0.06		?		
LIV-2-3 (Eocene clay)	3.61	0.39	1.37	0.46	0.24	0.05		?		
Godola (granite)	3.30	0.70	1.29	0.68	0.19	0.05	?	0.09	?	0.01

Table 1. Structural formulae of smectites from clayey to calcareous sediments of the late Cretaceous of tropical Northeastern Atlantic. Comparison with Paleogene detrital Al-Fe smectite (Wyoming) of the Mormoiron Basin, SE France, and with smectites from recent West African soils (After J. Duplay, H. Paquet, N. Trauth, H. Chamley et al. ; in Chamley 1989).

well-defined group of Al-Fe beidellites. The geochemical composition is very similar for flaky smectite particles and lathed overgrowths. This composition is comparable to that of smectites originating in continental soils from warm regions or reworked from such soils in pericontinental deposits (e.g., Table 1, and Paquet 1970). By constrast, the chemical composition of most Atlantic smectites differs greatly from that of smectites derived from the alteration of volcanic glass or basalt, of from hydrothermal or hydrogenous precipitation, that are much richer in magnesium or iron.

Rare earth elements composition. Sediments from the Eastern Atlantic Ocean rich in clay fraction (60-90 %) and containing abundant smectite within this fraction have been studied by neutron activation from both stratigraphic and geographic points of view (Bonnot-Courtois 1981). The vertical distribution of rare earth elements displays a very homogeneous pattern in most sediments, from early Cretaceous to Quaternary times. When normalized to continental shales, the curves appear rather flat. The rare earths are devoid of both enrichment in heavy elements (europium, gadolinium, terbium, tamarium, yterbium, lutetium) and depletion in cerium. A slight enrichment in light elements (lantanium, cerium, neodymium, samarium) is usually recorded. All these characters suggest a strong continental influence and the absence of noticeable impact of volcanic activity or sea-water influence on smectite formation.

Strontium Isotope Composition. Smectites of diverse grain sizes are studied by Clauer et al. (1990) from Cenomanian and Paleocene sediments of the North Atlantic Ocean. For instance, Paleocene oozes at DSDP site 386, Southeast of Bermuda, display $^{87}Sr/^{86}Sr$ ratios of 0.70860±0.0004 for the 0.2-0.4 µm fraction, and of 0.70899±0.00043 for the < 0.2 µm fraction. These values are much higher than those characteristic of Paleocene marine depositional environments (0.70783±0.00028), but slightly lower than those typical of pure detrital origin. The apparent age indicated by strontium isochrones and by K-Ar dating ranges between 167±5 and 200 ± 6.10^6 years, instead of 56 ± 3.10^6 years which corresponds to the Paleocene stage. These results indicate that Atlantic smectites were dominantly inherited from old, presumably terrigenous parent-rocks, and were only slightly modified by early-diagenetic changes.

Associated Minerals. Smectite in Atlantic sediments is often accompanied by clinoptilolite and opal CT (cristobalite-tridymite series), which could suggest a mineral paragenesis developed through diagenetic processes. In fact, the distribution of the three mineral groups is largely independent. If clinoptilolite- and/or opal-rich sediments frequently contain abundant smectite, smectite-rich sediments are often devoid of clinoptilolite and opal CT. Clinoptilolite and opal develop mainly in late Cretaceous to Paleogene sediments (Stonecipher 1976), while smectite-rich deposits occur from late Jurassic to late Paleogene. In addition, the smectite composition and relative abundance do not depend on the presence or absence of zeolite or opal.

Volcanic Constituents. Volcanic glass is rarely identified in significant amounts within late Jurassic to Cenozoic sediments of the Atlantic domain where smectite abundantly occurs. It is somewhat intriguing that marine sediments whose abundant smectite is attributed to the diagenetic transformation of volcanic glass are virtually devoid of undisputed volcanic remains (e.g, Jeans et al. 1982). Most authors supporting the hypothesis of a volcanic origin for smectite in common sediments imply that initial volcanic glass has been altered and destroyed through smectite formation. The often total absence of glass is nevertheless difficult to understand, knowing that the alteration potential of volcanic material greatly differs with size, shape, porosity and composition, and that non-altered volcanic debris usually still exist in deposits rich in unquestionably volcanogenic smectite. In addition, the sediments blanketing the basalt crust generally reveal a decrease in smectite abundance relatively to illite, chlorite, and kaolinite (e.g., Fig. 9 ; Robert 1987). Albian sediments deposited close to the North American continent contain abundant smectite, while contemporary deposits located close to the Atlantic ridge and in the volcanically active Cape Verde basin are depleted in this mineral (Chamley and Debrabant 1984).

Lithologic Distribution. In most Atlantic series, the relative abundance of smectite does not depend on the main lithologic successions (e.g., Fig. 9). The mineral content varies independently of the dominance of either reddish clay, organic-rich clay, limestone, chalk, chert, silt, pelagic or hemipelagic ooze. This general independence, combined with the absence of control of smectite abundance by the depth of burial, suggests that the mineral formation and evolution are not noticeably controlled by early diagenesis. Of course this general behavior does not exclude the possibility of minor lithologic control, such as local degradation in organic environment, or dependence on short-scale changes like limestone-marl alternations.

To summarize, the characters and distribution of most late Mesozoic to Cenozoic Atlantic smectites appear to depend little or not at all on the depth of burial, lithology, associated minerals and volcanic activity. On the contrary, these characters and distribution are largely compatible with land-derived minerals. This suggests that smectite is only little controlled by post-sedimentary processes and notably issues from land-masses adjacent to the Atlantic Ocean.

6. Geological applications of clay diagenesis data

6.1. ESTIMATION OF THE GEOTHERMAL GRADIENT

Comparison of present-day measurements and calculations from clay mineral data on sedimentary series buried in the course of the time may allow to reconstruct the actual temperatures and post-diagenetic history of a given region. Pollastro and Barker (1986) investigate the 1500-5500 meters sedimentary interval of the Green River basin, Wyoming. Late Cretaceous to early Tertiary rocks were drilled in the Pinedale anticline. Paleotemperatures calculated from changes in the composition and ordering of illite-smectite and in the mean vitrinite reflectance reached 200° C at 5500 meters. The present-day temperature measured at this depth is about 135-150° C. These data suggest that the maximum temperature responsible for the inorganic and organic evolution was about 50° C higher than present-day borehole temperature. The apparent temperature decrease can be explained by the erosion of about 1700 meters of the section, a fact documented by vitrinite reflectance data and implying that the geothermal gradient has remained constant since maximum burial. The erosion itself resulted from uplift of the Pinedale anticline during the Neogene.

6.2. TECTONIC OVERBURDEN

If sedimentary series are tectonically covered by thick geological formations, diagenetic effects may develop similarly to what occurs during progressive burial. The pressure and temperature due to the tectonic overurden may determine a clay mineral simplification and reorganization. If the superimposed tectonic units are ulteriorely removed by erosion, the autochthonous or para-autochthonous underlying formations may crop out and keep in their clay mineralogy the evidence of the former covering. Such a possibility is documented by Deconinck and Debrabant (1985) in the northwestern part of the French subalpine domain. Typical detrital clay assemblages occur in late Cretaceous limestones of the western sector, marked by dominant beidellitic-smectite associated with various amounts of illite and little random mixed-layers, chlorite or kaolinite. When crossing the Arcalod fault to the East, these assemblages in contemporaneous sediments are suddenly replaced by clay associations rich in well-crystallized illite and chlorite, with associated sub-regular mixed-layers. This change is thought to result from the overburden effect of former tectonic nappes, that issued from the inner Alps and invaded the eastern part of the subalpine range during Paleogene time. Further erosion determined the disappearance of the allochthonous formations, except in some local areas where isolated klippes are preserved. Clay mineral assemblages therefore appear as useful markers of past tectonic overburden in regions marked by active erosion.

6.3. LATERAL EXPRESSION OF THERMAL GRADIENT AND TECTONIC STRUCTURING

Tectonics and metamorphism are often associated. Intense deformations determine an increase in pressure and temperature that induces various chemical and mineralogical changes. Metamorphic zones develop in close connection with most active orogenic areas. Laterally the sedimentary formations involved in the tectonic structuring undergo attenuate thermodynamic efforts, and diagenetic changes can take place. Close to orogenic zones clay minerals tend to become less diversified and better crystallized, isotopic reequilibrations occur. This is particularly well documented in the subalpine and outer alpine range, Western Alps, where sediments present slight to moderate tectonic deformations.

Chlorite represents a particularly useful marker of post-sedimentary mineral changes in late Mesozoic calcareous sediments of the French sulpine range. As this mineral virtually lacks in non diagenetic sediments, its appearance obviously characterizes post-sedimentary modifications. Illite, smectite and kaolinite occur in variable amounts in both non diagenetic and diagenetic formations, and their percentage variations may reflect either changes in detrital sources and climate or thermodynamic constraints. Deconinck (1987) reports that chlorite abundance increases progressively in different early Cretaceous periods from the outermost (West) to innermost (East) parts of the subalpine range. The eastward increase of chlorite abundance correlates both with increasing temperature and sedimentary overburden. The chlorite amount increases eastwards much more rapidly in limestone beds than in marl interbeds, which is attributed to higher permeability and stronger heat transfer in calcareous levels. Marly beds appear to keep the paleoenvironmental signal of detrital clays much longer than calcareous beds. Diagenetic chlorites display an iron-rich type, which resembles the composition of chlorite deeply buried in Gulf Coast sediments (Hower et al. 1976) and could result from an iron release by pre-existing smectites.

Levert and Ferry (1988) confirm the eastward increase of Fe-chlorite abundance in four isochronous horizons of the Vocontian basin. Oxfordian, Kimmeridgian, Valanginian and Aptian bundles of limestone-marl alternations show a progressive enrichment of chlorite toward the inner Alps, where calcareous beds may contain up to 80 % of the mineral in the $< 2\ \mu m$ fraction (Fig. 10). The mineralogical change correlates with an increase in the maturation state of organic matter. In addition, local increases in more magnesian chlorite occur in the western part of the basin. The geographic situation of these anomalous chlorite enrichments varies according to the period, but preferentially characterizes some areas marked by fault systems, Triassic diapirs of brachyanticlines. Levert and Ferry (1988) suggest that the local Mg-chlorite augmentations, which sometimes correlate with a kaolinite enrichment, could result from hydrothermal mineralizations linked to fluid migrations along fault systems or above Triassic salt diapirs.

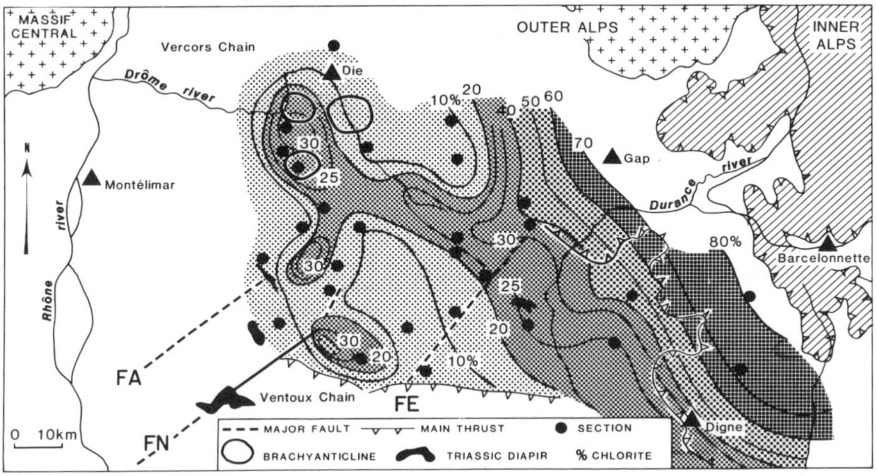

Figure 10. Percentage distribution of chlorite in the calcareous beds of Kimmeridgian isochronous limestone-marl alternations from the Vocontian basin, Southeastern France (After Levert and Ferry 1988).*FA* Eygalayes fault. *FM* Ménée fault. *BA* Brachyanticlines.

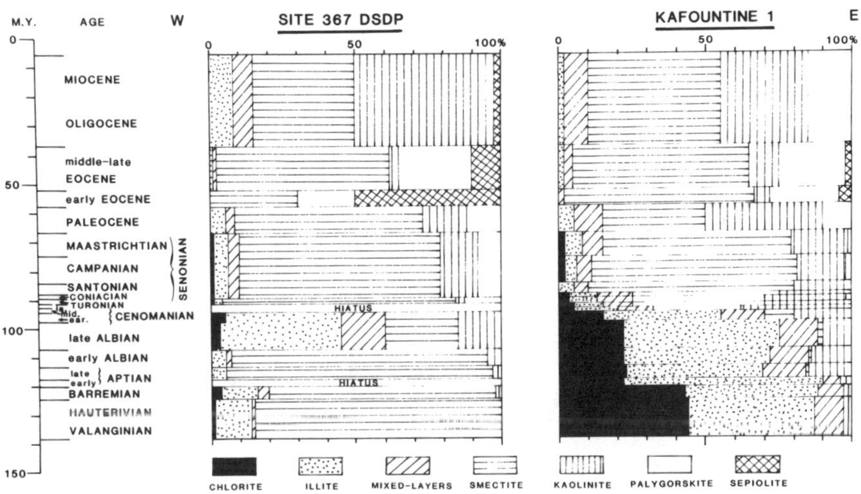

Figure 11. Comparative clay mineralogy at sites DSDP 367 and Kafountine 1. Results are given as average values for successive time intervals (After Chamley et al. 1988).

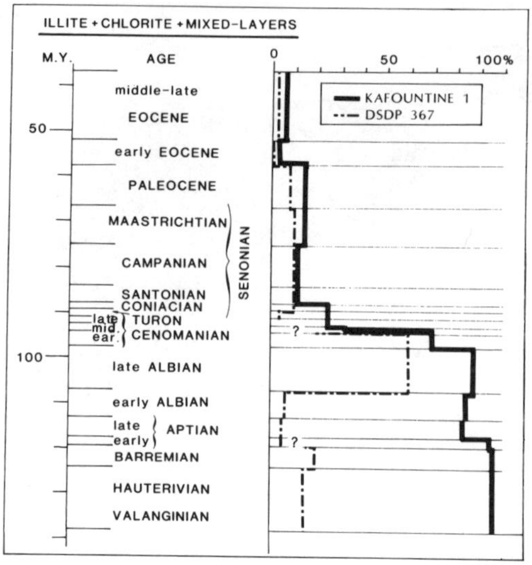

Figure 12. Relative abundance of illite + chlorite + random mixed-layers at sites 367 and Kafountine, during the Valanginian to Eocene time interval (After Chamley et al. 1988).

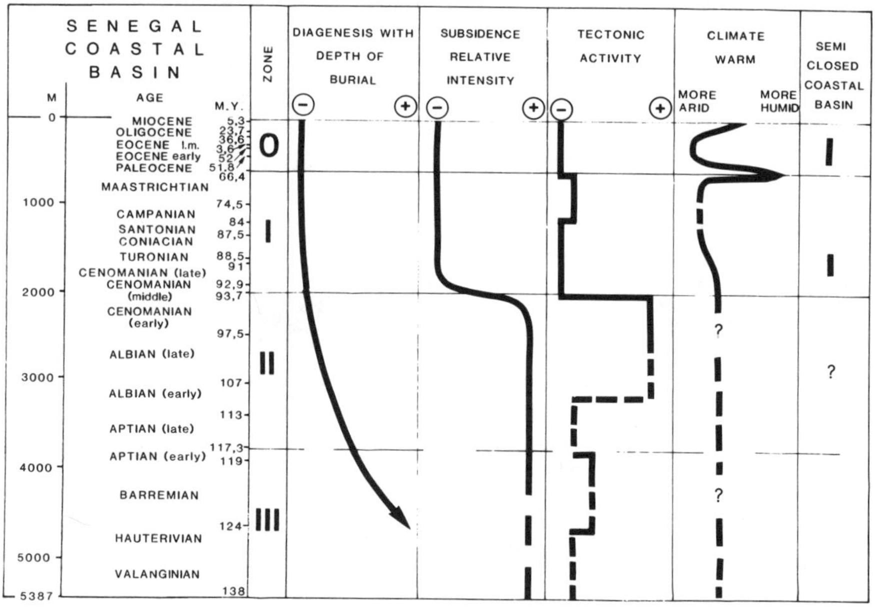

Figure 13. Cretacous to Paleogene sedimentological expression of diagenesis and paleoenvironment in the Senegal coastal basin (Kafountine borehole) based on comparisons with data from the Cape Verde deep-sea basin (After Chamley et al. 1988).

6.4. PROGRESSIVE OBLITERATION OF PALEOENVIRONMENTAL MESSAGE BY DIAGENESIS

Example of Cape Verde and Senegal Basins

A multidisciplinary approach including clay mineralogy, chemistry and morphology has been carried out on Cretaceous-Paleogene sediments at site Kafountine 1, drilled on land in the Senegal coastal basin, and at DSDP site 367 located in the adjacent Cape Verde deep-sea basin (NE Atlantic). It allows the identification of the respective controls on the sedimentary record of diagenesis relative to margin subsidence related to ocean opening, tectonic instability compared to climatic changes, sources, transport and depositional conditions (Chamley et al. 1988).

The diagenetic effects are mainly related to the depth of burial, and therefore occur only in the Senegal coastal basin, which experienced very high sedimentation rates (5387 m drilled at Kafountine site, Valanginian to Miocene). Diagenetic changes related to the depth are marked by an increase of clay grain size, increase of crystalline and perhaps magnesian chlorite, vermiculitic mixed-layers, Al-Mg illite, quartz and Na feldspar abundance, decrease of kaolinite, disappearance of smectite after a downward decrease of its interlayer charges deficit. Such changes do not exist at deep-sea site 367 (penetration 1153 m), which points to the importance of lithostatic pressure effects compared to hydrostatic ones (Fig. 11).The comparison at both sites of the downhole variation of the amount of illite, chlorite and mixed-layers clearly reflects the distinct behavior of clay assemblages (Fig. 12). Comparisons between coastal and deep-sea mineralogical and geochemical records indicate that the deepest diagenetic level studied in the coastal basin (5200 m) lies significantly above the top of the anchizone. Diagenetic effects appear to be low down to 2000 m, moderate down to 3500/3900 m, and stronger below 4500 m (Fig. 10). Clay diagenetic modifications related to sediment-interstitial water exchanges seem quantitatively unimportant. They mainly affect the deep-sea basin, where some clay-rich sediments show iso-mineralogical and nearly iso-geochemical micromorphological changes (recrystallization of flaky smectite into lathed smectites). Pore-water circulation did not significantly modify the sediments in either basin, which is demonstrated by the lack of correspondence between mineralogy or geochemistry and lithology, even in the permeable Late Cretaceous sandstones of western Senegal.

A strong mineralogical break identified in the Cretaceous series (diminution of the illite group. Fig. 9) suggests that tectonic activity dramatically decreased after the early-middle Cenomanian, close to the time at which the subsidence rate and crustal thinning decreased on the West African margin. After that time, moderate sedimentation rates developed in both coastal and deep-sea basins, which resulted in the disappearance of important diagenetic modifications of detrital sedimentary components. As a consequence, the comparative study of synchronous long-range sedimentary records in both coastal and deep-sea domains allows the identification of the combined effects of internal and surficial geodynamic modifications occurring at the continent-ocean boundary (Fig. 13).

The mineralogical modification produced by the tectonic instability prevented the identification of other palaeoenvironmental changes, which are more easily recognized during the intermediate and following relaxation stages. The comparison of coastal and deep-sea basin records confirms the warm and rather humid climatic conditions indicated by Cretaceous-Palaeogene soil-derived detrital assemblages (smectite). Minor changes in climate are better recognized in the proximal area (Senegal basin), suggesting drier conditions in the uppermost Cretaceous and parts of the Palaeocene-Eocene. The African provenance of the main terrigenous sources is confirmed by the presence of similar mineralogical and geochemical assemblages at both sites. The increased distance from sources is marked by a diminution of kaolinite supply, lesser compositional variations and stronger rerworking from evaporitic areas of semi-closed coastal basins (palygorskite). Thus the comparison between coastal and deep-sea sedimentary records points to both close genetic relationships in terrrigenous supply and significant differences in hydrodynamic and topographic conditions.

7. Prospects

Great progress arose the last ten years about the nature, processes and causes of clay mineral diagenesis in sedimentary deposits. Most new data concern diagenetic changes having affected the sandstones and other highly-porous rocks, and have been triggered by the need to better understand the post-sedimentary environment of possible hydrocarbon reservoirs. Studies on clay mineral diagenesis controlled by burial depth, which had been particularly active till the mid-seventies, have given way since a few years only to new investigations involving modern concepts and techniques. High-resolution electronmicroscopic and microgeochemical analyses provide undisputably the most high-performance data available presently.

Many basic questions on clay mineral diagenesis remain or arise presently. Some of these questions are the following : what are actually the more common mechanisms characterizing the diagenesis with increasing burial depth, and what are the corresponding mineralogical changes, chemical transfers and mass balance budgets ? Do there exist reliable ways to distinguish in a given sedimentary basin the respective influence of diagenetic constraints relative to mineralogical changes due to varying sources or depositional environments ? At which distance of hydrothermal vents are fluids able to influence the clay mineralogy of surrounding marine sediments ? Can the fluids migrating along fracture systems from subduction or fault zones determine any significant changes of the clay mineralogy ? What is the exact influence of volcanism and volcanic activity on the formation of recent and past sedimentary smectites, especially in open marine environments ? Were past hydrological and chemical marine conditions able to allow the syngenetic or diagenetic formation of beidellites and fibrous clay minerals ? What is the resistance of palygorskite and sepiolite to transportation processes and increasing burial depth, and what are the syngenetic or diagenetic conditions necessary for their in situ formation ? Finally what is the budget and the balance of ocean-derived and land-derived chemical elements responsible for the authigenic growth of sedimentary clays ? Answers to some of these questions will certainly determine great progress during the next decade.

REFERENCES

Ahn, J.H., Peacor, D.R. (1986) 'Transmission and analytical electron microscopy of the smectite-to-illite transition', Clays Clay Min., 34, 165-179.

Andrews, J.E. (1987) 'Jurassic clay mineral assemblages and their post-depositional alteration : upper Great Estuarine group', Scotland. Geol. Mag. (Tokyo), 124, 261-271.

Aronson, J.L.,,Hower, J. (1976) 'Mechanism of burial metamorphism of argillaceous sediment, 2 : Radiogenic argon evidence', Geol. Soc. Am. Bull, 87, 738-744.

Bjorlykke, K, Brensdal, A. ((1986) 'Diagenesis of the Brent sandstone in the Statfjord field, North Sea', Soc. Econ. Mineral. Paleontol., Spec. Publ., 38, 157-167.

Bonnot-Courtois, C. (1981) 'Géochimie des terres rares dans les principaux milieux de formation et de sédimentation des argiles', Thèse Sci. nat., Paris-Sud.

Brindley, G.W., Brown, G. (1980) Crystal structures of clay minerals and their X-ray identification. Mineral Soc., London.

Buatier, M., Honnorez J., Ehret G. (1989) 'Fe-smectite-glauconite transition in hydrothermal green clays from the Galapagos spreading center', Clays Clay Min, 37, 532-541.

Chamley, H. (1989) Clay Sedimentology, Springer-Verlag, Berlin.

Chamley, H., Debrabant, P. (1984) 'Paleoenvironmental history of the North Atlantic region from mineralogical and geochemical data', Sediment. Geol., 40, 151-167.

Chamley, H., Debrabant, P., Flicoteaux R. (1988) 'Comparative evolution of Senegal and eastern central Atlantic Basins, from mineralogical and geochemical investigations', Sedimentology, 35, 85-103.

Clauer, N., O'Neil, J.R., Bonnot-Courtois, C., Holtzapffel, T. (1990) 'Morphological, chemical, and isotopic evidence for an early diagenetic evolution of detrital smectite in marine sediments', Clays Clay Min., 38, 43-46.

Curtis, C.D., Ireland, B.J., Whiteman, J.A., Mulvaney, R., Whittle, C.K. (1984) 'Authigenic chlorites : problems with chemical analysis and structural formula calculation', Clay Min., 19, 471-481.

Debrabant, P., Delbart, S., Lemaguer, D. (1985) 'Micoranalyses géochimiques de minéraux argileux de sédiments prélevés en Atlantique Nord (forages du DSDP)', Clay Min., 20, 125-145.

Deconinck, J.F. (1987) 'Identification de l'origine détritique ou diagénétique des assemblages argileux : le cas des alternances marne-calcaire du Crétacé inférieur subalpin', Bull. Soc. géol. Fr., (8),3, 139-145.

Deconinck, J.F., Debrabant, P. (1985) 'Diagenèse des argiles dans le domaine subalpin : rôles respectifs de la lithologie, de l'enfouissement et de la surcharge tectonique', Rev. Géol. dyn. Géogr. phys., 26, 321-330.

Guthrie, J.M., Houseknecht, D.W., Johns, W.D. (1986) 'Relationships among vitrinite reflectance, illite crystallinity, and organic geochemistry in Carboniferous strata. Ouachita mountains, Oklahoma and Arkansas', Am. Assoc. Petrol. Geol. Bull., 70, 26-33.

Güven, N., Hower, W.F., Davies, D.K. (1980) 'Nature of authigenic illites in sandstone reservoirs', J. Sediment. Petrol., 50, 761-766.

Hower, J., Eslinger, E.V., Hower, M.E., Perry, E.A. (1976) 'Mechanisms of burial metamorphism of argillaceous sediment : 1. Mineralogical and chemical evidence', Geol. Soc. Am. Bull., 87, 725-737.

Huggett, J.M. (1984) 'Controls on mineral authigenesis in Coal Measures sandstones of the East Midlands, U.K.', Clay Min., 19, 343-357.

Jeans, C.V., Merriman, R.J., Mithell, J.G., Bland, D.J. (1982) 'Volcanic clays in the Cretaceous of Southern England and Northern Ireland', Clay Min., 17, 105-156.

Kisch, H.J. (1983) 'Mineralogy and petrology of burial diagenesis (burial metamorphism) and incipient metamorphism in clastic rocks', in G. Larsen, G.V. Chilingar (eds.), Diagenesis in sediments and sedimentary rocks, 2. Developments in sedimentology, Elsevier, Amsterdam.

Levert, J., Ferry, S. (1988) 'Diagenèse argileuse complexe dans le Mésozoïque subalpin révélée par cartographie des proportions relatives d'argiles selon des niveaux isochrones', Bull. Soc. géol. Fr., (8), 4, 1029-1038.

McHardy, W.J., Wilson, M.J., Tait, J.M. (1982) 'Electron microscope and X-ray diffraction studies of filamentous illitic clay from sandstones of the Magnus Field', Clay Min., 17, 23-29.

Morad, S., AlDahan, A.A. (1987a) 'A SEM study of diagenetic kaolinization and illitization of detrital feldspars in sandstones', Clay Min., 22, 237-243.

Morad, S., AlDahan, A.A. (1987b) 'Diagenetic chloritization of feldspars in sandstones', Sediment. Geol., 51, 155-164.

Nadeau, P.H., Tait, J.M., McHardy, W.J., Wilson, M.J. (1984) 'Interstratified XRD characteristics of physical mixtures of elementary clay particles', Clay Min., 19, 67-76.

Odin, G.S. (1988) Green marine clays. Developments in sedimentology, Elsevier, Amsterdam.

Paquet, H. (1970) Evolution géochimique des minéraux argileux dans les altérations et les sols des climats méditerranéens et tropicaux à saisons contrastées, Mém. Serv. Carte géol. Als.-Lorr., 30, Strasbourg.

Perry, E.A., Jr., Hower, J. (1972) 'Late-stage dehydration in deeply buried pelitic sediments', Am. Assoc. Petrol. Geol. Bull., 56, 2013-2021.

Pollastro, R.M. (1985) 'Mineralogical and morphological evidence for the formation of illite at the expense of illite/smectite', Clays Clay Min., 33, 265-274.

Pollastro, R.M., Barker, C.E. (1986) 'Application of clay-mineral, vitrinite reflectance, and fluid inclusion studies to the thermal and burial history of the Pinedale anticline, Green River basin, Wyoming', in Roles of organic matter in sediment diagenesis, Soc Econ. Pal. Min Spec. Publ., 38, 73-83.

Pye, K., Krinsley, D.H., Burton, J.H. (1986) 'Diagenesis of U.S. Gulf coast shales', Nature, 324, 557-559.

Ramseyer, K., Boles, J.R. (1986) 'Mixed-layer illite/smectite minerals in Tertiary sandstones and shales, San Joaquin basin, California', Clays Clay Min., 34, 115-124.

Reynolds, R.C., Jr, Hower, J. (1970) 'The nature of interlayering in mixed-layer illite-montmorillonites', Clays Clay Min., 18, 25-36.

Robert, C. (1987) 'Clay mineral associations and structural evolution of the South Atlantic : Jurassic to Eocene', Palaeogeogr., Climatol., Ecol., 50, 87-108.

Singer, A., Müller, G. (1983) 'Diagenesis in recent sediments', in G. Larsen, G.V. Chilingar (eds.), Diagenesis in sediments and sedimentary rocks, Developments in sedimentology, Elsevier, Amsterdam.

Stonecipher, S.A. (1976) 'Origin, distribution and diagenesis of phillipsite and clinoptilolite in deep-sea sediments', Chem. Geol., 17, 307-318.

Surdam, R.C., Crossey, L.J. (1987) 'Integrated diagenetic modeling : A process-oriented approach for clastic systems', Annu. Rev. Earth Planet. Sci., 15, 141-170.

Thomas, M. (1986) 'Diagenetic sequences and K/Ar dating in Jurassic sandstones, Central Viking graben : effects on reservoir properties', Clay Min., 21, 695-710.

Velde, B. (1985) Clay minerals. A physical-chemical explanation of their occurrence, Elsevier, Amsterdam.

Weaver, C.E. (1989) 'Clays, muds, and shales', Developments in Sedimentology, 44, Elsevier, Amsterdam.

Weaver, C.E. and associates (1984) Shale-slate metamorphism in Southern Appalachians, Elsevier, Amsterdam.

Wilson, M.D., Pittman, E.D. (1977) 'Authigenic clays in sandstones : recognition and influence on reservoir properties and paleoenvironmental analysis', J. Sedimentol. Petrol., 47, 3-31.

Wilson, M.J., Bain, D.C., McHardy, W.J., Berrow, M.L. (1972) 'Clay mineral studies on some Carboniferous sediments in Sotland', Sediment. Geol., 8, 137-150.

Yau, U.C., Peacor, D.R., McDowell, S.D. (1987) 'Smectite-to-illite reactions in Salton Sea shales : a transmission and analytical electron microscopy study', J. Sediment. Petrol., 57, 335-342.

Yeh, H.W., Savin, S.M. (1977) 'Mechanism of burial metamorphism of argillaceous sediments : I.O-isotope evidence', Geol. Soc. Am. Bull., 88, 1321-1330.

PORE-WATER FLOW AND MASS TRANSFER OF SOLIDS IN SOLUTION IN SEDIMENTARY BASINS

KNUT BJØRLYKKE
University of Oslo
Department of Geology
P.O. Box 1047
Blindern
0316 Oslo, Norway

ABSTRACT. Solids dissolved in pore water are transported by diffusion and advective flow through sedimentary basins. Diffusive transport is most important over shorter distances (< 100 m), but high-salinity gradients may facilitate significant diffusive transport over hundreds of metres. Flow of meteoric water may produce high fluxes, because rainwater is continuously supplied to the shallow parts of a basin from exposed land areas. Fluvial and shallow-marine sediments are more likely to be flushed by meteoric water than are sandstones of outer shelf facies or turbidites. Dissolution of K-feldspar or mica and precipitation of kaolinite requires constant removal of the released potassium and silica and cannot take place as an isochemical reaction. This reaction can, therefore, serve as a measure of the pore-ater flux, and indicate flow pathways and continuity of permeable sandstone facies.

Compaction-driven flow is limited by the volume of pore water present in the underlying sedimentary sequence. The flow rates and total fluxes produced by compaction are very small and generally are lower than the sedimentation rate, except on a very local scale. The potential for mass transport by such flow is therefore also very small, except for in some cases locally (i.e. along faults). The significance of faults as conducts for mass transport may be overestimated, however. Subsurface pressure data helps to indicate directions of pore-water flow, but the velocity and total flux over geological time are difficult to calculate.

The fact that the overall permeability on a large scale cannot be measured or estimated independently makes modelling for pore water flow very difficult. Significant "import" or "export" of major components like silica and carbonates in sandstones and limestones requires very high pore-water fluxes (10^7-10^8 cm^3/cm^2), which are difficult to attain except very locally by compaction-driven flow. Pore-water flow driven by thermal convection may reach very high fluxes, but the conditions required for Rayleigh convection are rarely met in sedimentary basins.

Non-Rayleigh convection will always occur if the isotherms are not completely horizontal, but only when relatively steep isotherms are formed, for example, by igneous intrusions or around salt domes, will the velocity be high enough to be significant in terms of mass transport of components like silica.

1. Introduction

Diagenesis includes all processes and reactions after deposition of sediments. In the marine environment, diagenetic reactions on the sea floor involve exchange with the sea water. Early marine cements are common in limestone but may also form in clastic sediments. In addition, biogenic components such as siliceous and calcareous fossils and organic matter are added to the clastic minerals by organisms on the sea floor and by pelagic sedimentation. Early marine diagenesis is therefore a very open system. Components from sea water can be transported downwards by diffusion 5-10 m below the sea floor (Berner, 1980). This is particularly the case for sulphate ions during sulphate reduction, and for uranium. These ions are highly soluble in the oxidized state (in sea water) but have low solubilities below the red ox boundary, so that high concentration gradients are set up, causing downwards-diffusion.

As sediments are buried more deeply, diagenetic reactions proceed in the direction of greater mechanical stability through compaction, and of greater thermodynamic stability through mineral reactions. If there is no introduction or removal of chemical components in the sediments (isochemical diagenesis), the compositions of the diagenetic minerals are constrained by the initial sediment composition at shallow burial and the reactivity of the mineral phases present. If, however, there is significant removal or introduction of chemical components during diagenesis, there is a higher degree of freedom in the system, and the distribution of diagenetic minerals is then more difficult to predict. The amount of mass transfer by pore-water flow is therefore one of the most important factors influencing diagenetic reactions. Only if the

A. Parker and B. W. Sellwood (eds.),
Quantitative Diagenesis: Recent Developments and Applications to Reservoir Geology, 189–221.

amounts of supply or removal of solids from a volume of sediments can be estimated can diagenetic reactions be predicted. There is thus a need to model fluid flow so that constraints can be put on the pore-water fluxes likely to flow through volumes of rocks in sedimentary basins.

In many cases simple calculations can be used to estimate orders of magnitude of fluid-flow rates and total fluxes, in order to evaluate their significance in terms of diagenesis and mass transfer.

A sedimentary basin does not consist of homogeneous sedimentary rocks, and fluid flow will be to a large extent controlled by permeable sandstones and limestones or faults. Diagenesis is not only a mineralogical and geochemical discipline; diagenetic studies should be integrated with basin analyses of depositional environment, stratigraphy and the structural history of the basin. This will help in understanding fluid flow in sedimentary basins. Reconstruction of the depositional environment based on outcrops, cores, cuttings, logs or seismic data requires an interpretation of the diagenetic processes so that the primary composition of the sediments can be interpreted. The interdependence between diagenesis, and the depositional and structural history of a basin, works both ways!

Because much of the fluid flow occurs along permeable sandstones and carbonate rocks, which define the basin plumbing system, different fluids will displace each other like fluids in a tube with a limited potential for effective mixing. This is particularly true when meteoric-water flow into sedimentary basins is channelled through deltaic or shallow-marine sandstones. The flow in porous rocks will nearly always be laminar, and two fluids with different densities, like meteoric and more-saline pore water, will mix mostly by diffusion. A better understanding of fluid flow in sedimentary basins is important for predicting the porosity and permeability of reservoir rocks. Migration of oil and gas involves two-, or in some cases three-phase flow, and these processes are still poorly understood.

Precipitation of ore minerals in sedimentary rocks also requires fluid flow, and permeable and porous host rocks. Ore-forming processes in sedimentary rocks must be seen in relation to other diagenetic processes, as ore-forming minerals compete with other types of cements to fill available pore space. Hot fluids introducing metals like Pb, Zn, and Cu will also in most cases precipitate large volumes of quartz. It is important to distinguish between fluid derived from the basin-fill and from the basement in connection with rifting and faulting.

2. The significance of diagenesis in sedimentology and sequence stratigraphy

Sequence-stratigraphic concepts (Van Wagoner et al., 1990) can provide a useful framework for interpretations of sand-body continuity, which determines the pathway for pore water flow.

Sedimentary rocks from outcrops,and from cores and cuttings from wells, have been through diagenetic changes from the time of deposition to their maximum burial and also during uplifts. The primary composition of the sediments can only be inferred from an interpretation of these diagenetic alterations. This is important in terms of interpreting the depositional environment.

The initial composition of a sediment is made of several different types of components: clastic minerals and rock fragments, amorphous phases (i.e. silica), biogenic minerals (i.e. carbonates) and adsorbed ions. A reconstruction of the primary content of feldspar, rock fragments and clay minerals is a good starting point for predicting burial diagenesis. Early diagenetic reactions on the sea floor and during meteoric water flushing may also strongly influence diagenetic reactions at greater burial depths. Evidence of early diagenetic reactions in clastic sediments, such as dissolution of feldspar and precipitation of kaolinite, may also help to provide evidence of depositional environment and proximity to a coastline and fresh-water lens.

The occurrence of carbonate-cemented layers in sandstones has been the subject of considerable interest, since they may form impermeable layers in reservoirs. Predictions of the lateral continuity of such carbonate-cemented layers depend very much on their mode of origin. If the carbonate cement is interpreted to have been "imported" from outside the sequence and precipitated by fluid flow, the lateral continuation should be predicted on the basis of models for such pore-water flow. If, on the other hand, the carbonate cements are interpreted to have formed by recrystallization of biogenic and clastic carbonate particles and early marine cement, then its distribution must be predicted on the basis of reconstructions of the depositional environment. Primary concentrations of early marine carbonate cement and biogenic carbonate may correspond to transgressive units where clastic input is held back, and these are very significant in terms of sequence stratigraphic interpretation.

In carbonate sequences, sea-level changes may cause meteoric-water flushing, which can be detected as fresh-water cementation above sea level and by karstic dissolution both above and below sea level. Thus, diagenetic studies may help in reconstructing sea-level changes.

3. Different types of pore water

For a full treatment of this subject the reader is referred to the short-course lecture notes (SEPM) by Hanor (1987).

Connate water:

The pore water buried with the sediments is often referred to as connate water. However, in a marine basin the initial sea water changes its composition drastically only a few metres below the sea floor by removal of sulphate and much of the magnesium content. The NaCl content of the sea water (10,770 ppm Na^+ and 19,350 ppm Cl^-) is not, however, removed by diagenetic processes, and the concentration in connate water can only be reduced by adding meteoric water and by the release of crystal-bound water during mineral reactions under burial. It is important to be aware of the fact that, when a layer of sediment is deposited on the sea floor, the pores will be filled partially or totally by pore water released by compaction of the underlying sediments. Where flushed by meteoric-water flow, the initial connate water is entirely removed (Clayton et al., 1966; Kharaka and Carothers, 1986; Egeberg and Aagaard, 1989).

Meteoric water:

Initially, meteoric-water is nearly pure water containing some carbon dioxide and other gases from the atmosphere. This composition becomes modified by reactions with minerals, amorphous materials, and organic matter in the soil horizon and in the basin.

Saline brines:

Saline brines have more than 100,000 ppm (10%) dissolved salt. These are highly soluble salts; mainly NaCl but also $CaCl_2$ often occur in very high concentrations. Evaporites will influence the composition of pore water by diffusion and advection.

Transport by diffusion:

Concentration gradients in pore water will set up diffusive transport as described by Fick's Law:

$$\partial C/\partial t = D_i \; \partial^2 C/\partial x^2$$

where Ci = concentration of component in mass per unit volume; Di = diffusion coefficient of i in area per unit time; x = direction of maximum concentration gradient.

The highest concentration gradients are normally found above and below the red ox boundary near the sea floor. When pore water at greater depths approaches equilibrium with the major mineral phases, such as quartz, feldspar etc., the concentration gradients are reduced for the elements present in these minerals. The gradients produced by the solubility/temperature functions are very small. The concentrations of components present in evaporite minerals are not buffered by the minerals present outside the evaporites and high gradients are therefore set up. Diffusion may be effective for dispersing salt hundreds of metres away from evaporites. In a non-dynamic system, a primary concentration gradient will be homogenized by diffusion. Primary differences in the composition of petroleum filling different parts of the reservoir will be homogenized at a rate dependent on the diffusion coefficient of the rocks.

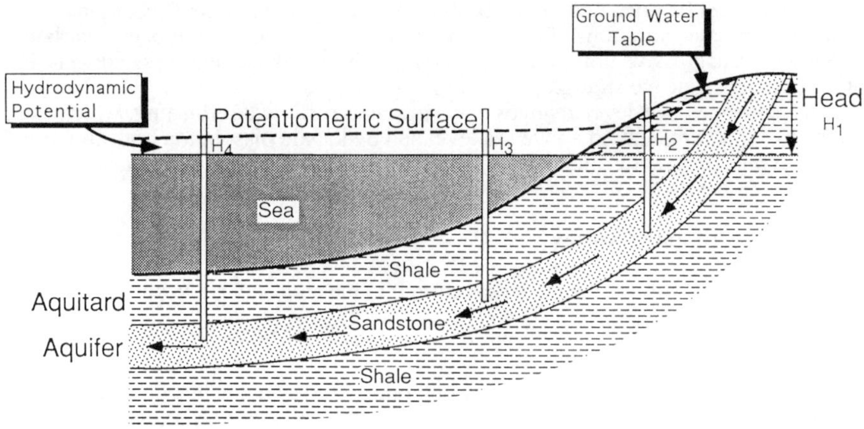

Fig. 1. Pore water flow in an aquifer from higher to lower hydrodynamic potential $H_1 > H_2 > H_3 > H_4$

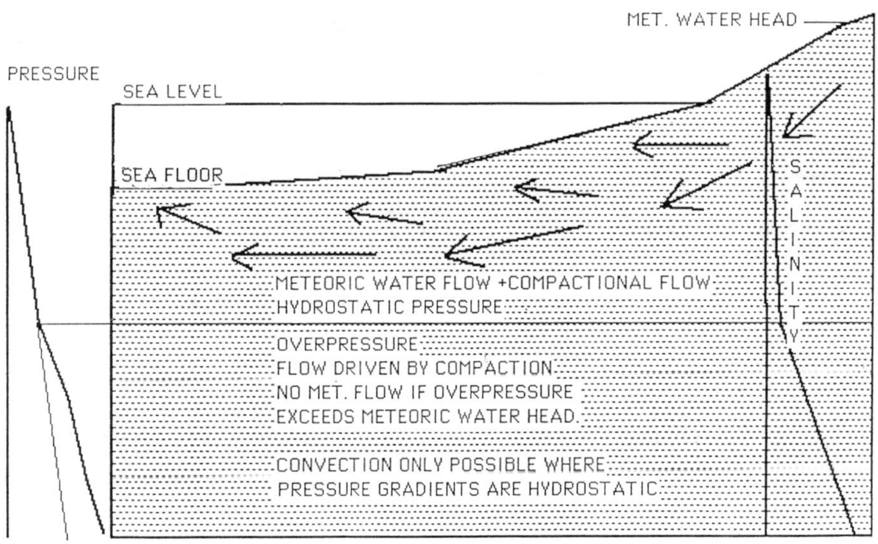

Fig. 2. Schematic representation of different types of pore water flow in sedimentary basins.

Mapping of compositional gradients in oil may help to depict barriers within the reservoir (Larter et al., 1990). Also in the water phase, abrupt compositional changes may indicate barriers and lack of communication by diffusion as well as by advective flow.

Fluid flow(advection) in sedimentary basins:

The flow of pore fluids in sedimentary basins is controlled by subsurface pressure gradients. However, increases in pressure due to the weight of the water column (hydrostatic pressure) will not cause pore-water flow. The hydrodynamic potential (Hp) as defined by Hubbert (1940) is:

$$Hp = Prgh$$

where P is the pore pressure, r the density of the fluid (water), g the level pressure and the actual pore pressure at any given point in the subsurface. This difference can also be expressed in terms of a potentiometric (or piezometric) surface H which is the height to which water would rise above sea level (or other reference levels like ground water level) in a pipe from a point in the subsurface (H= P/rg). Pore water with a higher potentiometric surface than sea level is defined as overpressured. Hydrodynamic flow will occur perpendicular to points of equal hydrodynamic potential.

The rate of flow (q) is controlled by Darcy's Law, where the pressure gradient is given in gradients of hydrodynamic potentials:

$$q=k.rg/h\ (\ H/L)$$

where k is permeability, r the density of the fluid, and h vicosity. H/L is the slope of the potentiometric surface. If the density of the fluid (r) varies, as it usually does, the pressure gradient cannot be expressed by the height of the potentionmetric surface H without making corrections for the variable density of the fluids in the pore water column.

When the viscosity of the fluid is known we can use the term transmissibility instead of permeability. Transmissibility is an expression of how much fluid ($cm^3/cm^2/s$) will be transmitted through a rock at a given pressure-gradient. The kinematic viscosity h varies with temperature. At about 20°C, the viscosity of water is 1.10-6 m^2/s and at 100°C it is 0.2.10-6 m^2/s (see Chapman, 1987).

Although pore-water flow in sedimentary sequences follow Darcy's equation, the initial pressure gradients are insufficient to predict fluid flow though time. Given the low compressibility of water, very small volumes of flow will cause large changes in pressure gradients. The persistence of overpressure over time is evidence of very limited fluid flow.

Another of the problems in applying Darcy's Law to predict fluid flow is that the average permeability on a large scale in a sedimentary basin cannot be measured or determined. Consequently, the flow rate cannot be calculated even if the pressure distribution is known.

Pore-water flow outside the influence of meteoric flow or thermal convection is mostly controlled by compaction. For further discussion of fluid flow in sedimentary basins, see Chapman (1987), Goff and Williams (1987), and Harrison (1989).

4. Transport of dissolved solids in pore water

Transport of ions dissolved in pore water is limited by the solubility and the volume of water through the different parts of the basin. An understanding of fluid flow in the subsurface is therefore essential.

Pore-water flow in sedimentary basins may be divided into three categories (Fig 2):

1) Meteoric water (rain water) flowing into a basin, driven by the head of the ground water table above sea level. This flow is limited by rainfall and its infiltration into the ground water.

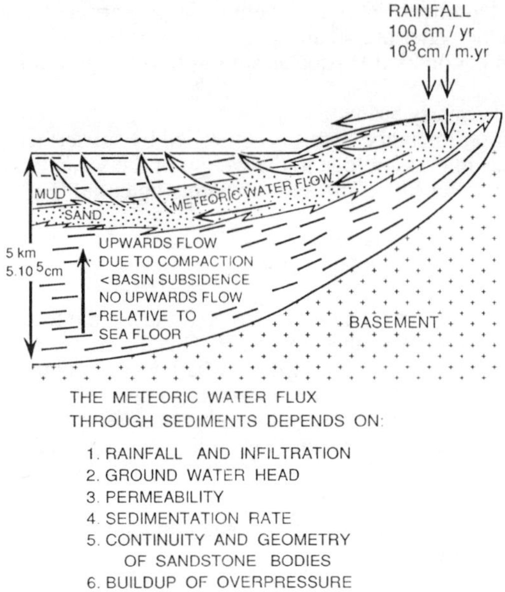

Fig. 3. Simple cross section through a sedimentary basin to show orders of magnitude of pore water flow. 1 m/yr of rainfall will amount to 10^8 cm^3/cm^2 in 1 M years, of which approximately 10^7 cm^3/cm^2 may infiltrate the ground water. The flux of meteoric water into the basin may then be 10^7-10^9 cm^3/cm^2 below the land surface (horizontal flow) and along the coast. the flux will be considerably lower in the more distal marine facies as most of the flow comes to the surface (or sea floor).

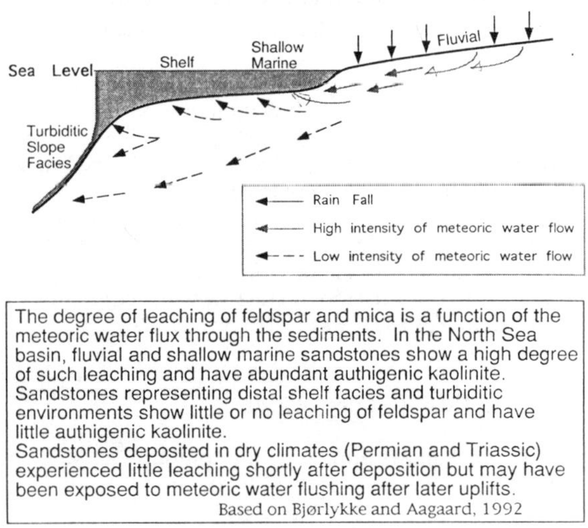

Fig. 4. Relationship between meteoric water flushing and depositional environment.

2) Compaction-driven flow, where fluids are driven by a reduction in the pore space as a function of burial. This flow is limited by the volume of pore water buried in the basin.
3) Pore-water flow driven by thermal convection. Here, the same volume of pore water is used in a closed cell so that high fluxes can be obtained. Meteoric-water flow will not extend into the overpressured parts of the basin. Thermal convection can theoretically occur inside overpressured compartments but the pressure gradients must be hydrostatic in the interval where convective flow can develop.

Although sedimentary basins contain large volumes of water, the amount of solids dissolved in the pore water is very small if chlorides and sulphates are excluded. The solubility of the silicate minerals is limited by low concentrations in silica (10-100 ppm) and aluminum (< 1 ppm) (Egeberg and Aagaard, 1989). The carbonate solubility is strongly influenced by the carbon dioxide pressure and salinity. Petrographic evidence also suggests limited mobility of aluminum (Hayes and Boles, 1992).

When pore fluids are moving at an angle to the isotherms, dissolution or precipitation will occur depending on the solubility functions of different minerals. Since the solubility gradient as a function of temperature is generally small, very large fluxes of pore water are needed to transport significant quantities of the common mineral phases in sandstones, limestones and shales.This can be quantified using this formula:

$$Vc= F.t .\sin b.(dT/dZ) .a .T/r$$

Here Vc is the volume of cement precipitated. F is the flux of pore water ($cm^3/cm^2.s$), t= time, dT/dZ is the geothermal gradient, a T is the solubility temperature function (transfer coefficient), r is the density of the mineral.

Flowing from higher to lower temperatures (upwards) silicate minerals which usually have a progressive solubility will precipitate, while carbonates which have regressive solubility functions will dissolve.

For quartz a T ranges from 1-3 ppm/^oC (1-3.10-6)Wood 1986). If the geothermal gradient is $30^oC/km$
($3.10^{-4}/^oC$), sinb=0,1) r is $2.7 g/cm^3$ and F is 100cm/yr, Vc =3. 10^{-9} cm^3/yr. For 1% quartz to precipitate(Vc=0.01) this flow rate has to continue for 3.10^6years). For vertical flow the total flux has to be 3. $10^7 cm^3/cm^2$. At low T (<100^oC) a T is1ppm/^oC and a flow of $108 cm^3/cm^2$ is required.This is a 1000 km-long water column.

5. Meteoric water flow

The flux of meteoric water into a sedimentary basin depends on the rainfall, the percentage of infiltration into the ground, the head of the ground-water table, and the permeability and continuity of the sandstone bodies (Fig. 3). In a moderately humid climate with an average annual rainfall of 1 m/yr, the total supply in 1 M yrs would be 108 cm^3/cm^2 of rain water on the land surface. Assuming that 10% of the rainfall infiltrates into the ground, the supply to the ground water will be $10^7 cm^3/cm^2$. Depending on the permeability, dip of the sedimentary strata, and the existence of confined aquifers, the meteoric water may flow below the sea floor far into marine sedimentary basins (Manheim and Paull, 1981; Bethke et al., 1988). Modelling suggests that the meteoric water lens may extend down to 3-4 km below the sea floor. Sea-level changes alter the head drastically and may act as a plumbing mechanism (Bethke et al., 1988). At low sea- level stand much of the shelf is exposed for meteoric water infiltration. In the case of relatively homogeneous sediments, many of the flow lines for meteoric water will reach the surface near the coastline. Sandstone beds separated by shales exposed on land and dipping into the basin represent good confined aquifers. Permian and Triassic sandstones in Britain are good examples of such aquifers and they may have a geothermal potential (Downing and Gray, 1986).

At low-to-intermediate sedimentation rates, coastal and deltaic sediments will spend several million years in the zone of meteoric flushing. In 10 M yrs the meteoric flux from infiltration to the ground water will be $10^8 cm^3/cm^2$, and even if only a small percent of this is flowing into the basin, rather high total fluxes (10^6-$10^7 cm^3/cm^2$) may penetrate each volume of permeable sediment. Since the leaching of feldspar at shallow burial and low temperatures is controlled by the kinetics of surface reactions (Aagaard and Helgeson, 1982), the amount of leaching will not only be a function of the total flux but also of time and temperature. The depth at which dissolution of feldspar and mica will continue depends on the flow rate of meteoric water and of kinetic reaction rates, which are difficult to estimate. Dissolution by meteoric water begins immediately after deposition and below unconformities produced by uplift. Below some of the unconformities in the Jurassic of the North Sea basin there is limited evidence of increased leaching.Sediments may be flushed by meteoric water both shortly after deposition and when uplifted and exposed on land. In the case of the Brent Group there is not always a clear increase in the degree of feldspar leaching and precipitation of kaolinite towards the Cimmerian unconformity (Bjørkum et al ., 1990). This may indicate that erosion kept pace with the uplifted and rotated fault blocks so that little subaerial relief was produced.

On present-day land surfaces meteoric-water seems to produce meteoric water leaching some times a few hundred metres below the land surface. In a study of the Milk River Aquifer in Alberta, Longstaffe(1984) found that the authigenic kaolinite down to 2-300 m below the surface had an isotopic compositiion which was in equilibrium with the present day meteoric water ($\partial^{18}O$ =-20) . Such low values for meteoric water are unlikely to have existed in Cretaceous times when the sands were deposted. This suggests therefore that kaolinite can continue to be formed along a relative long pathway before equilibrium with feldpar and mica is reached.In the Cretaceous aquifers in Missisippi, kaolinite is the dominant authigenic clay mineral down to about 300 m depth. At a greater depth the smectite content increases and the kaolinite content decreases (Hanor and Macmanus 1988). This pattern may also be a response to the present meteoric- water flow.

Meteoric water flowing into the basin through permeable aquifers must continue up to the surface for flow into the aquifer to continue (Fig. 3). In the case where a permeable sand bed pinches out into mud down-dip to the basin, pore water must be able to flow through the muddy sediments with lower permeability for the flow into the aquifer to continue. Recently-deposited and poorly- compacted clays, however, have considerable permeability and do not effectively confine the flow to the sandstone aquifers. They are therefore not good aquitards before they have been subjected to considerable compaction. At a burial depth of a few hundred metres, muddy sediments may have permeabilities of up to 1 mD (Rieke and Chillingarian, 1974). The flux of water through a layer of sand dipping into the basin can be distributed along the entire surface area of the sand sheet, which may be 10^3 or more times the area of the cross section. If the permeability is 10^3-10^4 times higher in the sand beds than in the mud, the flow will be balanced in the sense that the pressure gradients will be the same in the two lithologies. If the sediments have been buried, compacted and then uplifted, the permeability of the mudstones will be reduced and the flow into the subsurface will be less effective even if the sandstones remains permeable.

Given a 100 m head of the ground-water table, the flow rate is significant also in rather low-permeability rocks. Applying Darcy's Law and assuming an average permeability of 10 mD over a 10 km pathway though the mud, the flow will be 10-8cm^3/cm^2/s or 0.3 cm/yr and will take 3.10^5 yrs to flow along the pathway. This means that the low-permeability sediments would also be penetrated by meteoric water but at a very slow rate, insufficient to cause leaching, but which would replace the marine connate pore water. A flux of only a few pore volumes of meteoric pore water, however, is sufficient to change the $\partial^{18}O$ content of the pore water. Dissolution and reprecipitation of carbonate will therefore often be influenced by meteoric water also in low-permeability muddy sediments, even if the flux is too low for significant transport of solids.

The isotopic composition of the pore water in sedimentary basins also suggests that most of the marine connate water is flushed out by meteoric water flow (Clayton et al., 1966). At 100 m head the theoretical depth which a meteoric water lens surrounded by marine water (density 1.025 g/cm^3) will reach is 4 km. This assumes, however, a hydrostatic gradient. A build-up of a moderate degree of overpressure due to compaction will effectively prevent further

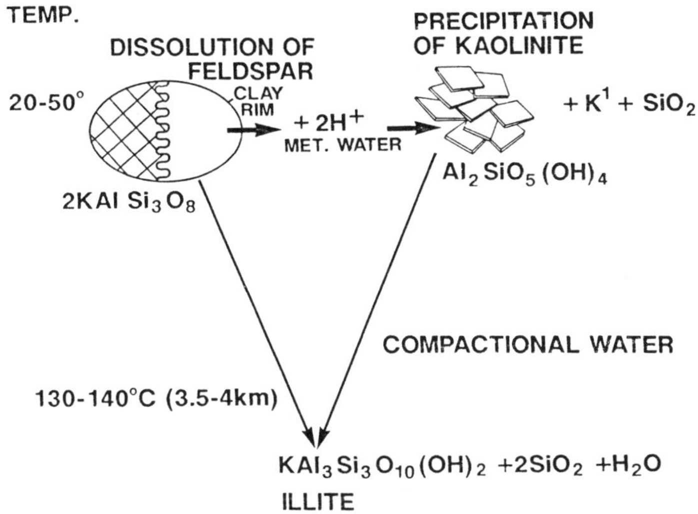

Fig. 5. Dissolution of K-feldspar and precipitation of kaolinite requires a continuous pore water flow through the sediments to remove potassium and silica. In the presence of both kaolinite and K-feldspar, illite will form at the expense of those two minerals at 3.5-4 km depth. If all the K-feldspar has been dissolved, kaolinite will be stable to much higher temperatures.

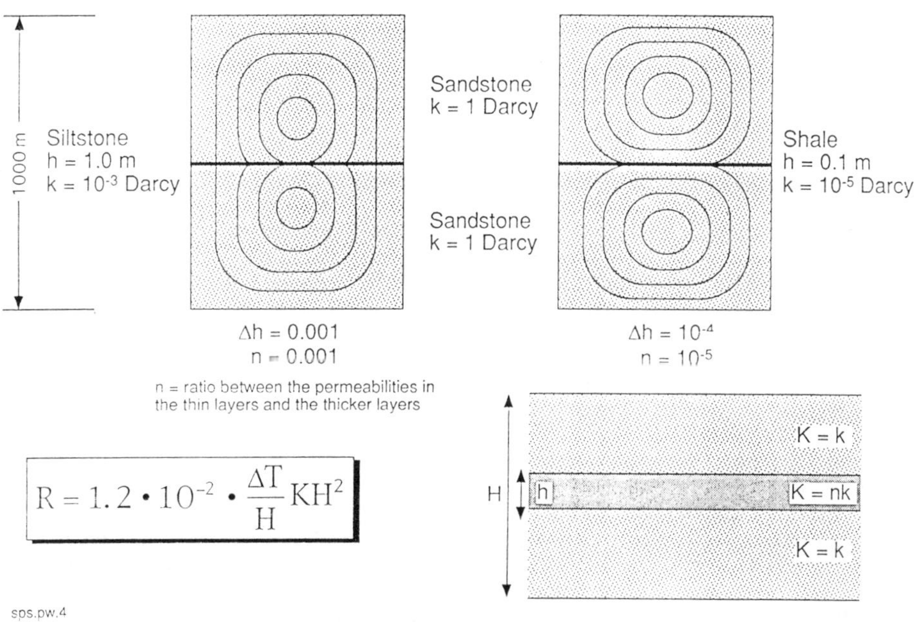

$$R = 1.2 \cdot 10^{-2} \cdot \frac{\Delta T}{H} K H^2$$

Fig. 6. Rayleigh convection in sedimentary basins (from Bjørlykke et al., 1988).

penetration of fresh water, and this is probably the main limitation on the extent of meteoric-water flow into many basins. In the Gulf of Mexico Basin, the present-day meteoric hydrologic regime is limited to the upper 2-3 km that are normally pressured (Harrison and Summa, 1991). Meteoric-water flushing in the Gulf Basin was probably more effective in Mesozoic and early Tertiary times, when sedimentation rates where lower. In Pliocene-Pleistocene times, when sedimentation rates were extremely high (> 1 mm/yr), overpressures due to compaction reduced the extent the extent of meteoric-water flushing. The high clay/shale content caused low overall permeability in these sequences. During sea-level lowstands in the Pleistocene, the ground water would have gained more than 100 m extra head. Harrison and Summa (1991) consider, however, that these low stands were of too short duration for the meteoric-water flux to have caused extensive diagenetic reactions. They may have caused bacterial degradations, however, according to the same authors.

In general, we may predict that coastal sediments deposited at low sedimentation rates will be subjected to much higher meteoric-water fluxes, both because they spend more time in the depth interval flushed by the meteoric water lens and because overpressures are much less likely to develop at slow sedimentation rates. Turbidites and shelf sandstones detached from the meteoric ground-water flow from land areas will not be subjected to extensive meteoric-water flushing. In the North Sea Basin, Upper Jurassic and Tertiary turbiditic sandstones show very little evidence of meteoric-water leaching or precipitation of authigenic kaolinite (Bjørlykke and Aagaard, 1992; Saigal et al., 1992). The Middle Jurassic Brent Group, on the other hand, which is one of the main North Sea reservoirs and was deposited in a shallow marine and fluvial environment, always shows evidence of such leaching (Bjørlykke et al., 1992 and Fig. 4).

Leaching of K-feldspar and precipitation of kaolinite at shallow depths have important consequences for diagenetic reactions at deeper burial (Fig. 5). If both kaolinite and K-feldspar are present, illite will form at 3.5-4.0 km burial depth (Bjørlykke, 1983; Bjørlykke et al., 1986, 1992; Ehrenberg, 1991). If all K-feldspar is dissolved, kaolinite will remain stable to much higher temperatures. Sandstones, not flushed by meteoric water, such as the Fulmar Sandstones and several other Upper Jurassic sandstones in the North Sea, do not have authigenic kaolinite as a precursor and develop less illite at depth (Bjørlykke and Aagaard, 1992; Saigal et al., 1992).

6. Convective pore water flow

Pore-water flow by convection has been suggested by several authors as a mechanism to move solids in solution within the subsurface (Wood and Hewett, 1982; Davis et al., 1985; Cassan et al., 1981; Haszeldine et al., 1984; Rabinowics et al., 1987). Mathematical modelling of Rayleigh convection has shown that low- permeability layers like thin shales (0.1 m) or cemented intervals within a sandstone sequence will effectively divide potentially large convection cells into smaller ones, which may then be too small to exceed the critical Rayleigh number (Bjørlykke et al., 1988 and Fig. 6). Assuming a permeability of 1 Darcy, a roughly 300 m-thick sandstone is required in order for thermal convection to take place. Since such thick homogeneous sandstones without any low permeability layers are rare, the conditions required for Rayleigh convection to occur are probably rarely met in sedimentary basins (Bjørlykke et al., 1988 and Fig. 6).

Calculations by Palm (1990) have shown that if the critical Rayleigh number is exceeded by only 10%, pore-water flow due to Rayleigh convection will be fast enough to dissolve 10% of the quartz in the downflowing parts of the convection cell and precipitate the same amount in the upflowing parts, in 10 Ma. This demonstrates that if Rayleigh convection is taking place, the effects on diagenesis would be fast enough to be observed by extensive quartz dissolution where the flow is upwards, and corresponding precipitation on the down-going part. Extensive quartz dissolution in sandstones is rarely observed except in connection with pressure solution and carbonate cements.

The convection cells if present are likely to follow high-permeability beds for longer periods of time until cementation severely reduces the permeability. Since carbonate minerals have the reverse temperature/solubility function, convective flow supplying silica would also remove a corresponding volume of carbonate. The ratio between silica supply and calcite removal by

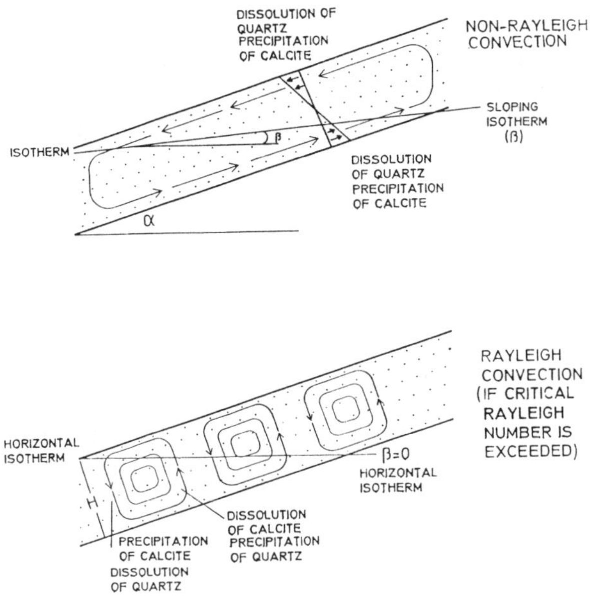

Fig. 7. Mass transport in thermal convection cells. Upflowing (cooling) pore water will precipitate quartz and dissolve calcite. The ratio between dissolved calcite and precipitated quartz, if the pore water is in equilibrium with both mineral phases, will depend on the temperature and pH. At pH 5 and 100°C, about 100 times more calcite will dissolve than the volume of quartz precipitated. The presence of calcite may then preclude introduction of significant volumes of quartz at such temperatures.

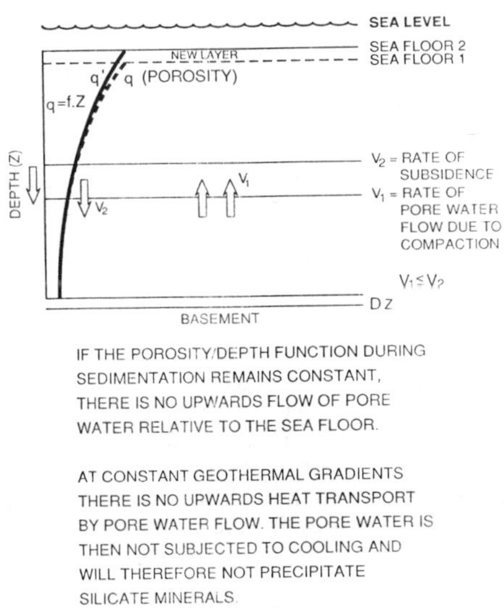

Fig. 8. Relationship between subsidence rate and pore water flow.

convective flow is 1:100 at pH 5 and 100 C (3 km) (Bjørlykke and Egeberg, 1993). For each volume of silica which is precipitated by circulating pore water, one hundred volumes of calcite will be dissolved if present. If calcite cements were still present and had initially made up 30% of the rock, only 0.3% of silica could have been introduced by convective flow.

At higher pH, the ratio between dissolved calcite and precipitated quartz is lower, but the pH is buffered by the silicate minerals (Hutcheon, 1989; Smith and Ehrenberg, 1989). An increase in the CO_2 content of pore water could result in the precipitation rather than dissolution of carbonate.

Upwards flow of pore water subjected to cooling would tend to dissolve carbonates (Fig. 7). The solubility is also a function of the CO_2 content, and where there is a rapid reduction in pressure so that CO_2 is released from solution, calcite will then precipitate. If the pore water is buffered by the silicate system, as suggested by Smith and Ehrenberg (1989) and Hutcheon (1989), the pH will remain relatively constant.

We have seen that in the case of horizontal isotherms, the pore water is stable unless a critical Rayleigh number is exceeded. Where the isotherms are not horizontal, there will always be some Non-Rayleigh convection. The velocities will be relatively low (10^{-4} m/yr), however, when the dip of the isotherm is less than 1-2° and the effective bed thickness is less than 10 m (Bjørlykke et al., 1988). Non-Rayleigh convection may be important near salt domes and around igneous and hydrothermal intrusions. Also in the case of Non-Rayleigh convection, the thickness of the bed is an important factor determining the fluid velocity. The effect of shales separating flow lines is less important than in the case of Rayleigh convection if the beds are long compared to the thickness (e.g. large aspect-ratio). Even if the cross-formational flow through each unit area of shale is small due to low permeabilities, the total flux through the shale may be considerable. Extensive thin shales will therefore not be as effective in terms of separating flow lines as in the case of Rayleigh convection. However, shales buried to more than 2-3 km depth have very low permeabilities, which means that interbedded shales in permeable sandstones have a strong influence on fluid flow.

The forces driving thermal convection are relatively weak and the pressure gradient must be close to hydrostatic. If not, the pore-water flow will be unidirectional from areas of higher potential to those of lower potential. Shales will often dewater both upwards and downwards into more permeable strata such as sand and siltstone beds. Thicker units of mudstones or shales will continue to dewater throughout burial, and therefore usually have higher pore pressures than extensive sandstones. This sets up a unidirectional flow out of the shale sequences, since pore water is less likely to flow through thick shales.

Many basins will also have an increase in salinity with depth, causing a density increase which may offset the density reduction caused by the thermal expansion of water (Bjørlykke et al., 1988). When the salinity of the isotopic composition of the pore water shows a systematic trend with respect to depth, large-scale convection can be excluded.

It is important to be aware that thermal convection is driven by the temperature distribution in the basin. The advective heat transport due to convection changes the temperature field, but only moderately, since most of the heat transport is still by diffusion (Ludviksen, 1992). In the case of Non-Rayleigh convection, the advective heat transport reduces the slope of the isotherms, producing a negative feedback.

Compaction-driven pore-water flow

When sediments are undergoing compaction in a subsiding sedimentary basin, pore water flows upwards to allow the reduction in porosity to take place. Let us assume, for simplicity, that there is constant sedimentation keeping pace with the subsidence and compaction so that the porosity/depth function remains constant. Theoretical calculations suggest that the average upwards pore-water flow relative to the sediments will not exceed the sedimentation rate, so that there is no net upwards-flow relative to the sea floor (Bjørlykke et al., 1989; Caritat, 1989; Figs. 8, 9). Compaction-driven flow is normally lower than the sedimentation rate (typically 0.1-0.01 mm/yr) and will not effect the temperature distribution in the basin (Bethke, 1985; Ungerer et al., 1990; Ludvigsen, 1992).

This means that the pore water does not flow upwards in relation to the sea floor (sediment/water interface), but that the sediments "sink through a column of pore water" (Bjørlykke et al., 1989). Since the pore water does not flow upwards relative to the sea floor, there is no cooling of the compactional water resulting from that flow if the geothermal

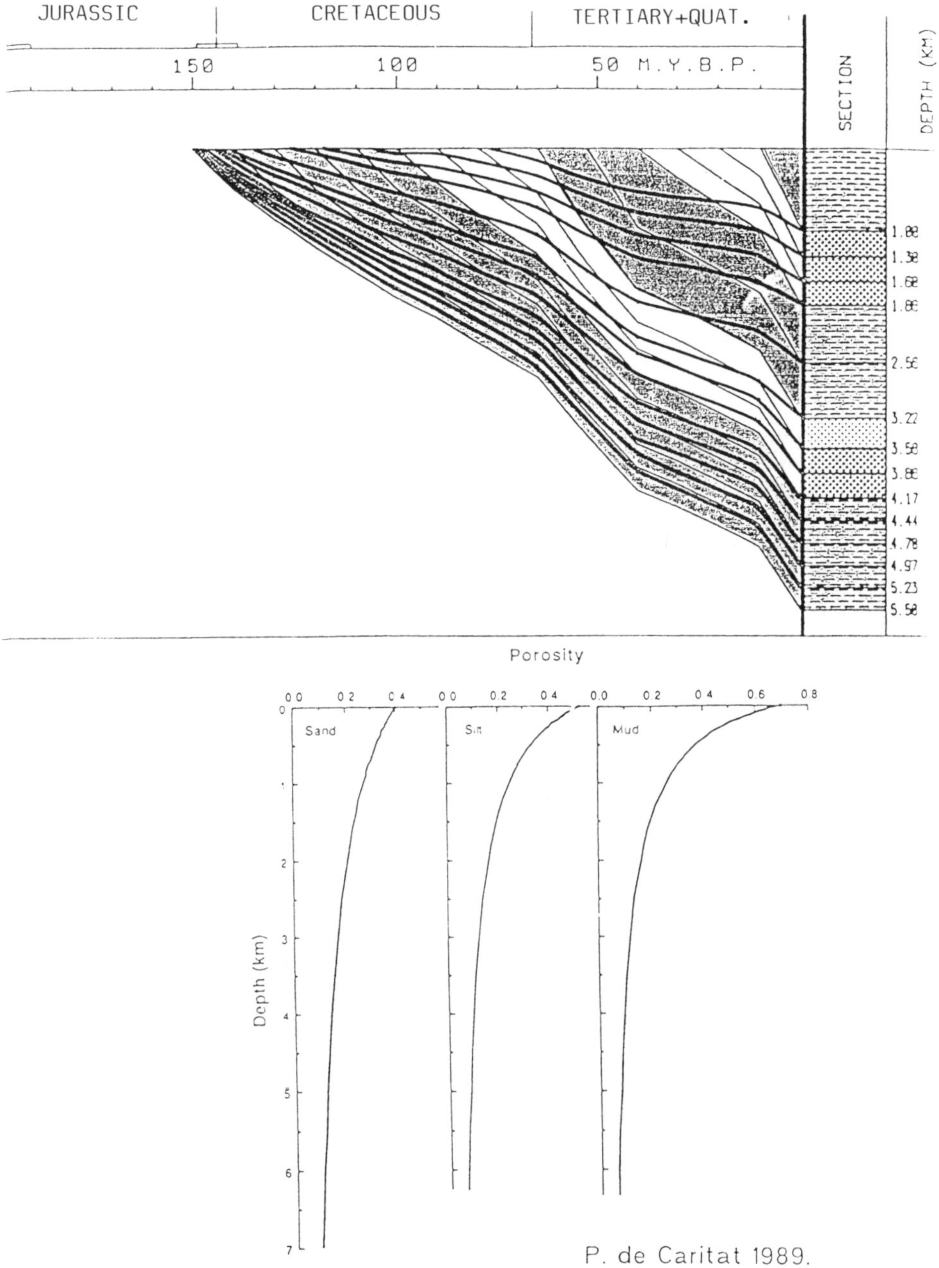

Fig. 9. Modelling of pore water flow relative to subsiding sediments and sea floor (Caritat, 1989).

gradient is constant. Consequently, compaction-driven pore water flow does not transmit heat upwards through the basin and will not cause precipitation of silicate minerals due to cooling of pore fluids, except locally where the flow is focused. When pore water flows upwards relative to the sea floor and cross the red ox boundary, any dissolved ferrous iron and reduced manganese ions dissolved in the pore water will precipitate near the sea floor as hydroxides, sulphides or other minerals. Modelling compaction-driven flow in the Gulf Basin, Harrison and Summa (1991) found that the vertical component of the compaction-driven flow is at maximum 0.2 cm/yr (2 km/My), which is approximately equivalent to the maximum sedimentation rate. The temperature distribution in the Gulf Basin at the present day does not reflect compaction- driven pore water flow nor do the temperatures derived from modelling of the basin during the Tertiary (Harrison and Summa, 1991; and Fig. 10).

Focused compactional flow may cause a significant heat flow by advection but only if the rate of pore-water flow is very high. The transport of heat, due to fluid flow (advection), is:

$$Q= P. C. F. T$$

Here, P is the fluid density, C the heat capacity of water and F the fluid flux. The temperature of the fluid is T, given in degrees Kelvin.

The heat capacity of water is 4200 J/kg K ($4.2\ 10^6$W.S./m^3/^oC). The heat capacity of the mineral matrix may be estimated to be about 8-900 J/kg K (Deming et al., 1990). At 20% porosity, about 2.5 times as much heat is stored in the mineral matrix (density 2.7) as in the water phase. During the first period of advective flow along a fault or through permeable sandstone beds, a high percentage of the advected heat will be lost to the mineral matrix. In most cases the advection can therefore be ignored and the heat flow will be a function of conduction mostly through the mineral matrix. The conductivity of the minerals in the matrix is 2 to 10 times higher than that of water. Most of the conduction thus takes place through the mineral matrix and the temperature of the pore water will, therefore, be close to that of the rock. Only if the pore water flow rate is very high will there be any significant difference. The most conductive minerals are quartz and evaporite minerals, such as halite and gypsum, between the temperature of pore water and mineral matrix.

During periods of zero sedimentation rate (hiatuses), compaction will continue and cause a net flow across the sediment water interface. This should be recorded by the precipitation of iron and manganese oxides on the sea floor due to oxidation of Fe^{2+} and Mn^{2+} in the pore water. The total flux across the sediment-water interface can be obtained from an integration of the difference in the porosity/depth function from start to the end of the time involved in the hiatus. Caritat (1989) has shown that, in the case of uplift and erosion, there is no upwards flow of pore water relative to the surface, since compaction will not occur during unloading. In a real sedimentary basin, pore water released by compaction does not flow upwards relative to the sediments in a uniform way, but may be focused through the more permeable sandstones or along faults. It is important to keep in mind, however, that the flux of pore water along such special pathways is limited by the total volume of water released by compaction and must be matched by a downward migration of pore water in other areas. The same continuity equation that applies for flow during oil production also applies for compactional flow (Caritat, 1989; and Fig. 9).

The total flux of pore water flowing through a sedimentary layer as it is subsiding to 5 km is on the order of 105 cm^3/cm^2 (average porosity 20%) (Bjørlykke, 1983; Bjørlykke et al., 1989).

This is 2-3 orders of magnitude lower than the fluxes that can be obtained during meteoric flow. Compactional pore water at deeper burial (> 80^oC) can be assumed to be in equilibrium with the main mineral phases present (Egeberg and Aagaard, 1989), and, since there is no upwards flow or cooling, little dissolution or precipitation will follow. At 3.5-4.0 km burial depth, most of the dewatering of the sediments has already occurred. The average porosity for the sequence at this depth may be 10-15% and the porosity gradient may be 3-4%/km, so that the pore water flux for 1 km burial will be only $3\text{-}4 \times 10^3\ cm^3/cm^2$.

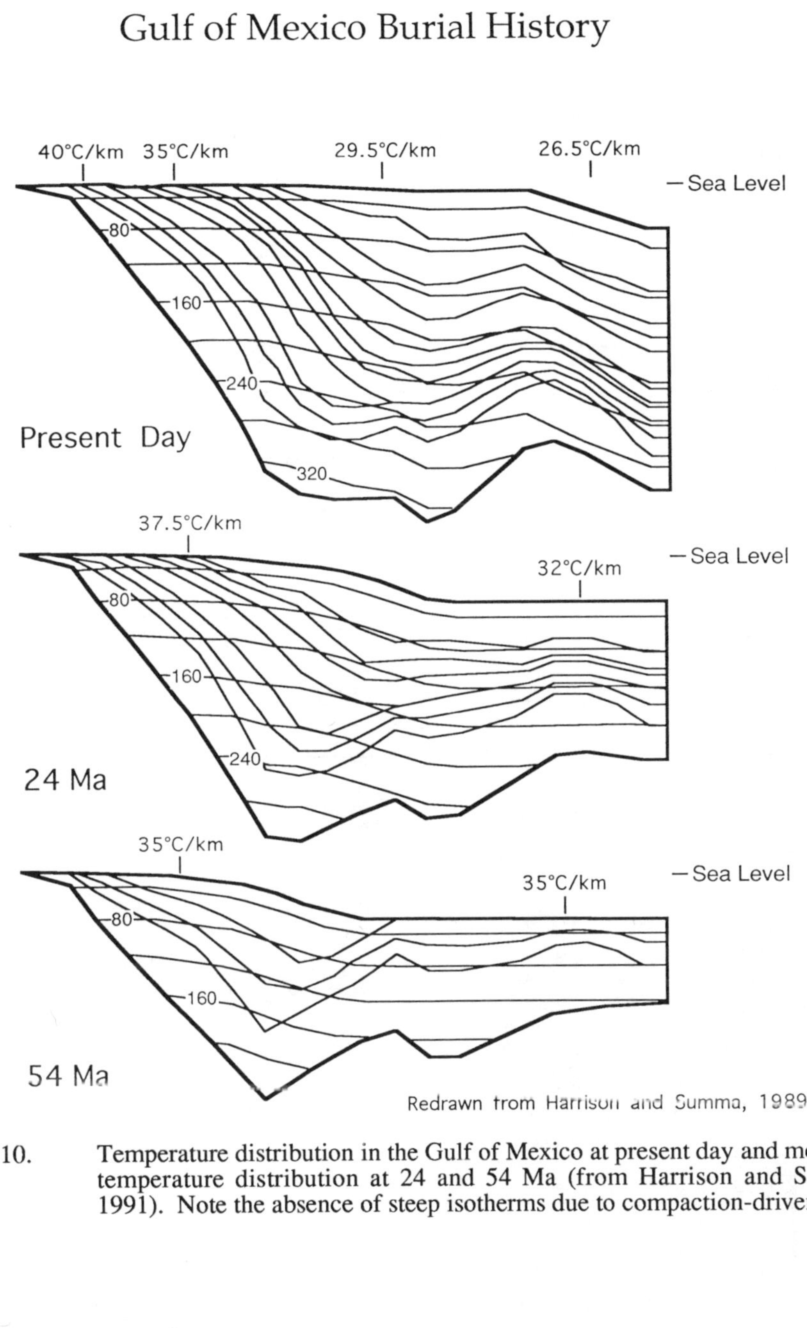

Fig. 10. Temperature distribution in the Gulf of Mexico at present day and modelled temperature distribution at 24 and 54 Ma (from Harrison and Summa, 1991). Note the absence of steep isotherms due to compaction-driven flow.

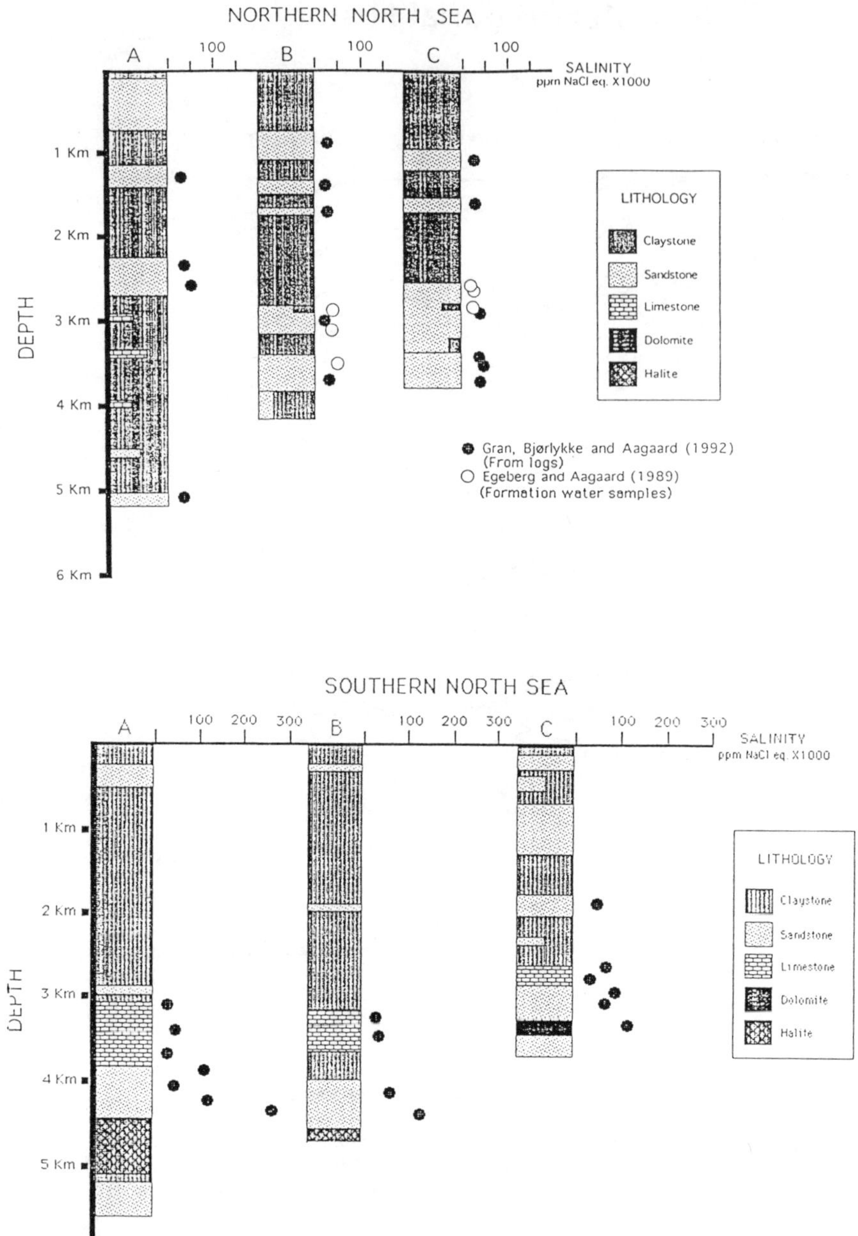

a

Fig. 11. Salinity/depth trends based on resistivity logs. Notice the steep increase in salinity towards the evaporites in the southern part of the Norwegian North Sea. In the northern North Sea, where evaporites are missing, highly saline brines are not recorded (after Gran et al., 1992).

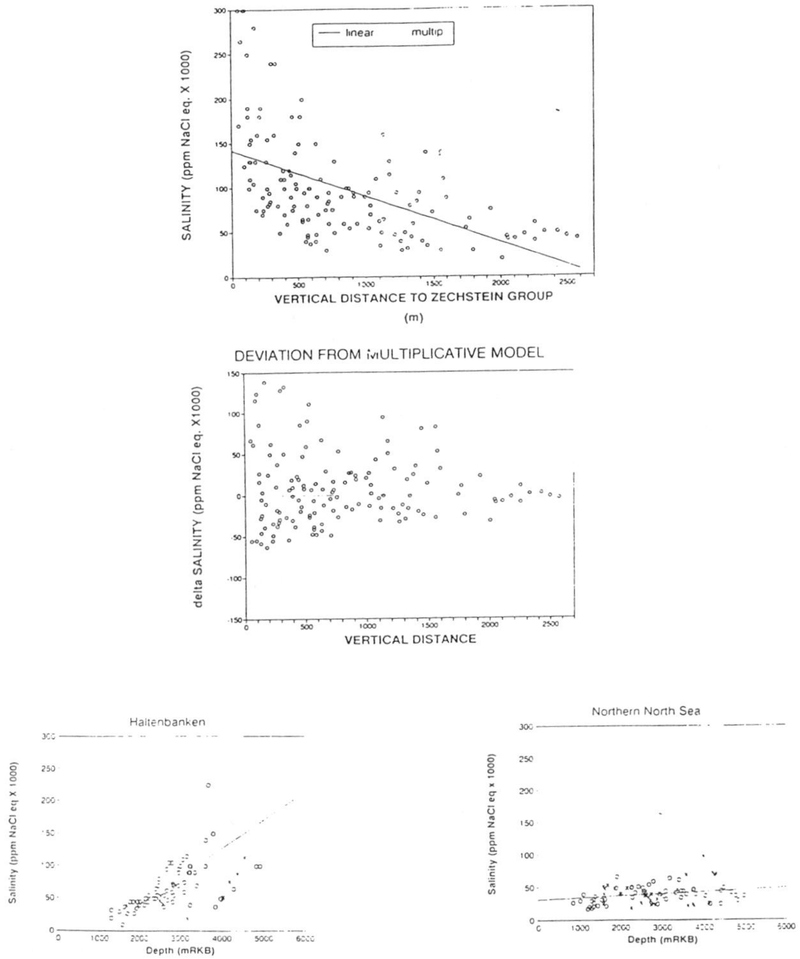

b

Fig. 11. Salinity/depth trends based on resistivity logs. Notice the steep increase in salinity towards the evaporites in the southern part of the Norwegian North Sea. In the northern North Sea, where evaporites are missing, highly saline brines are not recorded (after Gran et al., 1992).

If the depth to basement is about 6 km, the total reserve of pore water below a sandstone at 3.5 km is on the order of $2x10^4$ cm^3/cm^2. This is the depth at which quartz cementation starts to become very important. A flux of pore water of 10^8-10^9 cm^3/cm^2 is required to cement a 100 m-thick sandstone by vertical flow (Bjørlykke et al., 1992; Gluyas and Coleman, 1992). Focused pore-water flow along faults or permeable sandstone beds will have to drain all the pore-water in a volume of the basin which is 10^4 times larger than the volume of sandstone cemented. The pore-water flow along permeable dipping sandstone beds must be higher, since the vertical component (rate of cooling) will be lower. If the temperature gradient of the pore-water flowing upwards by focused flow does not deviate too much from the average geothermal gradient in the area, the precipitation will be distributed as a function of the flux and the solubility/temperature function (dc/dt). If the pore-water is flowing with a high velocity so that the heat supplied by advective transport dominates over heat diffusion to the rock matrix on a larger scale, the temperature profile along the flow line will deviate from that of the regional geothermal gradient. Even if there is a very rapid flow of pore water from the deeper parts of the basin, which is highly unlikely, the volume of pore water will be too small to cause cementation in overlying sandstones, except on a very local scale. In our example, the ratio between the volume of rocks drained of all compactional fluid (10%) and the volume of rocks receiving significant amounts of quartz cement (10%) is $3x10^4$. This assumes a temperature difference of 50°C and that the cooling takes place almost instantaneously within the sandstone. In the case of other elements with much lower solubility/temperature gradients, such as aluminium, the required flux to cement illite from outside sources is much higher. The solubility of quartz and other silicate minerals may be enhanced by complexing with organic acids (Surdam et al., 1984, Surdam, 1986). This complexing seems significant mainly at very low pH conditions (< pH4), which are not likely to occur in pore waters due to the buffering capacity of the silica and carbonate minerals (Hutcheon, 1989). Sandstones which are closely associated with source rocks do not show much evidence of enhanced dissolution (Bjørlykke and Aagaard, 1992).

Advective flow that is capable of supplying silica will also remove a corresponding volume of carbonate. This applies not only to pore water in a convection cell but also pore water driven upwards and cooled unless there is a rapid drop in the CO_2 pressure (Bjørlykke and Egeberg, 1992). This assumes, however, that the pore-water is in equilibrium with respect to quartz. This is not likely to be the case at temperatures lower than 70-80°C. If the pore water is buffered by the silicate minerals, as suggested by Smith and Ehrenberg (1989) and Hutcheon (1989), the temperature/solubility gradient is very much reduced and carbonate dissolution and precipitation could be slower during convective pore- water flow.

7. Constraints on fluid flow from compositional gradients (salinity gradients)

Chlorine is not consumed in diagenetic reactions and since there is no sink for this element, it is a good tracer for fluid flow. Cl^- is the main anion in saline water and its concentration, therefore, also determines the total cation equivalents.

In sedimentary basins such as the North Sea, steep salinity gradients are observed in the 300-500 m of sediments above evaporites (Gran et al., 1992; and Fig. 9).

Some of the highly-saline pore water overlying the Permian salt may represent primary pore water in the evaporites which has been squeezed out by compaction. In addition, there has been transport of salt dissolved from evaporites by diffusion and advection (Egeberg and Aagaard, 1989; Gran and Bjørlykke, 1992). The observed increase in salinity with depth and with decreasing distance to the evaporites is evidence of limited upwards-transport of salt by compaction-driven flow. This is to be expected, since there is limited supply of pore water from the deepest parts of the basin through the salt. This means that during the subsidence of the basin the pore water has moved downwards at almost the same rate as the basement, so that the upwards flow relative to the sedimentary sequence has been small (Fig. 11a,b, Gran et al., 1992). Salinity/depth gradients are also evidence that large-scale convection is not taking place (Bjørlykke et al., 1988).

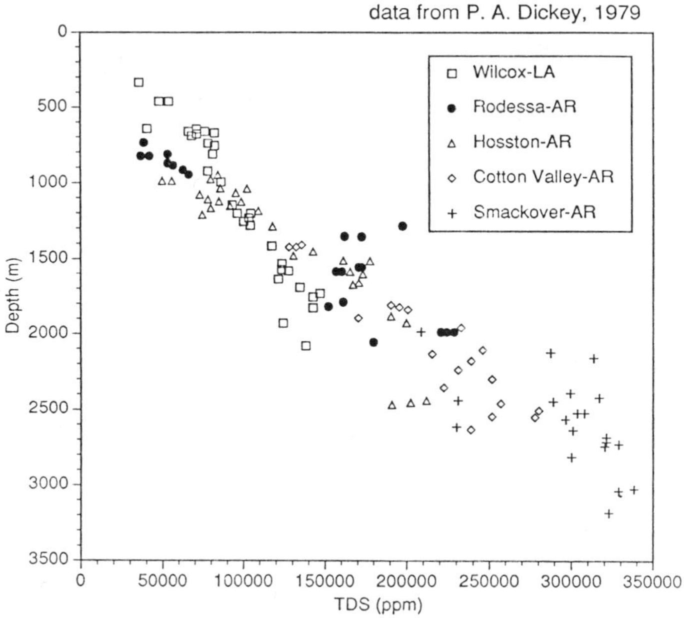

Fig. 12. Salinity/depth trends from Arkansas (Dickey, 1979).

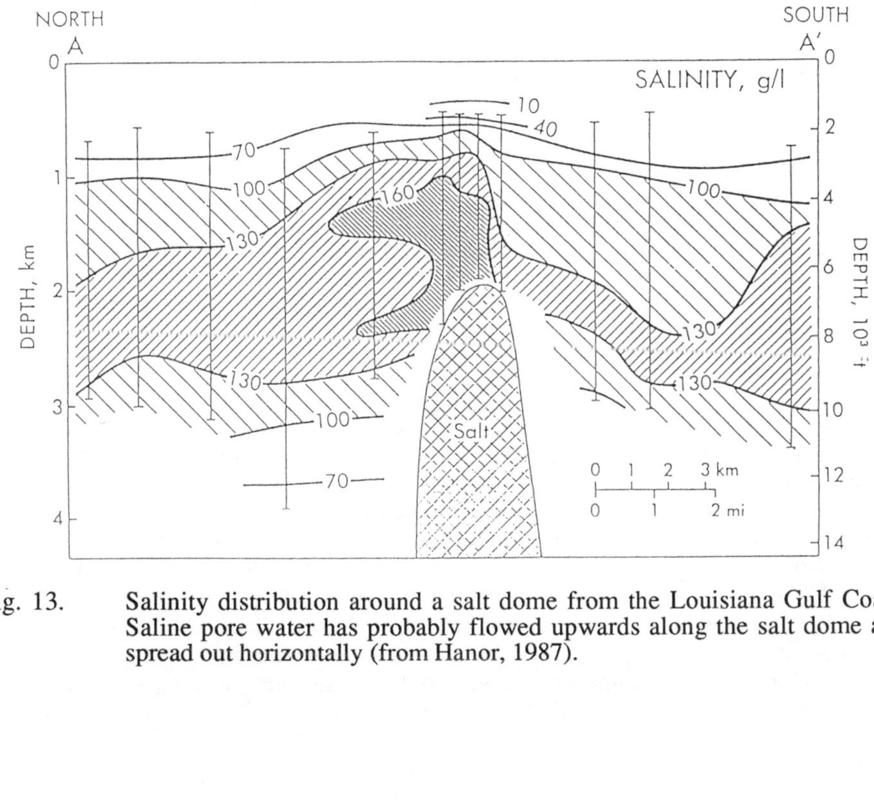

Fig. 13. Salinity distribution around a salt dome from the Louisiana Gulf Coast. Saline pore water has probably flowed upwards along the salt dome and spread out horizontally (from Hanor, 1987).

In the Gulf Coast Basin, an almost linear increase in salinity with depth is often observed (Dickey, 1979; Fig. 12). Around salt domes, however, saline pore water seems to rise and spread out horizontally in overlying sandstone beds, as described from the subsurface of Louisiana (Evans and Nunn, 1989; Bennet and Hanor, 1987; Hanor, 1987; Fig. 12). This results in a salinity inversion which will add to the downward decrease in density produced by the thermal expansion of water. Complete convection cells do not seem to develop, however (Fig. 13). The observed horizontal extension of saline brines between less-saline pore waters is very strong evidence of how limited the vertical movement of pore water is. This is probably due to the low permeability of the shales and to overpressures forcing pore water upwards. The driving force for thermal convection, even when reinforced by a reverse salinity gradient, is rather weak and will be counteracted by rather slight overpressure gradients. The result will then be unidirectional flow as defined by the potentiometric surface.

8. The importance of faults

Faults represent possible conduits for fluid flow in sedimentary basins. Fluids may be transmitted along the fault plane but faults are also important because fluids may flow across fault planes and connect different stratigraphically- juxtaposed sandstones. For a fault plane to be a highly permeable conduit for fluid flow, it must be kept open either by high pore pressures exceeding the geostatic pressure or by the wedging effect of coarse- grained breccias developed on the fault plane.

In crystalline rocks and well-cemented sedimentary rocks, tectonic shear may result in an extensional shear failure that opens up fractures which are then high-permeability conduits for fluids (Sibson, 1990). Such fractures are likely to be cemented up by carbonate or silica cement as a result of precipitation from fluids flowing through the fractures. Transport of cement into the fractures may also take place by diffusion. Since the fractures represent an area of minimum stress compared to the surrounding rocks, which are subjected to an overburden stress and a tectonic stress, there is likely to be a thermodynamic drive towards dissolution of the minerals under stress and precipitation into the pore space. Renewed fracturing is therefore required for the faults to remain permeable. As long as the pore pressure is less than the overburden pressure, there will be a net stress trying to close the fractures. A certain shear-strength of the rock matrix is therefore required in order for fractures along the fault plane to remain open. Fractures in soft clays and poorly cemented sandstones, therefore, cannot remain open for very long.

Growth faults, which develop slowly and gradually in soft sediments, will probably not develop open fractures (Chapman, 1987). Faults in the shallow parts of a basin (< 3 km) are often surrounded by poorly cemented and highly permeable sandstones. The fault plane is therefore only a well-defined pathway where there are shales on both sides of the fault. If overpressure causes a fault to open, pore water will probably be drawn from adjacent sandstones, not only from the deeper parts of the basin. The amount of fluid that can flow before the pressure is reduced will depend on the volume of rocks communicating with the fault system. However, the compressibility of water is low and the flow due to pressure release alone is limited (see below). The most significant flow is due to compaction caused by the increase in net stress when the overpressure is reduced. Seismic pumping of water along faults has been demonstrated to be an important process for driving fluid flow along seismically active faults, and water may then also be released from crystalline basement rocks (Sibson, 1990). Even if the total flux of water may be considerable, as demonstrated from the San Joaquin Valley where the surface effusion was estimated to be 10^7 m^3 (Sibson, 1981), the flux is only sufficient to explain very local diagenetic effects along the fault plane, which would be expected to gradually be cemented up by quartz cement. In two months, 10^7 m^3 of water flowed over a 35 km-long rupture at the surface; this is 300 m^3 per metre. For elements like silica, which have a solubility with respect to quartz in the area of 100 ppm at 150°C, only 0.003 m^3 of cement would precipitate along the fault plane for each metre-length along the fault. Near the surface, at temperatures lower than 60-70°C, precipitation would not take place in the form of quartz, however, but opal-CT. It is possible that this process could be repeated many times, but the volume of water available in the subsurface would represent another constraint.

At shallow depth, clays and poorly-indurated sandstones probably do not have sufficient shear strength to resist the effective stress against the fault plane. Clay smears along the fault plane tend to develop from adjacent clays and reduce permeability. When a porous sandstone is juxtaposed to a shale, the higher permeability will not be along the fault plane, but through the adjacent sandstone. When two thicker sandstones are juxtaposed, there will be little clay-smear. allowing flow through the fault plane connecting the two sandstones. This allows compactional fluid to "jump" stratigraphically to other high permeability sandstones. Faults may be an important part of the plumbing system in sedimentary basins for this reason.

Faulting may breach low-permeability barriers to the overpressured parts of the basin and cause an upwards discharge of fluids. This fault-valve model has been described by Sibson (1990). It is clear that the channelling of fluids along the fault plane will be most effective if the surrounding sediments have a low permeability. In the shallow section of sedimentary basins, where there is highly permeable loose sand, the water flow and the pressure will dissipate into the porous sands. Unless the fault plane is highly permeable right up to the sediment surface, the rate of fluid flow along the fault will be reduced by a temporary build-up of pressure in the shallow section. The rate of water supply to the fault from the overpressured section is also limited by the permeability of these sediments next to the fault plane.

9. Flow along faults and fractures

Faults may have a significant influence on fluid flow, in different ways, including: (i) the potential for fluid flow along the fault plane; (ii) offset of permeable rocks, such as sandstones or limestones, along the fault plane; and (iii) the effect of clay smear along fault planes when sandstones are juxtaposed on both sides of the fault.

The properties of fault planes are determined by the sediment lithology and the stress conditions at different depths. Unless there is an overpressure exceeding the lithostatic pressure, there will be a component of vertical stress acting on the fault plane to close it. This force has to be resisted by the shear strength of the rocks. Soft clays, mudstones and poorly cemented sandstones have low shear-strengths. Faults through such lithologies cannot remain open unless the overpressure comes close to equaling the component of lithostatic stress acting on the fault plane. Well-cemented sandstones, shales and limestones have high shear-strengths which can resist much higher mechanical stress. Brecciation developing along fault planes helps to wedge them open and maintain high permeability. A high flux of pore water on a fault plane will cause precipitation of cement, which reduces permeability. Silicates precipitate as the pore water cools during upwards flow, and carbonate minerals may precipitate due to degassing of carbon dioxide.

Open fractures will also tend to be slowly cemented by diffusion from the surrounding rocks. Carbonate and silicate minerals next to the faults are under lithostatic stress and therefore more soluble than unstressed minerals growing along the fracture. There is therefore a thermodynamic drive towards a diffusion of solids into the fracture. Inactive faults will therefore tend to be cemented up. It is generally difficult to explain total cementation of fractures by advective flow, since the permeability will reach a very low level before all the porosity is lost.

If the low permeability seal overlying or surrounding an overpressured part of the basin is broken, i.e, through faulting, a rapid pulse of pore water flowing upwards may follow. However, due to the low compressibility of water (4.3 x 10^{-10} Pa-1), the upwards flow of pore water necessary to reduce the pressure is relatively small. For an overpressure of 10^{7} Pa (potentiometric surface 1 km above sea level), the average expansion of water would be 4.3 x 10-3. In a vertical column though an overpressured sequence, this would result in an average upwards flow of 4.3 m for each km of sequence if all the overpressure were released at one time. If the flow were focused though a smaller cross-section, however, this figure would increase proportionally to the degree of focusing. It is very unlikely, however, that large volumes of shales would reduce their overpressure over a very short time, and the potential for episodes of rapid flow of compaction water is therefore limited due to the overpressure alone.

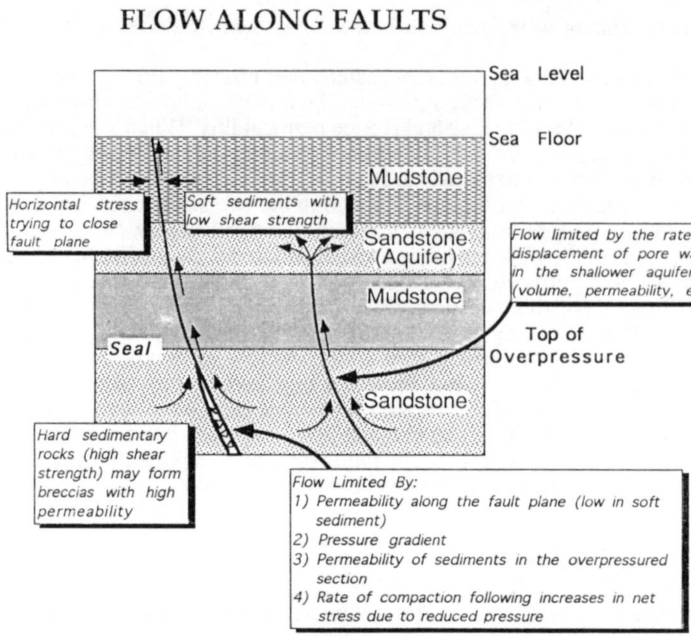

Fig. 14. Fluid flow along faults. Note the difference between faults extending all the way to the sea floor and faults dying out in the sequence.

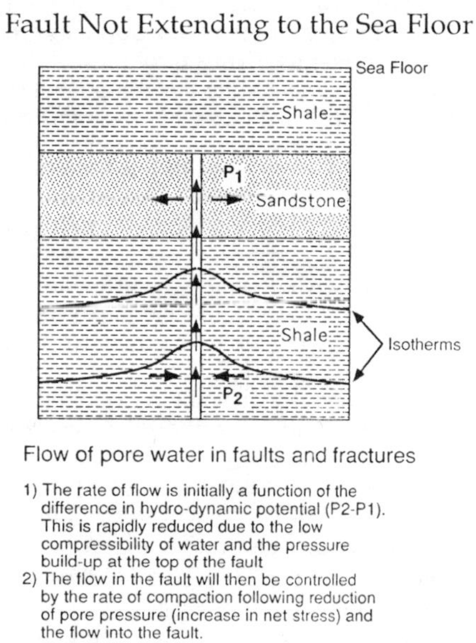

Fig. 15. Flow into shallow sandstones is dampened because of the necessary displacement of water.

If the water is saturated with respect to gases like carbon dioxide or methane, the compressibility of the pore fluid will increase significantly. A reduction in the pressure will cause gas to be released from solution and form a separate phase which has a high compressibility.

When the degree of overpressure is reduced by pore-water flow through a leaking seal, the net stress is increased and compaction will occur. This will probably not take place instantaneously, however, and compaction will slowly build up the pore pressure again unless there is continuous flow from the overpressured section. As we can see from these calculations, the flow resulting directly from the pressure release is relatively limited. The main expulsion of pore water is due to sediment compaction, which is an indirect consequence of the reduction in overpressure and increase in net stress (net stress = overburden stress - pore pressure). Compaction and expulsion of pore water resulting from increased net stress is a gradual and rather slow process, and this strongly influences the rate of water supply to faults from mudstones. The permeability of the surrounding mudstones limits the rate of flow into the fault zone.

If a fault plane does not extend up to the surface (or sea floor), the upwards- moving pore water will have to be accommodated in shallower strata (Fig. 14). Large volumes of pore water cannot suddenly be injected into shallower sandstones, even if they are normally pressured (Fig. 15). Flow into shallow aquifers of limited extent will result in a temporary pressure build-up before the water can be displaced, and this reduces the flow rate. The highest flow rates can be expected when the fault extends all the way to the surface and the pore water can flow into sea water or onto the land surface (Fig. 14). In the North Sea Basin, most of the faults extend only into the Cretaceous section and therefore were not very active during the Tertiary and Quarternary (Glennie, 1990). The timing of faulting must be taken into account when faults are called upon to explain fluid flow and diagenetic reactions in sedimentary basins (Burley, 1986).

In a sequences with a high clay/sand ratio, faults may act as a seal between two sandstones due to the clay smear on the fault plane.

10. Multiphase flow

The presence of hydrocarbons in sedimentary basins introduces the problems of two-and sometimes also three-phase flow. It becomes necessary, thus, to introduce the concepts of effective and relative permeability. Effective permeability is the permeability of one phase in combination with other phases. When pores are filled with two phases (i.e. oil and water), the permeability of each phase will be lower than it would have been in the case of only one phase. Relative permeability is the ratio between the permeability at a certain saturation of a phase and the permeability at 100% saturation. When the percentage of oil in the pore space is less than 20-30%, the relative permeability is close to zero, which means that only water will flow through the rocks. If we apply the principle of two -hase flow to sedimentary basins on a large scale, considering a large volume of rocks, the oil saturation will be smaller. Water will be the continuous phase and the relative permeability of oil and gas will be close to zero. Pressure differences within a sedimentary basin will essentially cause water flow and have a very limited effect on the movement of oil. We cannot, therefore, use the pressure regime (hydrodynamic potentials) in a sedimentary basin to predict the flow of petroleum. It is also important to remember that the pressure in oil and gas columns are different from those in the adjacent water phase, and that this difference is balanced by the capillary pressures.

While pore water will follow the potentiometric gradients, hydrocarbon flow is driven essentially by buoyancy (England et al., 1987). Pore-water flow out of compacting shales may have a locally-downwards component if draining into more permeable underlying sandstones. Hydrocarbons will not be influenced much by the potentiometric gradients. Hydrocarbon flow is principally driven by buoyancy and will therefore always be directed upwards (England et al., 1987), except during the primary migration when it is expelled from the shales (Leythaeuser et al., 1987). Since the driving forces are principally so different, the direction of pore-water flow and hydrocarbon flow usually will not coincide.

11. Pore-water flow in overpressured parts of sedimentary basins

The build-up of overpressure in a sedimentary basin is evidence that expulsion of pore water has not kept pace with subsidence. The pore pressure is balanced by low permeability rocks such as fine-grained calcareous or siliceous shales. Because of the interdependence of permeability, subsidence rate and overpressure, it is very difficult to estimate the rate of pore-water flow from pore pressure data. Bethke (1985) has shown that in intracratonic basins, the degree of overpressure is normally low and that pore water flow velocities are less than a few mm/yr. The North Sea Basin is an intracratonic basin with a high degree of overpressure. Overpressure is commonly associated with a rapid subsidence rate, and, although the flow relative to the sediments may be increased, the rate of upwards-flow relative to the sea floor is likely to be negative.

The total flux of pore water generated by compaction is limited by the total volume of water buried with the sediments. Under hydrostatic pressure and constant permeability/depth gradients, the upwards-flow of pore water may become close to the subsidence rate, so the rate of pore-water flow relative to the sea floor becomes close to zero. In overpressured sediments, the porosity is normally higher than normal for the depth and the pore-water flow will have a stronger net component downwards in relation to the sea floor (Fig. 16). When the overpressure is suddenly reduced due to breaking of tight shales or other seals, the expansion of the water column is rather small. A good introduction to the theory of subsurface pressures is given by Gretener (1977).

12. Diagenetic traps and seals

In many sedimentary basins, there are rather abrupt transitions from normally pressured to highly overpressured (or underpressured) rocks, suggesting the presence of nearly impermeable seals. Hunt (1990) suggested that the generation of oil and gas takes place when source rocks are inside seal-bounded compartments. Although a degree of overpressure can cause fracturing, which will help to facilitate oil expulsion, it is not a necessary requirement for oil generation. As pointed out by Larter et al. (1990), the transformation of kerogen to petroleum changes the fluid/rock ratio and causes the source rocks to undercompact, and local overpressure will result. The compartmentalization of the pore fluids, as suggested by examples referred to by Hunt (1990), puts very important constraints on the pore-water flow and thus on the transport of solids in solution.

13. Diagenesis and mass transfer

Estimates of the amount of mass transfer involved in diagenetic reactions are rather uncertain. Precipitation of a new mineral phase may have occurred while another phase was dissolving, and unless the composition and volume of the dissolving mineral are known, the "import" or "export" of ions cannot be determined.

Diagenetic minerals and possible precursor minerals:

Precursor phases	supply or removal of ions	diagenetic phase
Carbonate minerals:		
Aragonite	-Sr	Calcite
High Mg-calcite	-Mg	Low Mg Calcite
Low Mg calcite	-	Poikilotopic Calcite
Aragonite	+Mg	Dolomite
Dolomite (Fine-grained)	-	Dolomite (Coarse-grained)
Silicate minerals:		
Amorphous silica	(-Al)	Quartz

Opal CT	(-Al)	Quartz
Quartz		Quartz
K-Feldspar	-Na+, -K+	Kaolinite
Smectite	+K-H20(-Si)	Illite
Kaolinite	+Mg+++Fe++	Chlorite
K-feldspar + Kaolinite	-	Illite

Fluid flow and diagenetic reactions in sandstones

Theoretical modelling can be a significant help in predicting diagenetic reactions which strongly influence the reservoir properties. The most important factor in such predictions is the degree of transport of solids in pore water.

Early diagenetic reactions are more open because of the low kinetic reaction rate. The degree of leaching of feldspar, mica and rock fragments will depend largely on the flux of meteoric water. Petrographic studies of Jurassic reservoir rocks from the North Sea suggest that sandstones with a high degree of lateral continuity have been subjected to more feldspar leaching than equally permeable sandstones with a lower degree of continuity (Nedkvitne and Bjørlykke, 1992). Thus, the degree of feldspar leaching may provide some indication as to the continuity of reservoir rocks.

This study also showed that the most continuous sandstones, which had experienced the highest pore-water flux, had a lower content of kaolinite than would be expected from the volume of feldspar leaching. This suggests some degree of export of aluminium, but this may be on a relatively local scale. At low temperatures, the rate of mineral precipitation is very low, and the pore water may then be supersaturated or undersaturated with respect to mineral phases like kaolinite. The potential for transporting fluids which are supersaturated with respect to a mineral phase can be expressed as the half-saturation distances (Giles, 1987).

Shelf sandstones surrounded by shales, like the Upper Jurassic Fulmar sandstone from the North Sea, show very little feldspar leaching and contain little or no diagenetic kaolinite, probably because these sandstones received a low flux of meteoric water (Saigal and Bjørlykke, 1992). Also, turbiditic sandstones from the North Sea Basin show little or no evidence of feldspar leaching (Bjørlykke and Aagaard, 1992).

Calculation of fluid-flow rates and pore-water fluxes as discussed above suggests that they are insufficient to transport significant dissolved mass during burial diagenesis, except very locally. This means that diagenetic reactions are for the most part isochemical at a few metres scale. If this is accepted, rather detailed predictions can be made. If we know the mineralogy of a sandstone buried to a certain depth and temperature (i.e. 2 km and 80°C), the mineralogical changes this sandstone will undergo upon further burial (down-dip) can be predicted on the basis of thermodynamics and kinetic parameters. The association of kaolinite and K-feldspar is a good example. This association will be stable or at least metastable up to a temperature of 120-140°C, and in the North Sea Basin and Haltenbanken, Mid-Norway, a very significant increase in the concentration of illite is found below 3.7 km burial depth (Bjørlykke et al., 1992; Ehrenberg and Nadeau, 1989), which reduces the permeability significantly. In the absence of K-feldspar or kaolinite, we can predict that such a growth of illite would not occur and the reservoir properties would be better. The fact that kaolinite remains stable at much higher temperatures when K-feldspar is absent is an indication of the low degree of mobility of K^+ in the pore water. Potassium appears not to have been supplied from sandstones containing K-feldspar (Bjørlykke et al., 1992). A sandstone containing smectite at shallower depth will undergo alteration to mixed- layer minerals and illite with increasing burial in the temperature range of 70-100°C. As discussed above, due to the low solubility of aluminium, illite will not form unless suitable precursor minerals are present.

If quartz or carbonate cement are introduced into sandstones by advective flow of pore water from the outside, as suggested by Gluyas and Coleman (1992) and Robinson and Gluyas (1992a, b,), very high fluxes (10^7-10^8 cm^3/cm^2) would be required. Such pore-water flow will, for the most part, be channelled through the most permeable pathways, which may be porous sandstone, limestone or along faults. If the pore water is in equilibrium with these mineral phases, total precipitation (or dissolution) will be proportional to the flow rate. The most permeable beds, faults, and fractures will therefore receive most of the flow and be cemented up first, until the permeability is reduced to the level of the surrounding rocks.

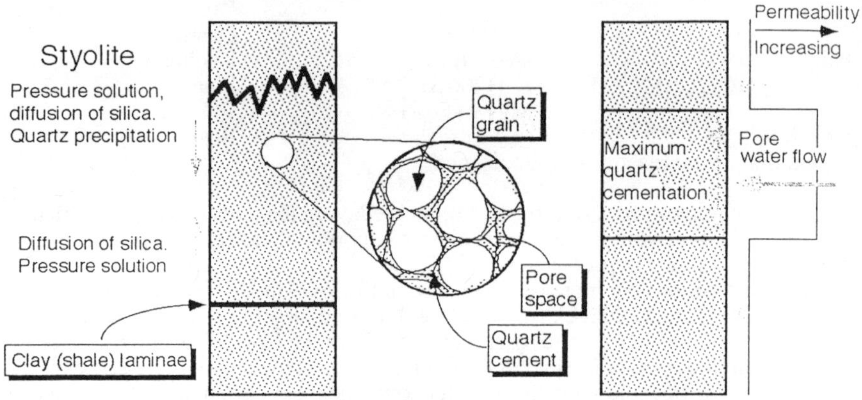

Fig. 16. Quartz cementation as a function of fluid flux. Import of silica through advective flow from outside a sandstone would imply that the rate of quartz cementation if proportional to the flux. This means that the most permeable pathways, like porous sandstones and faults and fractures, will receive most of the quartz cement.

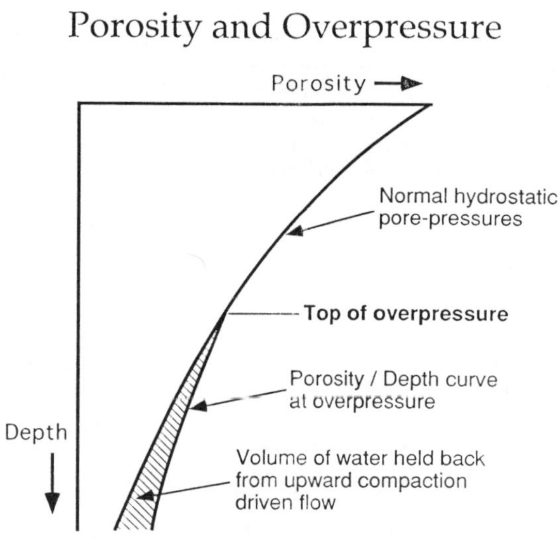

Fig. 17. Relationship between porosity and overpressure. Overpressured rocks will normally have excess porosity relative to hydrostatically pressured rocks at the same depth.

Introduction of cement by advective flow, therefore, predicts that a uniform permeability would develop. In reservoir sandstones and limestones, we nonetheless often observe abrupt changes in permeability and porosity (Fig. 17). This is easier to explain in the case of an internal source (i.e. pressure solution).

14. Mass transfer in mudstones and shales

At shallow depth before severe compaction, mudstones and shales are much more permeable than at 2-3 km burial. Meteoric water flowing into sedimentary basins will follow aquifers of sandstones or limestones, but cross-formational flow through mudstones or shales is often required for the pore-water flow to continue to the surface. Recently-deposited mud, which has not been subjected to much compaction, may have permeabilities exceeding 1 mD, while shales, i.e. from the Gulf Coast, may have only 10^{-10} mD (Reicke and Chillingarian, 1974).

The reduction of permeability in mudstones is due partly to physical compaction and partly to cementation. The content of calcareous fossils which may dissolve and form carbonate cement, and the amount of amorphous silica (diatoms, radiolari, volcanic glass, etc.), are important factors determining the amount of quartz cement. Well-cemented shales have very low permeabilities and are capable of maintaining overpressures for long periods of geological time.

The pore-water flow, resulting from compaction of shaly sequences will not cause sufficiently high pore-water fluxes to change the overall composition of shales significantly, at least not after 1-2 km of burial. As shown above, pore- water fluxes of 10^7-10^8 cm^3/cm^2 are needed, whereas less than 10^4-10^5 cm^3/cm^2 is likely to be experienced by shales as an average on a regional scale in a sedimentary basin. Locally, however, at contacts between shales (clays), sandstones, and along faults, higher fluxes may occur.

It is, nonetheless, very difficult to prove that changes in the composition of shales with increasing burial depth are due to diagenetic reactions and pore-water flow. Samples of the same stratigraphic horizon taken at different depths are usually also displaced horizontally by several 10's of kilometres. There is no reason to assume that mudstones should have a laterally-uniform composition. Changes in clay mineralogy, grain size and biogenic input (i.e. silica and carbonate) will influence the primary composition.

15. Mass transfer in carbonate rocks

The origin of carbonate cements in limestones has been addressed by many authors (see Bathurst, 1976). If calcite cements in limestones do not have a relatively local source, being transported by diffusion over distances smaller than (10 m), but are introduced by advective flow over longer distances, impossibly- high fluid fluxes are required. Each volume of pore water can be expected to precipitate only 10-100 ppm (Bathurst, 1976), so that 10^4-10^5 volumes are required in order to precipitate one volume of cement. To supply 50% calcite cement to a 100 m-thick limestone, a pore water flux of 5.10^7-10^8 cm^3/cm^2 is required. In the case of focused compaction-driven flow where the pore water has been cooled during flow, dissolution of calcite would occur instead of precipitation. There is considerable evidence to suggest, however, that the carbonate cement in limestones is derived from local dissolution and reprecipitation (Scholle and Halley, 1985).

The fact that $d^{13}C$ values in limestones often show variations that reflect changes in the primary contemporaneous sea water is evidence of a very closed system (Scholle and Halley, 1985). If the calcite cement was introduced from outside over longer distances, this $d^{13}C$ pattern would have been destroyed.

Meteoric water will initially be undersaturated with respect to calcite as well as other carbonate minerals (Fig. 8). Only dissolution will then take place and this may develop into a karstic dissolution pattern. However, the pore water will relatively rapidly become saturated with respect to calcite. Aragonite, high Mg-calcite, and other biominerals in fossils will dissolve, since the pore water is undersaturated with respect to these minerals, and low Mg-

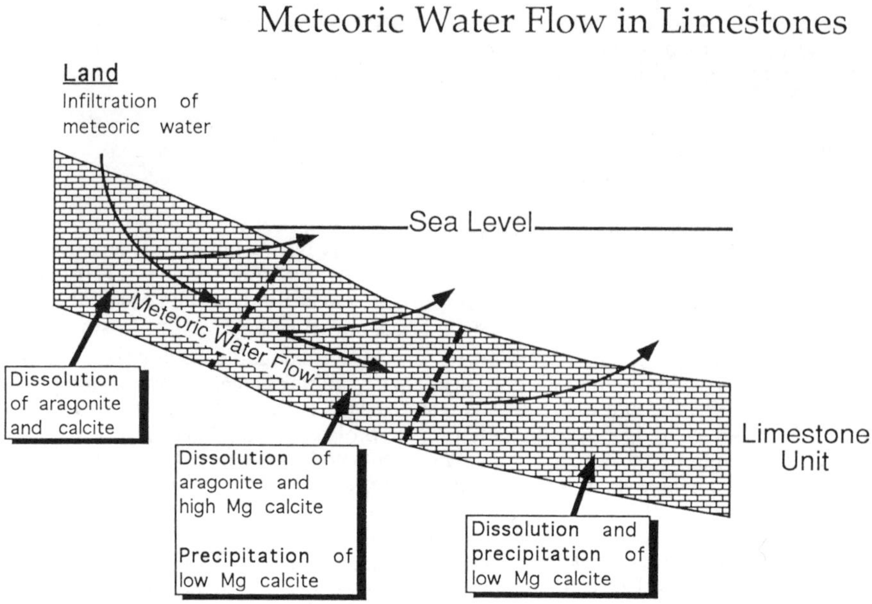

Fig. 18. Schematic representation of meteoric water flow in calcareous sediments. The pore water will gradually approach equilibrium with low Mg calcite.

calcite will precipitate. The rate of meteoric water flow into sedimentary basins is controlled to a large extent by sea-level changes, and it therefore influences near-surface carbonate diagenesis rather markedly (Matthews and Frohlich, 1987). As soon as saturation with respect to calcite is reached, meteoric water has a limited capacity for transporting carbonate in solution, although the solubility of calcite may vary somewhat with the local conditions. Most of the dissolved aragonite and high-Mg calcite must therefore precipitate relatively locally. Because of the small sample size it is difficult to reach firm conclusions about mass balance on textural evidence from thin-sections.

The most important diagenetic reaction which involves mass transfer on a large scale is dolomitization of calcite. Dolomitization of calcite or aragonite requires that calcite dissolves and dolomite precipitates as the most stable phase:

$$2CaCO_3+Mg^{2+}= Ca\,Mg(CO_3)_2 + Ca^{2+}.$$

Even if dolomite is the most stable phase in a certain type of pore water, this reaction cannot proceed without a continuous supply of Mg^{2+} and removal of Ca^{2+}. The Dorag model for dolomitisation (Badiozamani, 1973) assumes mixing of meteoric water and marine connate water, and producing a mixture in which dolomite is stable; however, it does not explain the large-scale supply of Mg^{2+} or the removal of Ca^{2+} which would be required. The concentrations of magnesium in subsurface pore waters are normally rather low (Egeberg and Aagaard, 1989), and very high fluxes of pore water would therefore be required to supply the necessary magnesium for dolomitization of large volumes of limestone.

Sea water, however, has a very high magnesium content (Mg/Ca ratio = 5) and may provide a source of Mg^{2+}. At low sulphate concentrations, dolomite will precipitate more easily (Kastner, 1984). As sulphate is removed from sea water by sulphate reduction, the pore water close to the sediment/water surface may be in the stability field for dolomite. Atolls and reefs form permeable limestones where sea water may circulate due to tidal forces and they thus form ideal settings for early diagenetic dolomitization (Saller, 1984). Sea water may then be both a source of Mg^{2+} and a sink for Ca^{2+}.

Dolomitization at greater burial depth is very difficult to understand in terms of magnesium supply. Water released by compaction of shale sequences will normally have a relatively low Mg content, because silicate phases like chlorite will compete for the available Mg^{2+}. Pore water in Cretaceous carbonates in South Texas is undersaturated with respect to dolomite at depths greater than 1 km (Land and Prezbindowsky, 1981).

When there is evidence from fluid inclusions or from stable isotopes that dolomite crystals have precipitated at high temperatures, it does not necessarily follow that dolomitization occurred at these temperatures and burial depths. Recrystallization of fine-grained early-diagenetic dolomite without a net supply of magnesium could produce the same geochemical and textural evidence. A useful introduction to the chemical and textural aspects of carbonate dissolution is presented as short course notes by Hutcheon et al., (1992).

16. Conclusion

Pore-water flow in sedimentary basins is driven by potentiometric gradients. However, since the overall permeability of the rocks on a large scale is difficult to determine independently, reliable flow velocities cannot be obtained using Darcy's Law. Flow resulting from overpressure is constrained by the low compressibility of water, and during burial, pore-water flow is limited by the rate of compaction.

Meteoric water flushing may cause pore-water fluxes that are several orders of magnitude higher than those resulting from unfocused compactional flow. The leaching of feldspar and rock fragments, and formation of secondary porosity, are to a very large extent controlled by depositional facies, climate, tectonics and sea level changes.

At the low temperatures that normally prevail during meteoric-water flushing, the pore water may be supersaturated or undersaturated with respect to mineral phases. This increases the potential for mass transport. While meteoric water flows from a source renewed by rainfall, the upwards flow of pore water is constrained by the volume of water buried in the sequence. During burial diagenesis, pore-water flow driven by compaction normally flows upward through the rock at a lower rate than the basin's subsidence and therefore has very little potential for mass transport. At higher temperatures (> 80-100°C), the pore water is in greater equilibrium with the minerals present. The mass transport can then be calculated from the temperature gradients and the solubility/temperature functions. Focused pore-water flow along permeable beds or along faults and fractures may precipitate significant amounts of quartz cement and other minerals. However, since fluxes on the order of 108 cm^3/cm^2 are needed in order to supply significant quartz, pore water must be derived from compaction of rocks (shales) with volumes 10^3-10^4 times larger than the volume of rocks to be cemented.

When pore water is circulating in a closed system driven by thermal convection, high pore-water fluxes can be attained. Because of the frequent occurrence of low-permeability layers, like shales or cemented intervals, the requirements for Rayleigh convection are probably rarely met in sedimentary basins. Non-Rayleigh convection is important where the isotherms are relatively steep, particularly where there is hydrothermal activity but also around salt domes. The pore-water salinity distribution around salt domes is often characterized by salinity gradients which would preclude large scale convection.

Compactional-water flow cannot explain the migration of hydrocarbons, which is driven by buoyancy. In a sedimentary basin the hydrocarbon saturation on a large scale is small and the relative permeability of oil is close to zero. A regional pressure gradient will therefore essentially cause flow only of the continous phase, which normally will be water. Fluid flow may be focused by faults, but at shallow depth in soft sediments, faults are more likely to cause sealing rather than be permeable, except when the degree of overpressure exceeds the overburden stress. Focused pore-water flow can only occur locally and such flow cannot be used to explain features which are commonly observed. Reservoirs are probably not the favoured pathway for compaction-driven flow, given the low permeability of the cap rocks.

When oil or gas start to accumulate, the underlying water is also trapped if the structure is not filled. Diffusion is important on small-to-intermediate scales in sedimentary basins. Many diagenetic reactions, including pressure solution, are dependent on a supply of elements by diffusion from the site of dissolution to the surface of the growing authigenic minerals.

Acknowledgements

The author greatfully acknowledges financial support from VISTA; a research cooperation between the Norwegian Academy of Science and Letters and Den Norske Stats Oljeselskap (Statoil), and NAVF; and the Norwegian Research Council. Scott Simmons and Amy Dale have provided useful technical assistance.

References

Aagaard, P. and Helgeson, H.C. (1982) Thermodynamic and kinetic constraints on reaction rates among minerals in aqueous solutions. I Theoretical Considerations 22. American Journal of Science 282, 237-285.

Badiozamani, K. (1973) The Dorag dolomitization model - application to the Middle Ordovician of Wisconsin. Journal of Sedimentary Petroleum 43, 965-984.

Bathurst, R.G.C. (1975) Carbonate sediments and their diageneis. Developments in Sedimentology 12. Diagenese. Elsevier, Amsterdam, Oxford, New York, 658 pp.

Bennett, S.C. and Hanor, J.S. (1987) Dynamics of subsurface salt dissolution at the Welsh Dome, Lousiana Gulf Coast. In Lerche, I. and O`Brien, J.J. (eds.), Dynamical Geolgy of Salt and Related Structures. Academic Press, N.Y., 653-677.

Berner, R.A. (1980) Early Diagenesis. A Theoretical Approach. Princeton University Press, 241 pp.

Bethke, C.M. (1985) A numerical model of compaction-driven groundwater flow and heat transfer and its application to the paleohydrology of intercratonic sedimentary basins. Journal of Geophysical Research 90, 6817-6828.

Bethke, C.M., Harrison, W.J., Upson, C., Altaner, S.P. (1988) Supercomputer analysis of sedimentary basins. Nature 239, 261-267.

Bjørkum, P.A., Mjøs, R., Walderhaug, 0. and Hurst, A. (1990) The role of the late Cimmerian unconformity for the distribution of kaolinite in the Gullfaks Field, northern North Sea. Sedimentology, 37, 396-406.

Bjørlykke, K. (1983) Diagenetic reactions in sandstones. In Sediment Diagenesis. Reidel Publication Company.

Bjørlykke, K., Aagaard, P., Dypvik, H., Hastings, D.S. and Harper, A.S. (1986) Diagenesis and reservoir properties of Jurassic sandstones from the Haltenbanken area. Offshore Mid Norway. In Spenser, A.M. et al. (eds.), Habitat of Hydrocarbons on the Norwegian Continental Shelf. Norwegian Petroleum Society, Graham and Trotman, 275-386.

Bjørlykke, K., Mo, A. and Palm, E. (1988) Modelling of thermal convection in sedimentar basins and its relevance to diagenetic reactions. Marine and Petroleum Geology 5, 338-351.

Bjørlykke, K., Ramm, M. and Saigal, G.C. (1989) Sandstone diagenesis and porosity modification during basin evolution. Geologische Rundschau 78, 243-268.

Bjørlykke, K. and Aagaard, P. (1992) Clay Minerals in North Sea Sandstones. In Houseknecht, D.W. and Pittman, E.D. (eds.), Origin, Diagenesis, and Petrophysics of Clay Minerals in Sandstones, SEPM Special Publication 47, 65-80.

Bjørlykke, K., Nedkvitne, T., Ramm, M. and Saigal, G. (1992) Diagenetic processes in the Brent Group (Middle Jurassic) reservoirs of the North Sea - an overview. In Morton, A.C., Haszeldine, R.S., Giles, M.R. and Brown, S. (eds.), Geology of the Brent Group. Geological Society Special Publication No. 61, 263-287.

Bjørlykke, K. and Egeberg, P.K. (1993) Quartz cementation in sedimentary basins. Submittted to AAPG Bulletin.

Burley, S.D. (1986) The development and destruction of porosity within Upper Jurassic reservoir sandstones of the Piper and Tartan Fields, Outer Moray Frith, North Sea. Clay Minerals 21, 649-694.

Cassan, J.P., Palacios, M.C.G., Fritz, B. and Tardy, B. (1981) Diagensis of sandstone reservoirs as shown by petrographical and geochemical analyses of oil bearing formations in the Gabon Basin. Bulletin Centre Rech. Elf-Aquitaine 5, 113-135 (Pau Cedex, France).

Caritat, P. de. (1989) Note on the maximum upward migration of pore water in response to sediment compaction. Sedimentary Geology 65, 371-377.

Chapman, R.E. (1987) Fluid flow in sedimentary basins: a geologist's perspective. Geological Society, Special Publication No. 34, 3-18.

Clayton, N.R., Friedman, I., Graf, D.L., Mayeda, T.K., Meents, U.F. and Shrimp, M.F. (1966) The origin of formation waters - I. Isotopic Composition. Journal of Geophysical Research 71, 3869-3882.

Davis, S.H., Rosenblat, S., Wood, J.R. and Hewett, T.A. (1985) Convective fluid flow and diagenetic patterns in domed sheets. American Journal of Science 285, 207-223.

Deming, D., Nunn, J.A. and Evans, D.G. (1990) Thermal effects of compaction driven ground water flow from overthrust belts. Journal of Geophysical Research 95, 6669-6683.

Dickey, P.A. (1979) Petroleum development geology. Petroleum Geology, PPC Books, Tulsa, OK, 398 pp.

Downing, R.A. and Gray, D.A. (1986) Geothermal energy - The Potential in the United Kingdom. British Geological Survey, Her Majesty's Stationary Office, 1986, 187 pp.

Egeberg, P.K. and Aagaard, P. (1989) Origin of formation waters from oil fields on the Norwegian Shelf. Applied Geochemistry 4, 131-143.

Ehrenberg, S.N. and Nadeau, P.H. (1989) Formation of diagenetic illite in sandstones of the Garn Formation, Haltenbanken area, mid-Norwegian Continental Shelf. Clay Minerals 24, 233-253.

Ehrenberg, S.N. (1991) Kaolonized, potassium-leached zones at the contacts of the Garn Formation, Haltenbanken, mid-Norwegian continental shelf. Marine and Petroleum Geology 8, 250-269.

England, W.A., MacKenzie, A.S., Mann, D.M. and Quigley, T.M. (1987) The movement and entrapment of petroleum fluids in the subsurface. Journal of Geological Society London 144, 327-347.

Evans, D.G. and Nunn, J.A. (1989) Free thermohaline convection in sediments surrounding a salt column. Journal of Geophysical Research 94, 12413-12422.

Giles, M.R. (1987) Mass transfer and problems of secondary porosity creation in deeply buried hydrocarbon reservoirs. Marine and Petroleum Geology 4, 188-201.

Glennie, E.W. (1990) Introduction to the petroleum geology of the North Sea. Blackwell, 402 pp.

Gluyas, J. and Coleman, M. (1992) Material flux and porosity changes during sediment diagenesis. Nature 356, 52-54.

Goff, J.C. and Williams, B.P.J. (1987) Fluid flow in sedimentary basins and aquifers. Geological Society Special Publication No. 34. Blackwell Scientific Publications, London. 230 pp.

Gran, K., Bjørlykke, K. and Aagaard, P. (1992) Fluid salinity and dynamics in the North Sea and Haltenbanken basins derived from well log data. In Hurst, A., Griffiths, C.M. and Worthington, P.F. (eds.), Geological Application of Wireline logs II. Geological Society, London Special Publication 66, 327-338.

Gretener, P.E. (1977) Pore pressures: Fundamentals, general ramification and implications for structural geology. AAPG Continuing Education Course Note Series No. 4, AAPG, Tulsa, OK. 87 pp.

Haszeldine, R.S., Samson, I.M. and Cornford, C. (1984) Quartz diagenesis and convective fluid movement: Beatrice Oilfield, UK, North Sea. Clay Mineralogy 19, 391-402.

Hanor, J.S. (1987) Origin and migration of subsurface sedimentary brines. SEPM Short Course No. 21.

Hanor,J. S. and Mcmanus, K. M. (1988) Sediment alteration and clay mineral diagenesis in a regional ground water flow system, Mississippii Gulf Coastal plain. Transactions-Gulf Coast Asss. of Geol Soc. 38, 495-502.

Harrison, W.J. (1989) Modelling Fluid/Rock Interactions in Sedimentary Basins. In Cross, T.A. (ed.), Quantitative Dynamic Stratigraphy, Prentice Hall, 195-231.

Harrison, W.J. and Summa, L.L. (1991) Paleohydrology of the Gulf of Mexico Basin. American Journal of Science 291, 109-176.

Hayes, J.M. and Boles, J.R. (1992) Volumetric relations between dissolved plagioclase and kaolinite in sandstones. Implications for aluminium mass transfer in the San Joaquin Basin, California. In Houseknecht, D.W. and Pittman, E.D. (eds.), Origin, Diagenesis, and Petrophysics of Clay Minerals in Sandstones. SEPM Special Publication 47, 111-123.

Hubbert, M.K. (1940) The theory of ground water motion. Journal of Geology 48, 785-944.

Hunt, J.H. (1990) Generation and migration of petroleum from abnormally pressured fluid compartments. American Association of Petroleum and Geology, Bulletin 74, 1-12.

Hutcheon, I.E. (1989) Application of chemical and isotopic analyses of fluids to problems in sandstone diagenesis. In Hutcheon, I.E. (ed.), Short course in burial diagenesis. Mineral Association of Canada, 270-310.

Hutcheon, I.E., Reime, J., Tearc, M., Dravis, J.J. and Wordlaw, N.C. (1992) Subsurface dissolution porosity in carbonates. Short course. AAPG Convention 1992, Calgary.

Karlsen, D.A. and Larter, S. (1989) A rapid correlation method for petroleum population mapping within individual petroleum reservoirs: applications to petroleum reservoir description. In Graham & Trotman (eds.), Correlation in hydrocarbon exploration. Norwegian Petroleum Society, 77-85.

Kastner, M. (1984) Control of dolomite formation. Nature 311, 410-411.

Kharaka, Y.K. and Carothers, W.W. (1986) Oxygen and hydrogen isotope geochemistry of deep basin brines. In Fritz, P. and Frape, J.Ch. (eds.), Handbook of environmental Isotope Geochemistry 2. The Terrestrial environment. Elsevier, Amsterdam, 305-360.

Land, L.S. and Prezbindowski, D.R. (1981) The origin and evolution of saline formation water, Lower Cretaceous carbonates, South Central Texas, USA. Journal of Hydrology 54, 51-57.

Larter, S.R., Bjørlykke, K., Karlsen, D.A., Nedkvitne, T., Eglinton, T., Johansen, P.E., Leythaeuser, D., Mason, P.C., Mitchel, A.W. and Newcombe, G.A. (1990) Determination of Petroleum Accumulation Histories: Examples from the Ula Field, Central Graben, Norwegian North Sea. North Sea Oil and Gas Reservoirs II. Norwegian Institue of Technology (Graham and Trotman). p 319-330.

Leythaeuser, D., Schaefer, R.G. and Radke, M. (1987) On the primary migration of petroleum. Proceedings, 12th World Petroleum Congress, Houston, 1987, Wiley, London, 227-236.

Longstaffe, F. J. (1984) The role of meteoric water in diagenesis of shallow sandstones : Stable isotope studies of the Milk River Aquifer and Gas Pool, southeastern Alberta. In Mcdonald, D.A. and Surdam,R.C. (eds) Clastic Diagenesis. American Association of Petroleum Geologsits . Memoir 37, 81-98.

Ludvigsen, A. (1992) Thermal convection and diagenetic processes in sedimentary basins. PhD Thesis, University of Oslo, 1992.

Manheim, F.T. and Paull, C.K. (1981) Patterns of ground water salinity changes in a deep continental-oceanic transect off the Southeastern Atlantic coast of the USA. Journal of Hydrology 54, 95-105.

Matthews, R.K. and Frohlich, C. (1987) Forward modelling of bank-margin carbonate diagenesis. Geology 15, 673-676

Nedkvitne, T. and Bjørlykke, K. (1992) A quantitative petrographic study in the Brent Group, Huldra Field, North Sea - Implications for sedimentary basins. Journal Sedimentary Petrology 61, 23-34.

Palm, E. (1990) Rayleigh convection, mass transport and change in porosity in layers of sandstone. Journal of Geophysical Research 95, 8675-8679.

Rabinowics, M., Danurand, J.L.,Jakubowski, M., Schott, J. and Cassan J.P. (1987) Convection in a North Sea oil reservoir: influences on diagenesis and hydrocarbon migration. Earth Planet. Science Letters 74, 387-404.

Rieke, H.H. and Chillingarian, G.V. (1974) Compaction or orgillaceous sediments. Developments in Sedimentology 16, Elsevier, 424 pp.

Robinson, A. and Gluyas, J. (1992a) Duration of quartz cementation in sandstones. Marine and Petroleum Geology 9, 324-327.

Robinson, A. and Gluyas, J. (1992b) Model calculations of loss of porosity in sandstones as a result of compaction and quartz cementation. Marine and Petroleum Geology 9, 319-322.

Saigal, G.C., Bjørlykke, K. and Larter, S. (1992) The effects of oil emplacement on diagenetic processes - examples from the Fulmar Reservoir Sandstones, Central North Sea. Bulletin American Association of Petroleum Geology 76, 1024-1033.

Saller, A.H. (1984) Petrologic and geochemical constraints on the origin of subsurface dolomite. Enewetak atoll: An example of dolomitization by normal sea water. Geology 12, 217-220.

Scholle, P.A. and Halley, R.B. (1985) Burial diagenesis: Out of sight, out of mind! SEPM, 309-333.

Sibson, R.H. (1981) Fluid flow accompanying faulting: Field evidence and models. In Simpson, D.W. and Richards, P.G. (eds.), Earthquake Prediction: an international review. Maurice Ewing Series 4, 593-603.

Sibson, R.H. (1990) Faulting and fluid flow. In Nesbitt (ed.), Fluids in tectonically active regimes of the continental crust. Short course, Mineralogical Association of Canada 18, 93-132.

Smith, J.T. and ehrenberg, S.N. (1989) Correlation of carbon dioxide abundance with temperature in clastic hydrocarbon reservoirs: relationship to inorganic chemical equilibrium: Marine and Petroleum Geology 6, 129-135.

Surdam, R.C., Boese, S.W. and Crossey, L.J. (1984) The chemistry of secondary porosity. In McDonald, D.A. and Surdam, R.C. (eds.), Clastic Diagenesis. AAPG, Memoirs 37, 127-150.

Ungerer, P., Burrus, J., Doliger, B., Doligez, P.Y., Chenet, P.Y. and Bessis, F. (1990) Basin evaluation by integrated two dimensional flow, hydrocarbon generation and migration. American Association of Petroleum Geology 74, 309-335.

van Wagoner, J.C., Mitchum, R.M., Campion, K.M. and Rahmanian, V.D. (1990) Siliciclastic sequence. Stratigraphy in well logs, cores and outcrops. Concepts for High-Resolution Correlation of Time and Facies. AAPG. Methods in exploration, Series No. 7.

Wood, J.R. and Hewett, T.A. (1982) Fluid convection and mass transfer in porous sandstones - a theoretical model. Geochimica et Cosmochimica Acta 46, 1707-1713.

Wood, J.R. (1986) Thermal transfer in systems containing quartz and calcite. In Gautier, D.L.(ed) Roles of Organic matter in Sediment Diagenesis. SEPM Special Publication ,38, 181-189.

STABLE ISOTOPIC CONSTRAINTS ON SANDSTONE DIAGENESIS IN THE WESTERN CANADA SEDIMENTARY BASIN

F.J. LONGSTAFFE
Department of Earth Sciences
University of Western Ontario
London, Ontario
Canada N6A 5B7

ABSTRACT. The oxygen and hydrogen isotope compositions of diagenetic minerals can be used to provide constraints on the origin and evolution of porewaters during the life cycle of a sedimentary basin. In the first part of this chapter, general controls on the oxygen and hydrogen isotope fractionations in diagenetic systems are discussed. In the second part, stable isotope results for diagenetic minerals from sandstones and conglomerates of the western Canada sedimentary basin are reviewed as an example of this approach. This review illustrates the importance of meteoric water throughout the diagenetic evolution of these rocks. For many units, the isotopic data show that meteoric water was abundant during early diagenesis, either because of depositional environment or because of early, post-depositional processes related to sea-level fluctuation. During burial, the oxygen isotope values of porewaters increased not only because of rock-water interaction, but also because of mixing with brines derived from underlying Paleozoic carbonates. The oxygen isotopic compositions of late diagenetic minerals indicate crystallization during regional, gravity-driven flow of meteoric water that was initiated in response to uplift of the western Canada sedimentary basin in early Eocene time.

The final part of the chapter describes Cretaceous oil sands from the Cold Lake and Primrose areas of Alberta, and shows how stable isotope measurements have provided information on both naturally occurring and artificially induced diagenesis. In the first instance, the oxygen isotope composition of early diagenetic berthierine (an unususal clay mineral abundant in some Clearwater Formation reservoirs in the Cold Lake area) suggests that early diagenesis occurred in brackish to fresh water. Ubiquitous crystallization of early diagenetic calcite was initially controlled by carbon dioxide of inorganic origin, but later calcite formation was driven by reactions in the methanogenic zone (reduction of carbon dioxide in particular). The association between berthierine and ^{13}C-rich calcite in these oil sands indicates that microbial activity played an important role in berthierine genesis. In the second instance, oxygen isotope geochemistry was used to examine the reactions that occurred during *in situ* steam injection of the Wabiskaw Member oil sands in the Primrose area. The results demonstrated that (i) calcite-cemented sandstones used to channel and confine steam and steam-condensate were not significantly affected by hot water-rock interaction, (ii) abundant smectitic clays formed at the expense of pore-filling kaolinite in certain sand intervals characterized by retention of bitumen, and (iii) neoformation of clays was not an important process in sand intervals that were swept of bitumen. Detrital clay minerals, however, had accumulated in these intervals because of migration of fines during steaming.

1. Introduction

The stable isotope compositions of diagenetic minerals can be used to provide constraints on the origin and modification of porewaters during the life cycle of a sedimentary basin. In most situations, the porewaters responsible for a particular product of diagenesis have long since disappeared from the system. However, a record of the water's passage may have been retained in the isotopic composition of minerals formed in equilibrium with the porewater at that time. In the best case, the variations among the isotopic compositions of diagenetic minerals that comprise a paragenetic sequence can be used to reconstruct the nature of porewater evolution throughout the geological history of the unit.

A. Parker and B. W. Sellwood (eds.),
Quantitative Diagenesis: Recent Developments and Applications to Reservoir Geology, 223–274.

In this paper, the principles of oxygen and hydrogen isotope fractionation that control the compositions of diagenetic minerals are reviewed first. Then, available isotopic data for Mesozoic sandstones and oil sands from the western Canada sedimentary basin (Fig. 1) are examined to illustrate how such information can be interpreted and utilized. The oil sand deposits, in particular, provide a special opportunity to demonstrate how stable isotopic studies can provide unique insight into the character of water-sediment-organic matter interaction, during both natural and artificial diagenesis. The latter process occurs during *in situ* steam injection of oil sands, and is a common method for heating and thus mobilizing the viscous hydrocarbons.

2. Oxygen and Hydrogen Isotopes as Tracers in Porewater Evolution

Stable isotopes are of special interest as tracers in sandstone diagenesis because at low temperatures they can be fractionated on a large scale by natural processes. The principles of stable isotope fractionation in low temperature systems have been reviewed by others (e.g., Faure, 1986; Hoefs, 1987; Kyser, 1987; O'Neil, 1987; Longstaffe, 1983, 1987, 1989; Coleman, this volume; Clayton, this volume). Only an overview of a few important features is given here, drawn in part from Longstaffe (1989).

2.1. DEFINITIONS AND PRINCIPLES OF FRACTIONATION

Oxygen has three stable isotopes, ^{18}O, ^{17}O and ^{16}O. In studies of terrestrial materials, it is conventional to measure only the ratio of ^{18}O to ^{16}O, given that the ^{17}O to ^{16}O ratio is systematically related to this parameter. Let us consider the typical behaviour of the oxygen isotopes in calcite. Calcite which has crystallized in equilibrium with porewater is enriched in ^{18}O relative to the water, the exact size of the fractionation depending upon the temperature at which the calcite formed. The oxygen isotopes have been distributed between the calcite and the porewater according to an isotope exchange reaction for which a temperature-dependent, equilibrium constant can be written. Equilibrium fractionations derive from small but significant differences in the thermodynamic properties between compounds containing an isotope with a higher (e.g., ^{18}O) versus lower mass number (e.g., ^{16}O). An important parameter controlling the size of equilibrium fractionations for a given element is the relative mass difference among the isotopes. Since this difference is greater for elements with lower atomic numbers (Z), the largest fractionations most commonly involve "light" elements such as hydrogen (Z = 1), carbon (Z = 6), oxygen (Z = 8) and sulphur (Z = 16).

A different yet equally important kind of stable isotope fractionation can be illustrated using carbon, which has two stable isotopes, ^{13}C and ^{12}C. During photosynthesis, $^{12}CO_2$ gas is preferentially selected during the production of organic matter (Park and Epstein, 1960; Deines, 1980). This fractionation of $^{12}CO_2$ relative to $^{13}CO_2$ gives organic matter a characteristic isotopic signature that points to its biological origin, and is controlled mostly by kinetic processes. The ultimate size and sign of the fractionation depends on the reaction rates and reaction pathways involved. Kinetic isotope fractionations associated with microbial activity are also of special importance, particularly in the production and destruction of methane and carbon dioxide. As both Clayton (this volume) and Coleman (this volume) point out, many early diagenetic processes are driven by microbially mediated reactions that produce highly variable carbon, hydrogen and sulphur isotope fractionations between the reactants and products.

The stable isotope fractionation factor between two phases, A and B, is defined as:

$$\alpha_{A\text{-}B} = R_A/R_B \tag{1}$$

where R_A is the ratio of the "heavy" isotope to the "light" isotope for phase A, and R_B is the ratio of the "heavy" isotope to the "light" isotope for phase B. Examples of R include D/H, $^{13}C/^{12}C$, $^{18}O/^{16}O$ and $^{34}S/^{32}S$. In almost every situation, the "light" (lower mass number) isotope has by far the highest abundance in nature.

Stable isotope results are normally reported using the delta (δ) notation in parts per thousand (‰

or permill), where (for phase A):

$$\delta_A = \frac{(R_A - R_{standard})}{R_{standard}} \times 1000. \tag{2}$$

Equation (2) illustrates that R for phase A is always compared to R for a standard material whose value is set at 0‰ exactly. For hydrogen, this standard is V-SMOW (Vienna Standard Mean Ocean Water; SMOW originally defined by Craig, 1961a; see also Hagemann et al., 1970). For oxygen, two standards are used, V-SMOW (Vienna Standard Mean Ocean Water; Craig, 1961a, as above; see also Baertschi, 1976), and V-PDB (Vienna Peedee Belemnite; PDB originally defined by Craig, 1957), which has a $\delta^{18}O$ value of +30.91‰ relative to V-SMOW. The use of PDB as the oxygen isotope reference should be restricted to ocean paleotemperature studies. The PDB standard is also used for carbon isotope measurements (Craig, 1957). Sulphur isotope ($\delta^{34}S$) values are reported relative to CDT (Canyon Diablo Troilite; Thode et al., 1961). Most δ-values can be determined with a precision of at least ±2‰ for H_2 gas and ±0.2‰ for other gases.

The relationship between α and δ is of special significance:

$$\alpha_{A\text{-}B} = \frac{1 + \delta_A/1000}{1 + \delta_B/1000} = \frac{\delta_A + 1000}{\delta_B + 1000}. \tag{3}$$

When isotopes of a given element are randomly distributed among the available sites in phases A and B (equilibrium condition), the fractionation factor (α) can be related to the equilibrium constant (K) by

$$\alpha = K^{1/n} \tag{4}$$

where n is the number of atoms exchanged. In an oxygen isotope exchange reaction between CO_2 and H_2O involving the exchange of only one atom (O'Neil, 1979),

$$1/2C^{16}O_2 + H_2{}^{18}O = 1/2C^{18}O_2 + H_2{}^{16}O, \tag{5}$$

the equilibrium constant (K) will be:

$$K = \frac{(aC^{18}O_2)^{1/2}(aH_2{}^{16}O)}{(aC^{16}O_2)^{1/2}(aH_2{}^{18}O)}. \tag{6}$$

Since the activities (a) of isotopic phases are equivalent to the concentration of a given isotope within that phase, then

$$K = \frac{({}^{18}O/{}^{16}O)CO_2}{({}^{18}O/{}^{16}O)H_2O} = \alpha. \tag{7}$$

Values of α are normally very close to unity, typically 1.00x or 1.0x. For example, the oxygen isotope fractionation factor for CO_2-H_2O at 25°C is 1.0412 (Friedman and O'Neil, 1977).

Being a true equilibrium constant, the equilibrium isotope fractionation factor (α) should be temperature dependent. For an ideal gas, lnα varies as a function of $1/T^2$ (°K) at high temperatures (generally >273.15°K for all elements except H) and 1/T (°K) at low temperatures. In oxygen isotope systems, experimentally determined α-values for mineral-mineral or mineral-water pairs ($\alpha_{A\text{-}B}$) normally fall on smooth curves when $10^3\ln\alpha$ (or permill fractionation) is plotted versus $10^6/T^2$ (°K), with permill fractionation decreasing with increasing temperature. The temperature dependence of α in equilibrium systems forms the basis of oxygen isotope geothermometry (Urey, 1947; Bigeleisen and Mayer, 1947).

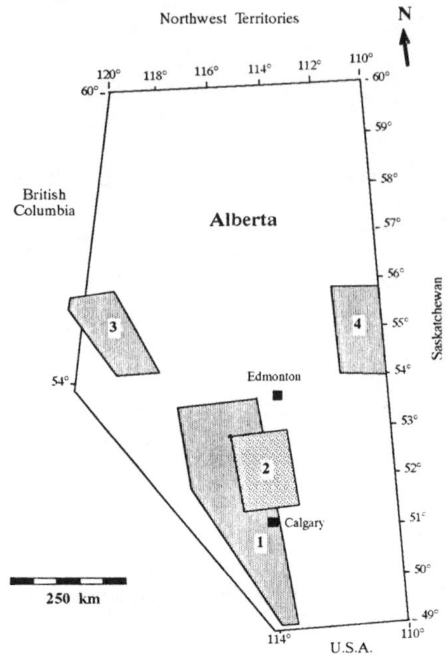

Figure 1. Locations of Mesozoic clastic sedimentary rocks described in this study of the western Canada sedimentary basin: 1 - Belly River/Brazeau Group; 2 - Cretaceous Viking Formation; 3 - Alberta Deep Basin; 4 - Clearwater Formation and Wabiskaw Member (oil sands).

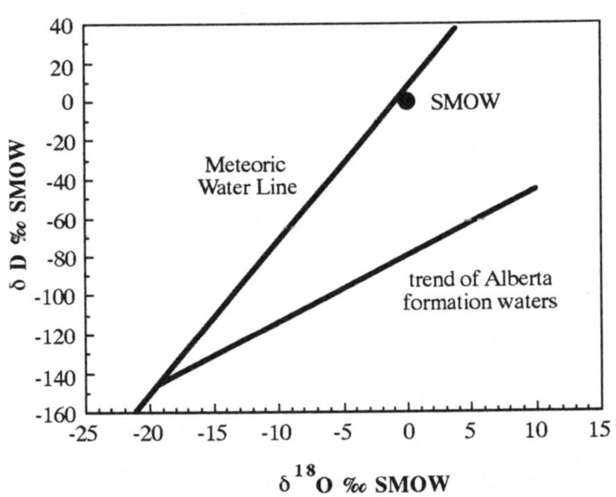

Figure 2. General relationship between the Meteoric Water Line (after Craig, 1961b) and formation waters in the western Canada sedimentary basin. The data for the formation waters are from Clayton et al. (1966), Hitchon and Friedman (1969) and Connolly et al. (1990b).

A useful mathematical approximation is that for two phases, A and B, $10^3\ln(1.00x) \approx x$. Because of this behaviour, $10^3\ln(1.00x)$ can be approximated by the Δ-value (big delta):

$$\Delta_{A\text{-}B} = \delta_A - \delta_B \approx 10^3\ln\alpha_{A-B}. \tag{8}$$

However, for values of permill fractionation >10, which are common in low temperature systems, the permill fractionation should be calculated directly.

2.2. OXYGEN AND HYDROGEN ISOTOPE VARIATIONS IN NATURAL WATERS

If the oxygen and hydrogen isotope compositions of authigenic minerals are to be used as tracers of porewater origin and evolution, the possible variation in natural waters must be understood. Further, while meteoric and seawater initially have characteristic oxygen and hydrogen isotope compositions, these signatures can be substantially modified by rock-water interaction and mixing.

2.2.1. *Ocean Water*. Unmodified ocean water has $\delta^{18}O$ and δD values ≈ 0‰ (SMOW; Craig, 1961a). These values may vary slightly because of dilution with fresh water. For instance, a high latitude, shallow inland sea that has received an anomalously high influx of surface runoff relative to the body's normal volume could acquire an isotopic composition < 0‰, especially in its surface layers. Depending on the rate of mixing within the water column, low $\delta^{18}O$ and δD porewaters derived from such a sea might be trapped in shoreline sands. Based on evidence from fossil shell material, the model of a brackish inland sea has been invoked to explain periods of low $\delta^{18}O$ values for the Cretaceous Interior Seaway in western Canada (Kyser et al., in press).

Evaporation can also cause some variation in the isotopic composition of seawater. Because ^{16}O and H are preferentially partitioned into the vapour phase during evaporation, remaining surface waters within a restricted basin become progressively enriched in ^{18}O and D, at least until extreme levels of solute concentration are reached (Sofer and Gat, 1975). The isotopic composition of evaporated waters can be described by:

$$\delta D = m\delta^{18}O. \tag{9}$$

The slope (m) decreases with increasing evaporation/precipitation ratios, and is generally characteristic of a given region (Faure, 1986). Normally, the enrichment in ^{18}O is less than 5 to 6‰. Isotopic variations correlate well with changes in salinity that accompany evaporation.

Besides evaporation and mixing with fresh waters, other processes, such as continental glaciation, can also affect the isotopic composition of seawater. Large ice caps represent major reservoirs of ^{16}O- and H-rich water. Melting of the present Greenland and Antarctic ice caps would cause a change in the $\delta^{18}O$ value of the ocean of -0.5 to -0.9‰; studies of ice-cover fluctuations in the Pleistocene suggest that the variation was no larger than 0.8 to 1.3‰ (Craig, 1965; Shackleton, 1967; Dansgaard and Tauber, 1969, Savin and Yeh, 1981).

Whether the oxygen isotopic composition of the ocean has changed with time on a much larger scale is more difficult to evaluate: The $\delta^{18}O$ values of both marine limestones and cherts decrease with increasing geological age (Degens and Epstein, 1962; Keith and Weber, 1964; Perry, 1967; Veizer and Hoefs, 1976; Knauth and Lowe, 1978; Perry et al., 1978). The explanation that carbonate rocks have undergone progressive isotopic exchange with meteoric waters since their deposition (Degens and Epstein, 1962; Keith and Weber, 1964) is apparently untenable for chert, which seems not to be greatly affected by post-depositional exchange (Knauth and Epstein, 1976; Perry, 1967; Clayton et al., 1968). The proposal that marine cherts crystallized from seawater which had become progressively richer in ^{18}O towards present time because of seawater cycling through the mantle (Perry, 1967; Perry and Tan, 1972; Chase and Perry, 1972; Perry et al., 1978) has been challenged by Muehlenbachs and Clayton (1976) and Gregory and Taylor (1981). They have proposed that reactions between oceanic rocks and seawater should have long buffered the $\delta^{18}O$ value of the ocean to about its present value of ≈0‰, basing their conclusion largely on oxygen isotope data obtained for ophiolite sequences.

Knauth and Epstein (1976) and Knauth and Lowe (1978) suggested that the variation in the $\delta^{18}O$ values of chert reflects crystallization from seawater of constant isotopic composition, but at higher temperatures in the past. Karhu and Epstein (1986) reported results for Precambrian and Phanerozoic chert-phosphate assemblages that apparently support this idea. Using chert-phosphate mineral pairs as a geothermometer, Karhu and Epstein (1986) proposed that temperatures on the earth's surface had declined substantially since Precambrian time, but that the oxygen-isotope composition of the oceans has not varied more than a few permill over the last three billion years. However, Perry (1990) has questioned the selection of samples analyzed by Karhu and Epstein (1986), and others have challenged the geological and biological viability of hot Precambrian oceans in general (e.g., Veizer et al., 1982). An additional complication is that some cherts may be of hydrothermal origin, deposited on the seafloor to temperatures as low as 30-60°C ("white smokers") (e.g., Crerar et al., 1982; Herzig et al., 1988).

The debate continues without reconciliation of divergent observations. The accumulating data for "least altered" shell materials (mostly brachiopods) and marine cements continue to show a trend towards ^{18}O depletion with age (to -7‰ for the Cambrian; see Wadleigh and Veizer, 1992), while data from ophiolites consistently support the concept of a 0‰ ocean throughout much of geologic time. To further complicate matters, the oxygen isotope compositions of very early (≈3900 Ma) Archean granitoid rocks have even been interpreted to hint that the very early Archean oceans had high $\delta^{18}O$ values (≈+6‰), which were the product of high temperature outgassing of the Earth's mantle during early crustal segregation (Muehlenbachs and Bowring, 1992).

The first step to resolving this dilemma lies with a better understanding of critical differences between the specific rock-water environments involved in precipitation and preservation of both shell materials and cherts on one hand, and halmyrolysis and alteration of ocean crust on the other. In the interim, though, the lack of certainty about the oxygen isotope composition of ancient seawater must be considered in isotopically based, paleoporewater interpretations.

2.2.2. *Meteoric Water.* Meteoric water shows a large but systematic variation in isotopic composition ($\delta^{18}O \approx -55$ to 0‰; $\delta D \approx -400$ to +10‰; Fig. 2). Its behaviour is described by the meteoric water line (Craig, 1961b):

$$\delta D = 8\ \delta^{18}O + 10\ (‰). \tag{10}$$

This regularity is the product of evaporation and condensation cycles that can be modelled by the expression for Rayleigh distillation:

$$R/R_o = f^{(\alpha-1)} \tag{11}$$

where f = fraction of the original vapour remaining in a cloud mass, R_o = the $^{18}O/^{16}O$ (or D/H) of the water vapour just before condensation begins, R = $^{18}O/^{16}O$ (or D/H) of the remaining water vapour for a given value of f, and α = liquid-vapour, isotope fractionation factor for the temperature at which condensation is occurring.

Water initially evaporated from the ocean becomes progressively depleted in ^{18}O and D as it participates in successive evaporation-condensation cycles while travelling across the continents. This continuous distillation results in a geographically controlled distribution of oxygen and hydrogen isotope compositions, fresh water in general becoming progressively depleted in the heavier isotope at higher altitudes and latitudes (Craig, 1961b; Taylor, 1974; Gat, 1980; Yurtsever and Gat, 1981). The depletion is further accentuated by lower average air-mass temperatures and larger α (liquid-vapour) towards higher latitudes and altitudes, and by the re-evaporation of meteoric waters from the continental surface.

Because the oxygen isotope composition of meteoric water is strongly controlled by climatic parameters such as temperature and the precipitation/evaporation regime, it has become useful to analyze neoformed minerals from regolith deposits to deduce the $\delta^{18}O$ value of ancient meteoric waters. Such estimates should provide insight into continental paleoclimate during regolith formation (e.g., Bird et al., 1992a,b). It is important, however, not to overinterpret such results, given the complexities found in many modern hydrologic systems. For example, low $\delta^{18}O$ values are not uncommon in tropical regions influenced by monsoonal climatic patterns (Yurtsever and Gat, 1981; Aharon, 1983; Bird, 1988). Weathering products formed primarily during such episodes of intense rainfall might have much lower oxygen isotope compositions than otherwise expected for a tropical climate (Bird et al., 1992a,b).

2.2.3. *Porewaters in Oceanic Sediments.* The complexities in interpreting oxygen and hydrogen isotope compositions of ancient porewaters are well illustrated by results for ocean sediments. The $\delta^{18}O$ and δD values of porewaters trapped in Cenozoic oceanic sediments commonly decrease with depth to values <0‰ (Savin and Yeh, 1981), yet it is unlikely that meteoric water ever penetrated these units.

Friedman and Hardcastle (1973) proposed that variations in the δD values obtained for porewaters from sediments of the Caribbean and Red Seas reflected primary differences in bottom water compositions resulting from glacial and evaporative episodes. Good correlation between the hydrogen isotope data of Friedman and Hardcastle (1973, 1974) and the oxygen isotope data of Lawrence (1973, 1974) for the same porewaters can be used to support this conclusion; a cross-plot gave a slope typical of seawater for regions of high evaporation (Savin and Yeh, 1981). Lawrence and Taviani (1988), however, have noted that diffusion would substantially attenuate such isotopic signals over periods greater than 10 my. Further, if primary porewater compositions had been preserved, a more oscillatory character with depth to the coupled variation in hydrogen and oxygen isotope values might have been expected.

Lawrence et al. (1975, 1979) and Gieskes and Lawrence (1981) suggested that the lower values (up to 4‰ for oxygen) arise from chemical reactions between porewater and volcanic ash in the sediment or underlying basalt (e.g., neoformation of smectite). Low temperature crystallization of ^{18}O-rich authigenic clay minerals would measurably deplete the porewater reservoir of ^{18}O in a quasi-closed system with a high mineral/water ratio. However, such an explanation cannot account for similar trends in the δD values of these porewaters, which can decrease by more than 10‰. Low temperature neoformation of smectitic clays in a closed system should cause D-enrichment in the remaining porewater (assuming that the clays acquired all of their hydrogen from the water), since structural OH in smectite should have a δD value up to 70‰ lower than the porewater. The tendency to enrichment should be attenuated by fixation of some porewater into the smectite interlayer. The hydrogen isotope fractionation between smectite interlayer water and free water is not known, but a value of +20‰ has been suggested by Lawrence and Taviani (1988) based on the ice-like structure of interlayer water, and O'Neil's (1968) measurement of the ice-water hydrogen isotope fractionation. The net result of smectite neoformation, according to Lawrence and Taviani (1988), is that virtually no change in δD values should accompany a 3‰ decrease in the $\delta^{18}O$ value of porewater.

Some of the uncertainty associated with the isotopic behaviour of porewaters in quasi-closed, reducing environments such as oceanic sediments derives from our relative state of ignorance concerning the behaviour of hydrogen isotopes during clay-organic-water interactions. That some clay minerals can exchange hydrogen isotopes with porewaters independently from oxygen isotopes has been proposed by Bird and Chivas (1988) and Longstaffe and Ayalon (1990), among others. Such exchange with pre-existing clays could cause a lowering of porewater δD values, depending on the porewater/clay mineral ratio for hydrogen, the initial hydrogen isotopic composition of the clays, and the temperature of exchange.

Another possibility is that oxidation of organic matter, methane and hydrogen may be responsible for lowered δD values in ocean sediment porewaters. Water produced in all of these reactions can be expected to have very low δD values (≤ -200‰) (Clayton, this volume). Lawrence and Taviani (1988) examined such possibilities to explain porewater δD values as low as -14‰ in the Norwegian Sea. They concluded that the apparent abundance of organic matter and biogenic methane in these sediments was insufficent (by a factor of three) to account for the depletions in δD, and suggested instead that contribution of methane or hydrogen from the mantle provided the additional flux needed for oxidation (by sulphate and ferric iron reduction) to carbon dioxide and water. Better isotopic constraints (both hydrogen and oxygen) are still needed to fully assess the mass balance calculations upon which these conclusions are based. Nevertheless, Lawrence and Taviani (1988) have demonstrated that contribution of water produced during oxidation of organic matter, methane and hydrogen may be important in quasi-closed, sediment-dominated systems.

2.2.4. *Formation Waters.* In a review of available chemical data, White (1965) concluded that most saline formation waters were derived from original (connate) seawater trapped during deposition, and subsequently modified by rock-water interaction. This view was challenged soon after by Clayton et al. (1966) who used oxygen and hydrogen isotope data to argue that many oil-field brines had evolved from meteoric water. This opinion was reinforced by several subsequent studies (e.g., Hitchon and Friedman, 1969; Kharaka et al., 1973). As a result, research into the origin of saline formation waters began to focus less on connate waters (i.e., evaporite brines) and more on mechanisms of rock/water interaction which would permit fresh waters which had invaded sedimentary basins to evolve into saline brines (e.g., membrane filtration, Graf et al., 1965; dissolution of evaporites and albitization of feldspar, Land and Prezbindowski, 1981). More recently, Knauth and Beeunas (1986) and Knauth (1988) proposed that complete flushing of sedimentary basins by meteoric water was not required to interpret oxygen and hydrogen isotope data like those of Clayton et al. (1966). Instead, such result can reflect mixing between meteoric water and the remnants of connate waters, such as evaporitic brines.

Some typical isotopic data for formation waters from the western Canada sedimentary basin are illustrated in Figure 2. Many other basins show similar trends. The most important observation is that the $\delta^{18}O$ and δD values of these waters do not plot along the meteoric water line. Instead, they verge from it, trending to higher $\delta^{18}O$ and δD values along a lower slope than the meteoric water line. Formation waters with the highest temperatures (greatest depths) and salinities are generally richer in ^{18}O and D than cooler, less saline waters. The trend line for formation waters generally intersects the meteoric water line at an isotopic composition typical of surface and shallow groundwater at that locality. For example, recent results for formation waters from central Alberta follow the same general trend shown in Figure 2, intersecting the meteoric water line at $\delta^{18}O$ = -16.3‰ and δD = -121‰, virtually identical to the current weighted mean annual value for rainfall at that locality (Edmonton, Alberta) (Connolly et al., 1990b). Such correlations caused Clayton et al. (1966) to propose that meteoric water comprised an important fraction of these formation waters.

Interaction and isotopic exchange between minerals in the pore system (e.g., ^{18}O-rich phases such as calcite) and porewaters have been invoked to explain the deviation of formation waters from the meteoric water line, and in particular the trend to higher $\delta^{18}O$ values. However, the extent of ^{18}O-enrichment depends on the isotopic composition of the exchanging solid phase, the effective mineral/water ratio (i.e., how easily does the mineral in contact with the water undergo isotopic re-equilibration, section 2.3) and the temperature of the system. Extensive exchange may be plausible in carbonate-dominated sequences, but seems less likely in siliciclastic units (section 3.2).

Transfer of porewaters from shales to sandstones in systems where the smectite to illite reaction controls shale porewater chemistry has also been proposed to explain the ^{18}O-rich nature of some sandstone porewaters (e.g., Suchecki and Land, 1983; Kharaka and Carothers, 1986). Isotopic re-equilibration between shale porewaters and illitic clay, facilitated by a dissolution-reprecipitation style smectite to illite reaction, should produce ^{18}O-rich shale porewaters at high clay/water ratios (sections 2.3 and 3.3). One of the great debates in sandstone diagenesis has revolved about the importance of the smectite-illite reaction in generating cations for transport (via shale porewaters) to sandstones (Hower et al., 1976; Boles, 1978, 1981; Boles and Franks, 1979). However, it appears that large-scale, cyclic movement of water is required if this process is to have played a significant role in sandstone cementation (Land, 1984; Longstaffe, 1988). Except under special conditions of very high (originally smectite-rich) shale to sand ratios, it seems unlikely that clay mineral dehydration in shales can control porewater chemistry or isotopic composition in adjacent sandstones.

Explaining the hydrogen isotope composition of formation waters is even more difficult. The role of interaction with hydrocarbons or clays is difficult to evaluate. As mentioned earlier, insufficient data are available on the nature of hydrogen isotopic exchange between formation water and hydrogen gas, hydrogen sulphide gas and/or liquid and gaseous hydrocarbons. However, because of the potentially large fractionation factors at diagenetic temperatures, such interaction might be very important (Hitchon and Friedman, 1969). Whether formation waters are enriched or depleted in D during hydrogen isotope exchange with clay minerals depends on the effective clay/water ratio, temperature, the initial hydrogen isotopic composition of both detrital and diagenetic clay minerals in the system, and the mechanism of hydrogen isotope exchange (section 2.3). Increasing δD values with depth are most likely in diagenetic systems where all hydrogen fixed in diagenetic clay minerals has been derived exclusively from the water, when the contribution of hydrogen to the water from exchange with pre-existing clay minerals is minimal, and where the effective clay/water ratio for hydrogen is high (see Kharaka et al., 1979; Kharaka and Carothers, 1986; Yeh, 1980). In sandstones of the western Canada sedimentary basin, for example, where effective clay/water ratios are quite low, the water is controlling the hydrogen isotope composition of the clay, rather than the other way around (Longstaffe and Ayalon, 1990).

Membrane filtration has also been proposed to explain the oxygen and hydrogen isotope compositions of formation waters. In this hypothesis, movement of dissolved constituents in water is attenuated by passage through shale or mudstones, gradually increasing the concentration of solute until a highly saline water is formed. Isotopic fractionation should accompany diffusion of water through this micropore system (Graf et al., 1965, 1966; Coplen and Hanshaw, 1973; Graf, 1982; Benzel and Graf, 1984). However, experimental results are contradictory. Hadyen and Graf (1986) reported that the $\delta^{18}O$ value of water passed through a smectite membrane varied from -0.80 to -0.96‰ at 20°C to +0.15 and +0.05‰ at 140° and 180°C, respectively. A significantly smaller membrane-induced fractionation of -0.25‰ at 21°C and pressures equivalent to 1.5 km depth has been reported by Demir (1988). Whether isotopic exchange between brine and clay minerals at elevated temperatures has significantly contributed to these apparent membrane-effects also remains to be evaluated. Isotopic effects induced by clay membranes remain potentially significant, if formation water movement involving significant, fracture-independent, cross-formational flow through shale sequences can be demonstrated.

Knauth and Beeunas (1986) and Knauth (1988) attempted to reconcile (i) the chemical data of Carpenter (1978), which suggest that many deep basin brines are evaporated seawater which later mixed with new fluxes of seawater or meteoric water, and/or reacted with the sedimentary host rocks, and (ii) the isotopic data of Clayton et al. (1966) which has been interpreted to indicate that many brines are derived entirely from meteoric water and then modified during water/rock interaction. Knauth's study of Permian salt deposits from the Palo Duro Basin showed that connate water trapped in halite fluid-inclusions was enriched in ^{18}O relative to the meteoric water line but depleted in D relative to ocean water, much like many formation waters (Fig. 3). If these fluid inclusions were dominantly of evaporative origin, how could such isotopic compositions be explained? Knauth and Beeunas (1986) noted that early in seawater evaporation, residual water becomes enriched in ^{18}O and D, the exact amount depending upon climatic variables (Craig and Gordon, 1965; Lloyd, 1966). However, under conditions of extreme evaporation (e.g., halite saturation), progressive enrichment of the heavier isotopes ceases, and with further evaporation, the residual liquids become depleted in ^{18}O and D (Lloyd, 1966; Sofer and Gat, 1975; Holser, 1979; Pierre et al., 1984). Knauth and Beeunas (1986) concluded that the isotopic compositions of fluid inclusions in halite could represent such liquids. They also showed how the composition of other fluid inclusions which were still more depleted in ^{18}O and D could have formed by mixing between hyper-evaporated waters and meteoric water subsequently introduced into the system.

This interpretation explains the isotopic trends observed for many formation waters, and obviates the need for complex chemical processes to create concentrated brines from fresh waters. However, others have noted that the process required to reverse the isotopic evaporation trend is rather complicated (and by inference, uncommon). O'Neil et al. (1986) offered different explanations of similar isotopic results for brines and fluid inclusions from salt beds in the Delaware Basin, located near to the Palo Duro study area. Isotopic results for fluid inclusions and brine from one site in particular were interpreted as a mixture between the evaporative water released by complete dehydration of gypsum, and meteoric water, which may have entered the evaporite system early or late in its geological history, possibly during recrystallization of halite. Another possibility was that the fluid inclusions were samples of evaporated meteoric water, introduced after the marine basin had dried up.

No single model can account for the oxygen and hydrogen isotope evolution of formation waters in sedimentary basins, given the complicated interplay between fluid mixing and rock-organic-water interaction. What we can attempt, however, is more quantitative assessment of the relative role played by each process within the specific geologic and hydrodynamic setting of a particular sedimentary basin.

2.3. OXYGEN AND HYDROGEN ISOTOPE EXCHANGE

The oxygen and/or hydrogen isotope compositions of diagenetic phases such as quartz, calcite and

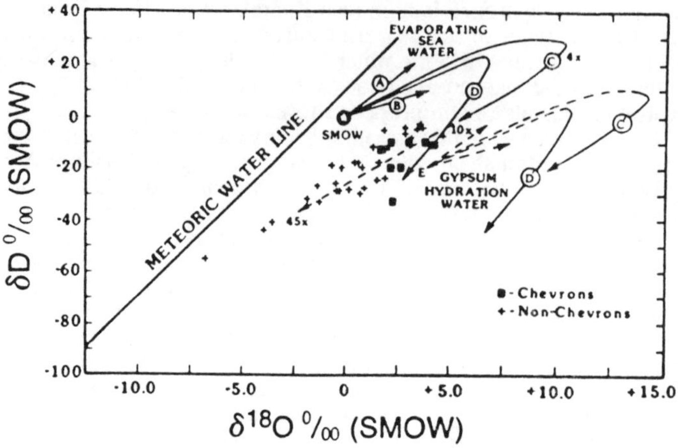

Figure 3. Isotopic fluid-evolution paths for seawater undergoing evaporation compared with fluid inclusion data (chevron and non-chevron data points) for Permian salt deposits (from Knauth and Beeunas, 1986). Curve A describes initial evaporation under relatively humid conditions. Curve B describes evaporation under arid conditions. Curve C is an estimate of evaporating seawater to 10X concentration (Holser, 1979); the dashed line extrapolation is to 45X concentration. Similar data are given in curve D after Pierre et al. (1984). Curves C' and D' represent the isotopic composition of hydration water in gypsum crystallized at any point along the evaporation curves C and D, respectively.

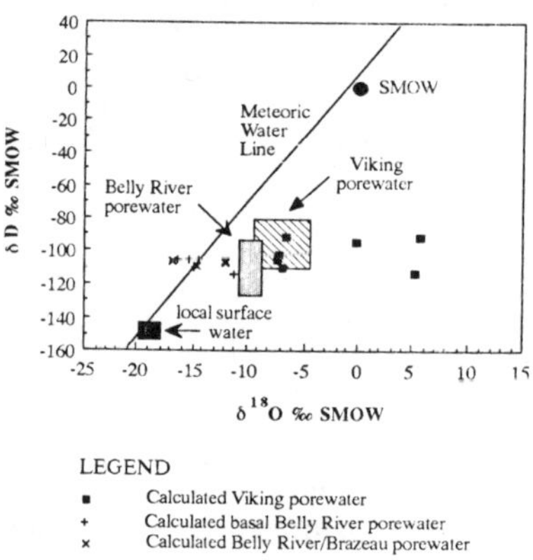

Figure 4. Comparison of measured hydrogen and oxygen isotope compositions for formation waters from the Viking Formation and Belly River/Brazeau Group with those calculated using the isotopic compositions of diagenetic kaolinite. The range of measured compositions for formation waters and surface water are shown by the shaded boxes. Calculated formation water compositions (individual data points) were deduced by assuming that oxygen and hydrogen isotopic equilibrium was attained between kaolinite and formation water at present subsurface conditions. The hydrogen isotope, kaolinite-water geothermometer of Lambert and Epstein (1980) and the oxygen isotope kaolinite-water geothermometer of Land and Dutton (1978) were used to perform the calculations. After Longstaffe and Ayalon (1990).

clay minerals are commonly used to deduce the nature and origin of ancient porewaters. This approach assumes that each diagenetic mineral in a pore system retains an isotopic composition characteristic of the conditions that prevailed during its crystallization. In this section, we wish to evaluate this assumption, and consider the extent to which isotopic exchange and re-equilibriation can occur once a diagenetic phase has initially precipitated.

Temperature, grain-size, surface area and crystal chemistry are the important factors controlling oxygen isotopic exchange between minerals and water. Given the very low rates of solid-state diffusion at sedimentary temperatures, isotopic exchange between stable minerals and water requires solution and recrystallization of the solid phase - that is, chemical bonds in the structure are broken and then reformed (Longstaffe, 1983; O'Neil, 1987). Consequently, isotopic exchange between many minerals and water is insignificant at surface temperatures. No evidence exists for oxygen isotope exchange between quartz and water up to at least 170°C, even for the finest size-fractions (Syers et al., 1968; Yeh and Savin, 1977). In contrast, because of its much greater ease of recrystallization, oxygen isotope exchange affects calcite (and to a lesser extent, dolomite) at much lower temperatures than silicate minerals (Clayton, 1959). The possibility of such exchange must be recognized in interpretation of oxygen isotope data for diagenetic carbonate minerals. For example, the oxygen isotope compositions of texturally early calcite cements may have been reset by re-equilibration with porewaters during subsequent, hotter stages of diagenesis. In such situations, though, the carbon isotope composition of the original carbonate is commonly preserved, since the original carbonate remains the major reservoir of carbon in the system.

2.3.1. *Clay Minerals.* The susceptibility of clay minerals to isotopic exchange is more complex. Oxygen and hydrogen in interlayer water of swelling clay minerals re-equilibrate with external water vapour at room temperature within hours or days (Savin and Epstein, 1970a). Accordingly, during conventional procedures, clay samples are heated under vacuum to remove absorbed and interlayer waters prior to isotopic analysis. However, what is less commonly recognized (or monitored during analysis) is that the temperature for complete removal of interlayer water is quite variable among clays. In some smectites, dehydration is not complete even at 300°C. The residual interlayer water, which is then extracted with structural water during a conventional analysis, can be very rich in D (δD values as high as +29‰, Marumo et al., in review), seriously contaminating measurements presumed to reflect structural water alone.

Exchange of oxygen isotopes between water and structural sites in clay minerals is normally very slow in low temperature environments (O'Neil and Kharaka, 1976) unless the mineral has been artificially expanded (James and Baker, 1976; Savin and Lee, 1988). Savin and Epstein (1970a,b) showed that detrital oceanic clay minerals had not re-equilibrated oxygen isotopes with seawater (< 5°C) over at least hundreds of thousands of years; these results were further confirmed by Yeh and Savin (1976), Eslinger and Yeh (1981) and Yeh and Eslinger (1986). Without neoformation of new clay, oxygen isotope exchange seems to be insignificant at temperatures typical of ocean sedimentation, even for very fine (<0.3 to <0.1 μm), smectite-rich material.

The rate of oxygen isotope exchange between clay minerals and water increases as temperatures rise, with significant exchange occurring by 300°C. Smectite is affected more readily than illite or kaolinite (O'Neil and Kharaka, 1976). Most evidence suggests that mineralogical reactions, such as the smectite to illite conversion, are required to facilitate isotopic exchange. In argillaceous rocks from the Texas Gulf Coast, for example, various clay size-fractions (dominated by illite-smectite with a high percentage of smectite layers) are out of internal isotopic equilibrium at shallow depths (1300 m), but by maximum burial depths (5500 m), near-homogeneous $\delta^{18}O$ values (<1‰ variation) are obtained for highly illitized illite-smectite (Yeh and Savin, 1977; Hower et al., 1976).

While the extent of isotopic exchange between illite-smectite and porewater can be correlated with the degree of conversion from smectite to illite, the details of this dependency are complex. Whitney and Northrop (1988) suggested from experimental results that ≈65% oxygen isotope exchange occurs during the first stage of illitization, which is controlled by clay transformation reactions yielding randomly ordered illite-smectite. The final (deepest and hottest) stage of illitization, during which regularly ordered illite-smectite is formed, is dominated by neoformation of illite, and characterized by 100% oxygen isotope exchange.

Hydrogen isotope exchange between clay minerals and water is less well understood. Experiments by O'Neil and Kharaka (1976) show that while such exchange occurs very slowly at low temperatures, the rate is nevertheless faster than for oxygen. Significant hydrogen isotope exchange can occur at temperatures below 100°C, smectitic clays being most readily affected. At temperatures as high as 200°C, extensive hydrogen isotope exchange can occur unaccompanied by major chemical or oxygen isotopic shifts. Such behaviour suggests that replacement of H^+ rather than $(OH)^-$ controls hydrogen isotope exchange between clay minerals and water (O'Neil and Kharaka, 1976; Kharaka and Carothers, 1986) at or below this temperature. At higher temperatures (specific to each clay mineral family), replacement of $(OH)^-$ should become much more important in controlling hydrogen (and oxygen) isotope exchange with water. Notably, the α (kaolinite-water) for hydrogen shows a rather sharp reversal from decreasing values with increasing temperature to increasing values with increasing temperature at about 200°C (Kyser, 1987).

Regardless of the experimental data, conclusions differ on the extent to which hydrogen isotope exchange between clays and water actually occurs during sedimentary processes. Yeh and Epstein (1978) and Eslinger and Yeh (1981) reported that hydrogen isotope exchange between seawater and detrital clay minerals from two to three million year old oceanic sediments was very limited, and confined to the <0.1 μm fraction, if it occurred at all. In contrast, Lawrence and Taylor (1972) noted that the expandable clay mineral halloysite rapidly exchanged hydrogen isotopes with groundwater at surface temperatures, and Wilson et al. (1987), Bird and Chivas (1988) and Longstaffe and Ayalon (1990) have reported significant post-formational hydrogen isotope exchange between clays and sandstone porewaters at <80°C without a corresponding effect on the oxygen isotope composition of the clay.

Longstaffe and Ayalon's (1990) conclusions rest mainly on the remarkably small range of δD values obtained for Cretaceous sandstones from the western Canada sedimentary basin, specifically early and late diagenetic kaolinites from the Viking Formation (-132 to -112‰) and Belly River/Brazeau Group (-137 to -128‰). Kaolinite from both units is in, or close to, hydrogen isotopic equilibrium with current formation waters at present subsurface temperatures (Fig. 4). By comparison, a proportionally much larger range in kaolinite $\delta^{18}O$ values was obtained for each unit (+13.8 to +26.9‰, Viking Formation; +5.9 to +12.7‰, Belly River/Brazeau Groups); a general state of oxygen isotope equilibrium does not exist between the kaolinite and formation waters (Fig. 4). Similar observations were made for some diagenetic illitic clays from the same units, albeit that α (illite-water) for hydrogen is not well known. Other illitic clays, however, are in both hydrogen and oxygen isotope equilibrium with present formation waters, perhaps implying neoformation at relatively low temperatures, a possibility which is also not without controversy.

The general implication is that in some situations, diagenetic clay minerals have exchanged hydrogen isotopes with current formation waters to temperatures as low as 40°C. Because this exchange was unaccompanied by significant changes in the original oxygen isotope composition of the clay, it most probably occurred in the absence of dissolution-precipitation reactions. Instead, proton exchange reactions may control hydrogen isotope compositions of clay minerals in water-dominated (low rock/water ratio) systems at low temperatures. Separate consideration is required in the interpretation of oxygen and hydrogen isotope ratios for clay minerals from such environments.

Rock-dominated systems, however, can behave quite differently. For example, shallow argillaceous rocks from the Texas Gulf Coast show a large variation among the δD values of different clay size-fractions, but these differences disappear at greater depths in a fashion similar to that obtained for the oxygen isotope compositions of these minerals (Yeh, 1980). The finest size fractions (almost entirely illite-smectite) and shale porewaters have reached isotopic equilibrium. In this case, a mineralogical reaction (smectite to illite conversion) seems to have been critical in facilitating extensive hydrogen and oxygen isotopic exchange. A depth dependent increase in δD values is shown by the clay minerals, and is accompanied by a decrease in the δD values of formation waters from the same area. This pattern can be predicted using a simple, material-balance calculation in which the clay minerals dominate the system (Kharaka and Carothers,1986).

2.3.2. *Isotopic Material Balance.* Comparison of the hydrogen isotope behaviour of clays in water- versus rock-dominated systems illustrates the importance of isotopic material balance considerations in diagenesis, in particular the extent to which a given phase participates in an isotope exchange reaction. Using the Law of Conservation of Mass, the isotopic shifts during porewater-mineral interaction can be quantified for simple systems by:

$$\delta_{system} = \Sigma X_i \delta_i \quad (12)$$

where X_i = mole fraction of phase i for the element (e.g., oxygen, hydrogen) of interest. This approximation can be very accurate for oxygen and hydrogen isotopes (Gregory and Criss, 1986). The net result is that if mineral/water ratios are very low, precipitation of diagenetic minerals has little effect on the isotopic composition of coexisting porewater. If mineral/water ratios are high, pore-lining and pore-filling minerals may ultimately determine the isotopic composition of entrapped porewaters in the system, depending on the nature and extent of dissolution-precipitation and /or proton-exchange processes.

Simple equations that describe mineral/water ratios during diagenetic and hydrothermal processes have been described by Suchecki and Land (1983), Lawrence et al. (1976), Dickinson (1987) and Nabelek (1987), and are discussed more fully in Longstaffe (1989). Only a few general points need to be reviewed here. First, such calculations commonly assume closed-system behaviour (no leakage in or out of the system, only one pore volume of water interacting with each pore). A first order approximation of open-system behaviour during diagenesis can be made by allowing multiple pore volumes of water to react with the mineral assemblage, and then move out of the system. The equations can also be modified to account for the porosity and permeability of the rock, and to permit more than one mineral to be involved in the exchange reaction.

Second, most models for calculating mineral/water ratios assume the absence of kinetic isotope effects during mineral neoformation. O'Neil (1987) suggested that most minerals crystallize in or very close to isotopic equilibrium, and that kinetic effects are secondary, except during (i) very rapid crystallization or loss of volatiles (Gonfiantini et al., 1968) or (ii) special situations associated with biological activity (Keith and Weber, 1965). Data for neoformed phases from soils and weathering profiles also indicate that most minerals precipitate in isotopic equilibrium at low temperatures (Lawrence and Taylor, 1971, 1972; Bird et al., 1992a,b). However, the potential for kinetic isotope fractionation under certain conditions of crystal growth should not be ignored (Bird et al., 1993).

Finally, complete isotopic re-equilibration between solids and water in the system is commonly assumed in calculations of mineral/water ratio, but as discussed in section 2.3.2, this is not always the case. The degree of isotopic exchange between a pre-existing mineral and coexisting porewaters must be evaluated (i.e., the "effective water/mineral" ratio). For oxygen, pore-lining calcite could be highly exchangeable whereas quartz cements in contact with porewaters would probably have minimal involvement. Likewise, effective mineral/water ratios for hydrogen in sandstones like those from the western Canada sedimentary basin are probably very small; water composition and temperature decide the isotopic composition of the diagenetic clays with little effect on the waters themselves. In comparison, because large amounts of hydrogen are contained in clay minerals in argillaceous sequences, and because of the ease of hydrogen isotope exchange (relative to oxygen), high effective mineral/water ratios can be expected, especially for compacted, deeper sequences (Kharaka and Carothers, 1986). In such cases, the initial δD values of pre-existing clay minerals (commonly -80 to -25‰) will strongly influence the final isotopic composition of the porewater.

2.3.3. *Equilibrium isotope fractionation factors.* As the previous discussions have demonstrated, $\delta^{18}O$ and δD values of diagenetic minerals are controlled by several parameters, including the isotopic composition of water present during crystallization, the material balance of oxygen or hydrogen between the mineral and porewater during crystallization, the degree to which isotopic equilibrium was attained during crystallization, the mode of diagenetic mineral formation (i.e., neoformation versus transformation), and the extent of mineral-water isotopic exchange following crystallization. Under equilibrium conditions, however, temperature is the most important parameter that determines the "instantaneous" isotopic fractionation ($\alpha_{mineral\text{-}water}$) between the diagenetic mineral and water.

Fractionation factors have been determined using naturally occurring samples, laboratory equilibration experiments, and theoretical and empirical calculations. Many commonly employed oxygen isotope geothermometers useful in studies of diagenesis have been summarized by Longstaffe (1989). All need to be applied with due caution. For example, conclusive experimental determinations of α (mineral-water) are uncommon because of the difficulties in demonstrating reversibility in low temperature systems. As a result, many equations available for higher temperature systems have been extrapolated to lower temperatures, presuming linearity for $1000 \ln \alpha$ versus $1/T^2$ (T in °K). This assumption can lead to considerable error. In addition, complex chemical substitutions in phases such as clay minerals ensure a significant compositional as well as thermal dependence of α (mineral-water) (i.e., Si-O bonds concentrate ^{18}O relative to Mg-O or Fe-O bonds; Taylor and Epstein, 1961). Hence, additional uncertainty is associated with application of α-values determined for a specific mineral composition to the range of compositions possible for that mineral. In clay minerals, the situation is even more complex; cation-hydroxyl bonded oxygen (e.g., Al-OH) has a different $\delta^{18}O$ value than cation-oxygen bonded oxygen (e.g., Al-O-Si). This variation may even have potential as a single mineral geothermometer (Savin, 1967; Hamza and Epstein, 1980; Savin and Lee, 1988; Girard et al., 1991).

Because of these difficulties, serious attempts have been made to calculate α (mineral-water) values using a combination of theoretical approaches (statistical mechanics) or empirical bond length measurements (e.g., Savin and Lee, 1988; Zheng, 1991). Recent results are encouraging; Bird et al. (1993), for example, have demonstrated excellent agreement between the values for α (rutile-water) obtained from synthesis experiments at 22° and 50°C, and those calculated following Zheng (1991).

Hydrogen isotope equilibrium fractionations between hydrous phases and water, or for that matter, hydrous phases and organic liquids or gases (hydrocarbons), are not well characterized. Most of the limited data available for low temperatures have been reviewed by Suzuoki and Epstein (1976), Kyser (1987) and Savin and Lee (1988). Additional data for illite-smectite, mica-smectite and chlorite-smectite have been presented recently by Capuano (1992) and Marumo et al. (in review). In almost all cases, hydrous phases are depleted in D relative to water, but in general, α (mineral-water) for hydrogen, especially for clay minerals, is poorly known. Kaolinite has been studied the most; those results indicate that the near-linearity of $1000\ln\alpha$ versus $10^6/T^2$ that characterizes oxygen isotope fractionation for most mineral-water pairs should not be assumed (section 2.3.1; Lambert and Epstein, 1980; Liu and Epstein, 1984). Somewhat conflicting results for α (illitic clay-water) are reported by Yeh (1980) and Capuano (1992). Little is known for chlorite formed at low temperatures. Graham et al. (1987) and Savin and Lee (1988) have presented some information for the temperature range 130-700°C, and Marumo et al. (1980, in review) have suggested that Fe/Mg ratios in chlorite may influence α (mineral-water) for hydrogen.

3. Meteoric Diagenesis in Sandstones, West-Central Alberta

In this section, we illustrate how oxygen isotope data for diagenetic minerals have been used to constrain the nature of porewater evolution in Mesozoic sandstones from western and central Alberta. These sandstones represent the more deeply buried portions of the clastic sedimentary pile in the western Canada sedimentary basin. Most results presented here are summarized from Longstaffe et al. (1992a) and Longstaffe (in press). The most significant conclusion is that meteoric water played an important, even dominant role throughout diagenesis.

3.1. GENERAL GEOLOGIC AND HYDRODYNAMIC SETTING

The western Canada sedimentary basin consists of a thick sequence of Paleozoic carbonate rocks and evaporites overlain by sandstones and shales of mostly Cretaceous to Tertiary age. Along its western margin, the basin can be up to 5,000 m thick, but its thickness and elevation decrease towards the east-northeast until the Precambrian basement rocks are exposed in northeastern Alberta, and northern Saskatchewan and Manitoba. Because of this configuration, this basin provides a good opportunity to evaluate the role played by meteoric water in sandstone diagenesis.

The presence of active meteoric flow systems in the basin, with recharge in major highlands and discharge in major lowlands, has been demonstrated both on regional and local scales by Hitchon (1969a,b) and Tóth (1978, 1980) (Fig. 5). Development of the gravity-driven flow regime is related to the formation of the Cordillera, and probably began in the early Eocene, when uplift produced substantial topography. The first phase of the Laramide Orogeny caused thrusting and

CONCEPTUAL MODEL FOR REGIONAL TRANSPORT

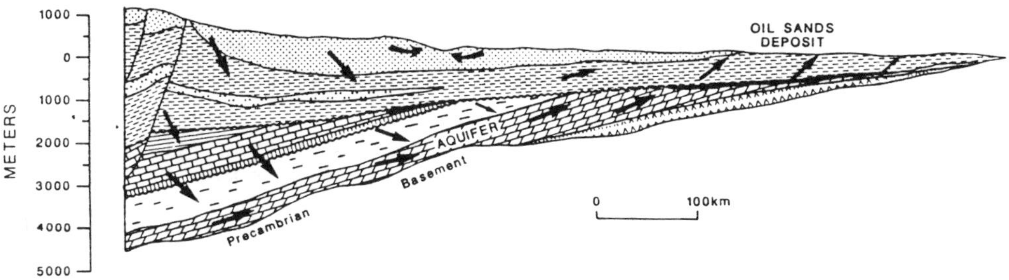

Figure 5. Schematic diagram of gravity-driven groundwater flow in the uplifted western Canada sedimentary basin (from Garven, 1989). According to Garven (1989), descending groundwater leached soluble oil from deeply buried source rocks, and was focused into regional aquifers. Discrete-phase oil was also added to these aquifers from below. The shale acted as an aquitard which was impermeable to discrete oil droplets but not to groundwater. The oil sand deposits accumulated at the discharge end of the regional flow system.

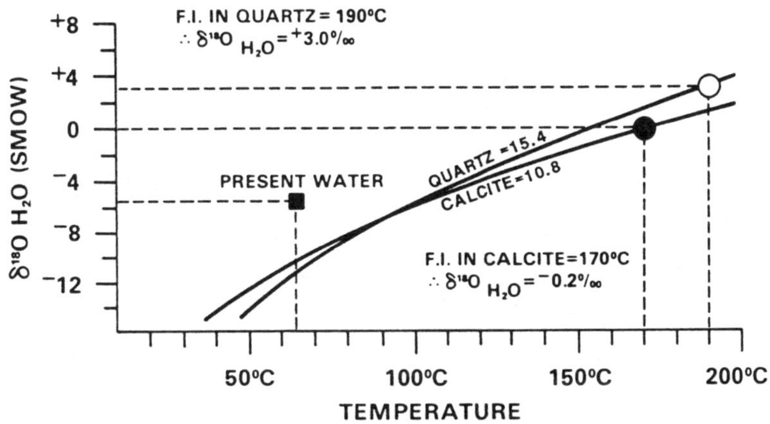

Figure 6. In Figures 6a-e, development of a porewater evolution diagram (porewater $\delta^{18}O$ value versus temperature) is illustrated for a conglomerate from the Fahler Member in the western portions of the Alberta Deep Basin. All diagrams are from Tilley and Longstaffe (1989).

Figure 6a. The $\delta^{18}O$ value of porewater versus temperature for diagenetic quartz and calcite from conglomerates of the Fahler Member. The quartz and calcite curves have been determined using the equations presented in Friedman and O'Neil (1977), after Clayton et al. (1972) for quartz and O'Neil et al. (1969) for calcite. Present formation water is indicated by the black square. The white and black circles on the mineral curves represent fluid inclusion temperatures for quartz and calcite, respectively.

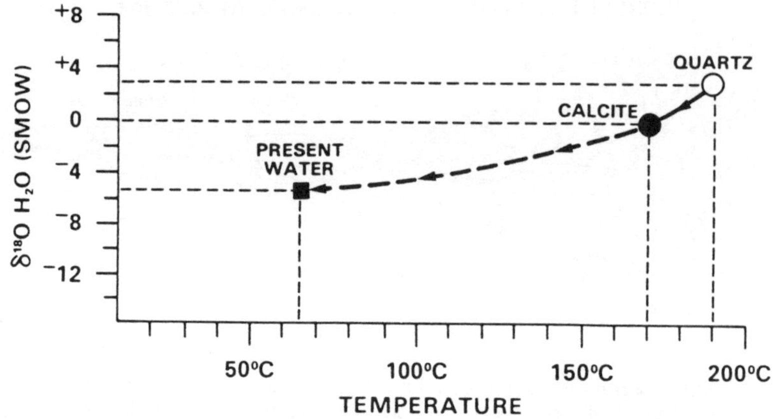

Figure 6b. Idealized porewater evolution pathway for late-stage diagenesis in conglomerates of the Fahler Member. White (quartz) and black (calcite) circles are points determined using fluid inclusion temperatures and mineral oxygen-isotope compositions, whereas the black square represents present formation waters (see Fig. 6a). Therefore the dashed line represents the pathway of porewater evolution in water-saturated portions of the Fahler Member.

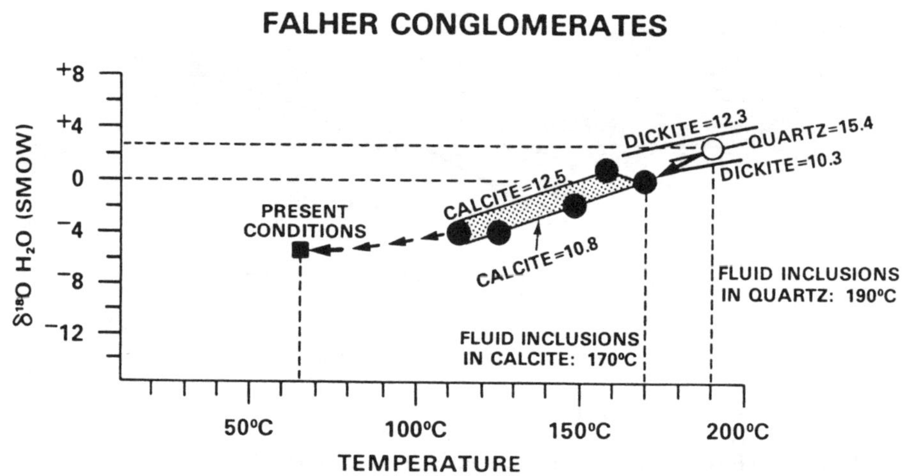

Figure 6c. Idealized porewater evolution pathway for late-stage diagenesis in conglomerates of the Fahler Member. Higher temperature parts of the mineral curves for dickite and calcite are shown for the range of $\delta^{18}O$ values obtained for each phase. The dickite curves were calculated using the geothermometer of Land and Dutton (1978). The shaded band from 113 to 170°C indicates the range of conditions in which calcite precipitated. Black dots within this band represent individual samples for which both stable isotope compositions and fluid-inclusion temperature were obtained.

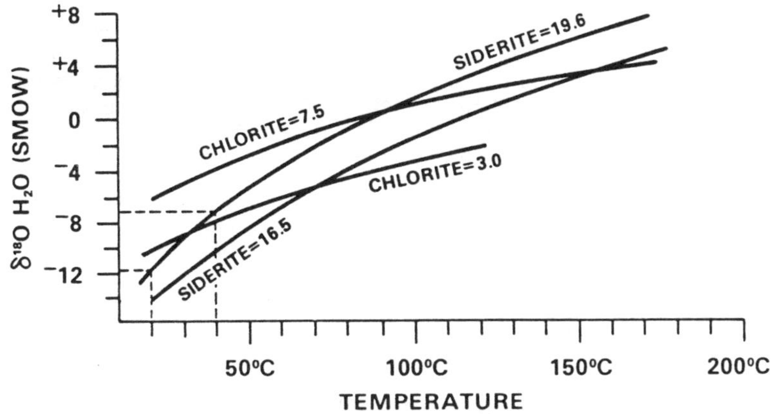

Figure 6d. The $\delta^{18}O$ value of porewater versus temperature for the range in oxygen isotope compositions obtained for diagenetic siderite and chlorite from conglomerates of the Fahler and Cadotte Members. The siderite and chlorite curves have been determined using the equations of Becker and Clayton (1976) and Wenner and Taylor (1971), respectively. Both equations have since been superseded by Carothers et al. (1988) and Savin and Lee (1988). Only the new equation for chlorite significantly changes the result shown here. If the chlorite results were replotted according to Savin and Lee (1988), the new curves would track very closely to those obtained for siderite.

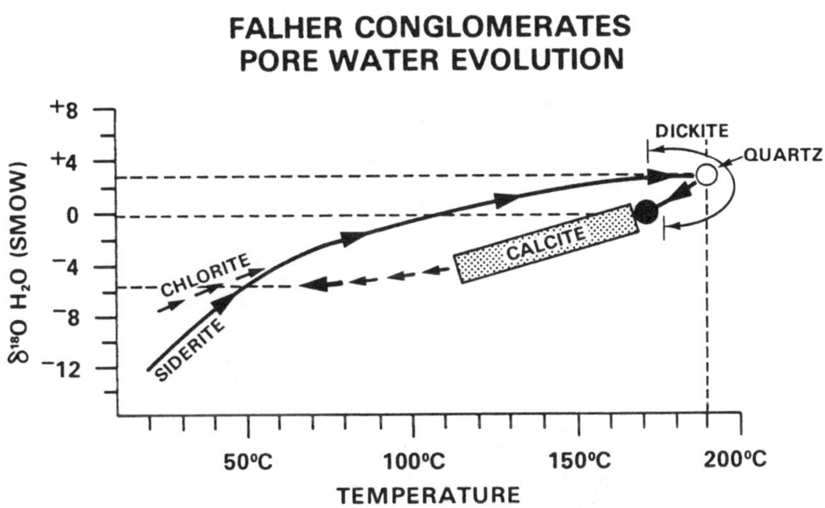

Figure 6e. Idealized porewater evolution pathway during diagenesis of conglomerates of the Fahler Member, summarized from results presented in Figures 6a-d. The discontinuous arrows at the end of the curve represent late-stage porewater evolution in portions of the study area that remained water saturated (rather than filled by methane gas). The porewater pathway shown here for chlorite was deduced using the isotope geothermometer of Wenner and Taylor (1971). If the revised chlorite geothermometer of Savin and Lee (1988) is used, both chlorite and siderite track along similar paths.

uplift along the eastern Cordillera in late Mesozoic time. To the east, continental deposition dominated throughout the Late Cretaceous and into the Early Tertiary as sediment was shed eastward into the downwarped sedimentary basin. Maximum burial probably occurred in the late Paleocene or early Eocene (Taylor et al., 1964; Hitchon, 1984). In the early Eocene, the major pulse of the Laramide Orogeny resulted in extensive deformation of the eastern Cordillera and substantial uplift of the sedimentary basin in western Alberta. Uplift created a high potentiometric surface, causing large-scale, meteoric recharge into permeable sedimentary rocks in the basin (Hitchon, 1984). This process continued, albeit gradually diminishing, as erosion of accumulated Upper Cretaceous and Tertiary rocks proceeded (Taylor et al., 1964; Beaumont, 1981).

Whether meteoric water introduced in this fashion mixed with, rather than entirely replaced, earlier porewaters is an issue of some importance (Domenico and Robbins, 1985). How should oxygen and hydrogen isotope data for formation waters be interpreted, given that the same results have been used to infer both processes (section 2.2.4)? Either way, however, introduction of meteoric water should have strongly influenced the chemical reactions and mass transport that produce sandstone diagenesis (Schwartz and Longstaffe, 1988). Further consequences have also been suggested. For example, some of the northeastward flow of groundwater initiated by Laramide uplift may have been focused in highly permeable Paleozoic carbonates (e.g., Mountjoy, this volume). This "regional aquifer" may have drained source rocks, and transported the hydrocarbons that now comprise in degraded form, the bitumen and heavy oil accumulations of eastern and northeastern Alberta (Fig. 5; Hitchon, 1969b, Garven, 1989) (section 4).

3.2. METEORIC DIAGENESIS AND THE LARAMIDE OROGENY (EARLY EOCENE)

It has long been recognized that formation waters in the western Canada sedimentary basin contain a sizeable fraction of meteoric water or its evolved equivalent (Clayton et al., 1966; Hitchon and Friedman, 1969). More recently, the impact of meteoric water on diagenesis in Cretaceous sandstones throughout the basin was demonstrated by the low $\delta^{18}O$ values (≈+5 to +15‰) of most late diagenetic clay, quartz and/or carbonate minerals (Longstaffe, 1983, 1984, 1986; Longstaffe and Ayalon, 1987; Ayalon and Longstaffe, 1988; Tilley and Longstaffe, 1989; Longstaffe and Ayalon, 1990, 1991). The oxygen isotope compositions of most diagenetic phases formed at or following maximum burial can only be explained by a signficant lowering of porewater $\delta^{18}O$ values that began at, or shortly following, maximum burial. Lowering of porewater $\delta^{18}O$ values during late diagenesis reflects the gravity-driven influx of meteoric water that began in the early Eocene as the result of uplift, and which has continued in gradually subdued fashion as erosion of the uplifted sediments proceeded. An example of how this pattern for porewater evolution was deduced is illustrated in Figure 6 for the Lower Cretaceous Fahler Member conglomerate from the Alberta Deep Basin (after Tilley and Longstaffe, 1989).

The post-Laramide meteoric recharge represents one very important end-member in the evolution of formation waters from the western Canada sedimentary basin. We must not forget, however, that the formation waters are the cumulative result of both mixing and rock/water interaction. Hitchon and Friedman (1969) invoked an important role for mixing between evolved (connate) seawater and surface-derived meteoric water to explain the isotopic compositions of these formation waters (Fig. 2). Connolly et al. (1990a,b) developed this idea further, recognizing two distinct fluid regimes in central Alberta.

Connolly et al. (1990a,b) concluded that brines of the lower regime formed by evaporation of seawater beyond the point of halite saturation, followed by dilution with 50 to 80% of (post-Laramide) meteoric water. The lower regime contains two groups of formation waters, between which some mixing has occurred (see Connolly et al., 1990a for plots of Cl vs Br, K vs Br, etc.). One group ($\delta D \approx -85‰$; $\delta^{18}O \approx -5‰$; TDS ≈ 120 g/l) is more saline, and dominates Devonian, Mississippian and lowermost Jurassic carbonate rocks. The other group ($\delta D \approx -85‰$; $\delta^{18}O \approx 7‰$; TDS ≈ 75 g/l) is generally hosted by Middle Jurassic and Lower Cretaceous sandstones. Waters from the lowermost fluid regime have $^{87}Sr/^{86}Sr$ ratios of 0.7076 to 0.7129; those contained in Devonian carbonate rocks generally have the most radiogenic compositions. Most samples are more radiogenic than predicted for connate seawater, and have higher $^{87}Sr/^{86}Sr$ ratios than coexisting carbonate cements. Only Precambrian basement rocks, and some Cambrian and Devonian shales have $^{87}Sr/^{86}Sr$ ratios sufficiently high to account for such compositions, thus providing further evidence for fluid communication between Paleozoic carbonate and Mesozoic clastic rocks.

The upper fluid regime is situated mostly in Upper Cretaceous sandstones, and is characterized by dilute, meteoric waters (δD ≈ -110‰; $\delta^{18}O$ ≈ -12‰; TDS ≈ 12 g/l). The formation waters have very low $^{87}Sr/^{86}Sr$ ratios (0.7058-0.7089) (as do coexisting late diagenetic carbonate and clay cements, Ayalon and Longstaffe, unpublished data; Connolly et al., 1990b). These values are significantly lower than Upper Cretaceous seawater. The lowest values occur in the uppermost unit sampled (Belly River/Brazeau Group, $^{87}Sr/^{86}Sr$ = 0.7058-0.7063), and probably reflect the emerging influence of sediments derived from a volcanic provenance during Cretaceous time (Connolly et al., 1990b).

The upper and lower fluid regimes are chemically and isotopically decoupled. Connolly et al. (1990b) concluded that only the upper regime is now hydrodynamically connected to the post-Laramide recharge of meteoric water; hydrochemical isolation of the lower regime began in the Pliocene. Garven (1989) proposed that the post-Laramide, gravity-driven flow system began to dissipate in Miocene-Pliocene time. Gradual waning of the regional hydrodynamic drive may have permitted upward diffusional flow and density stratification of formation waters within the lower hydrological regime (Connolly et al., 1990a,b).

Mixing has clearly been important in determining the composition of formation waters in sandstones and carbonates from the western Canada sedimentary basin. The importance of oxygen isotope exchange with carbonate cements in modifying the composition of formation waters (section 2.2.4) may have been overemphasized in the past, especially for waters contained in clastic units. For any locality in central Alberta, for example, formation waters generally have $\delta^{18}O$ values either too low or too high for equilibrium with coexisting carbonates. Instead, Cl and Na versus $\delta^{18}O$ relationships showed mixing to be the dominant control on the oxygen isotope composition of formation water (Connolly et al., 1990b). Nevertheless, reactions involving carbonates, clay minerals and feldspars still play an important role. The $^{87}Sr/^{86}Sr$ ratios of many formation water samples, for example, lie outside the range known for seawater of any age, and must be explained by rock/water interaction with feldspars, micas and clay minerals.

3.3. POREWATER EVOLUTION DURING BURIAL DIAGENESIS

As discussed in sections 2.2 and 2.3, porewaters in sedimentary basins commonly (but not always) evolve towards ^{18}O-rich compositions during burial. In the shale-dominated Texas Gulf Coast and the Great Valley Sequence of California, sandstone porewaters have acquired $\delta^{18}O$ values as high as +8‰, a result commonly attributed to conversion of smectite to illite in interbedded shales (Yeh and Savin, 1977; Suchecki and Land, 1983; Taylor, 1990).

Tilley (1988) and Longstaffe et al. (1992a) addressed this issue in the Alberta Deep Basin (Fig. 1). A succession of Permian to Upper Cretaceous sandstones was examined from the thick sandstone-shale sequence that overlies the Paleozoic carbonate rocks. The lowermost unit studied (the Permian Belloy sandstone) is located within the uppermost portion of the Paleozoic carbonate sequence. All sandstones showed a similar diagenetic evolution: siderite or chlorite formed early; quartz, kaolinite and then illitic clays crystallized as temperature and burial depth increased; quartz druse, dickite/late kaolinite, and then ankerite and calcite precipitated during the post-Laramide influx of meteoric water (Tilley, 1988). Maximum burial temperatures in each unit were determined from a combination of organic maturation and fluid inclusion data (Tilley, 1988; Tilley et al., 1989; Tilley and Longstaffe, 1989). The oxygen isotopic compositions of minerals formed at these temperatures were used to calculate maximum porewater $\delta^{18}O$ values prior to the post-Laramide dilution by meteoric water (e.g., Fig. 6).

This investigation verified earlier observations (Longstaffe and Ayalon, 1987; Ayalon and Longstaffe, 1988; Tilley and Longstaffe, 1989) that sandstone porewater $\delta^{18}O$ values attained by maximum burial in the western Canada sedimentary basin were generally no higher than +2 to +3‰, and sometimes much lower. Only with depth, as the Paleozoic carbonate platform was approached, were higher values (>+3‰) obtained.

One implication of these observations is that significant contribution of ^{18}O-rich shale porewaters to the sandstones (via the smectite to illite reaction) cannot be convincingly demonstrated.
Diagenetic illitic clays in typical Texas Gulf Coast shales ($\delta^{18}O$ ≈ +16 to +19‰; Yeh and Savin,

1977) are in isotopic equilibrium with porewaters of $\delta^{18}O \approx +5$ to +10‰ at 130-170°C. Because porewaters in associated sandstones have similar $\delta^{18}O$ values, substantial transfer of shale porewaters to the sandstones has been presumed (Kharaka and Carothers, 1986; Taylor, 1990). In the Mesozoic section examined in the western Canada sedimentary basin, shale is also very abundant, has been heated to equivalent temperatures, and generally contains abundant illitic clay. Why then are maximum $\delta^{18}O$ values deduced for sandstone porewaters generally <+3‰. In a few cases, the oxygen isotope signal resulting from the smectite-illite reaction may have been lower (e.g., mudstones from the continental Belly River/Brazeau Group; Longstaffe and Ayalon, 1991). But most illitic clays from shales in the western Canada sedimentary basin have $\delta^{18}O$ values similar to the Texas Gulf Coast (Longstaffe et al., 1992a). Perhaps ^{18}O-rich porewaters derived from shales were volumetrically insufficient to displace pre-existing sandstone porewaters, and instead mixed with, and were diluted by, lower ^{18}O connate waters (section 3.4). But it also may be that the smectite-illite reaction was unimportant, at least in some shales. The detrital versus diagenetic nature of illitic clays in Mesozoic shales in the western Canada sedimentary basin has not yet been demonstrated in any great detail; an origin via the smectite to illite reaction in quantities sufficient to control shale porewater compositions should not be an automatic assumption.

Another observation arising from the Deep Basin study is that the highest porewater $\delta^{18}O$ values occur closest to the Paleozoic unconformity, and decrease upwards in the siliciclastic section: Permian Belloy Formation, +9‰; Triassic Halfway Formation, +7‰, Jurassic-Lower Cretaceous Cadomin/Nikanassin formations, +5‰, and overlying Cretaceous units, +2 to 0‰ (Longstaffe et al., 1992a). This distribution suggests mixing between pre-existing, lower ^{18}O porewaters in the sandstones (section 3.4) and ^{18}O-rich waters from underlying Paleozoic carbonates, similar to that which has modified the composition of some current formation waters.

The ^{18}O-rich character of waters originating from the Paleozoic units probably arose both from evaporative brine formation and rock/water interaction. Trapping and evaporation of marine connate waters would be anticipated during formation of the Paleozoic interbedded carbonates, calcareous shales and evaporites; residual waters with $\delta^{18}O$ values ≈+7‰ can be expected by 4X seawater concentration, assuming that seawater was 0‰ (Fig. 3). The higher concentrations required to reverse the evaporative trend and produce $\delta^{18}O$ values < 0‰ (section 2.2.4) were probably not attained by most brines (Connolly et al., 1990a). In addition, regional dolomitization of some Paleozoic carbonates indicates that extensive rock-water interaction has occurred; isotopic re-equilibration with the average Paleozoic carbonate at 125-175°C would produce porewaters with $\delta^{18}O$ values of +3 to +11‰.

3.4. POREWATER EVOLUTION DURING EARLY DIAGENESIS

Figure 6 illustrates that low-^{18}O porewaters were present during formation of early chlorite and siderite in the Fahler Member conglomerates. This feature is typical of early diagenetic minerals from many sandstones and conglomerates throughout Alberta (Longstaffe and Ayalon, 1987; Ayalon and Longstaffe, 1988; Machemer and Hutcheon, 1988; Tilley and Longstaffe, 1989; McKay et al., 1989; Wickert et al., 1989; Bloch, 1990; Longstaffe and Ayalon, 1991; Hart et al., 1992; Longstaffe et al., 1992a,b; McKay, 1992). Some units were deposited in continental settings, or deltaic and coastal plain environments; hence a significant meteoric contribution early in diagenesis is expected. For instance, comparison of the continental Brazeau Group sandstones, underlying deltaic, shoreline and nearshore basal Belly River Group sandstones, and dominantly marine sandstones of the Lower Cretaceous Viking Formation showed that the average $\delta^{18}O$ value of a given diagenetic mineral, whether formed early or late is lowest in the Brazeau Group and highest in the Viking Formation (Fig. 7). Some memory of differences in connate porewater composition has apparently persisted throughout diagenesis; post-Laramide meteoric displacement of (evolved) connate waters, whether they be brines, seawater or freshwater, has not been complete (Longstaffe and Ayalon, 1991).

However, even dominantly 'marine' sandstones, such as Viking Formation in central Alberta and the Paddy Member in northwestern Alberta, commonly contain early diagenetic minerals with lower $\delta^{18}O$ values than predicted for crystallization from open marine seawater (Longstaffe and Ayalon, 1987; Longstaffe et al., 1992a). One explanation is that the seaway may have been brackish. Kyser et al. (in press) have suggested that the continental seaway in western Canada was decoupled

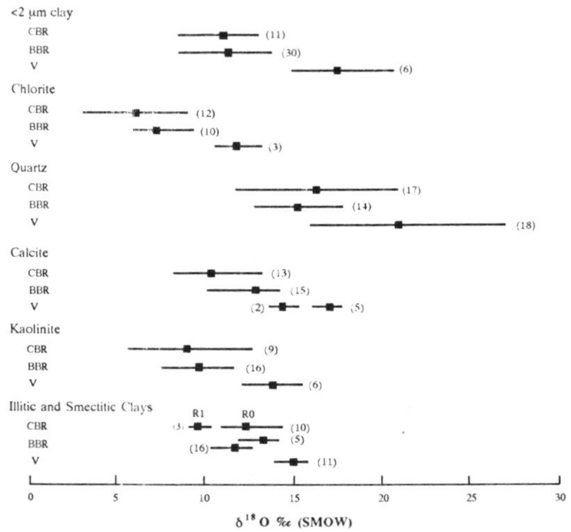

Figure 7. Range (bars) and mean (squares) $\delta^{18}O$ values for diagenetic minerals from continental sandstones of the Upper Cretaceous, Belly River/Brazeau Group (CBR), mixed meteoric-marine sandstones of the basal Belly River Group (BBR) and marine sandstones of the Lower Cretaceous, Viking Formation (V). Diagenetic minerals from the Viking Formation sandstones are generally enriched in ^{18}O relative to the same phase from the Belly River/Brazeau Group sandstones. The number of samples analyzed is given in parentheses. After Longstaffe and Ayalon (1991).

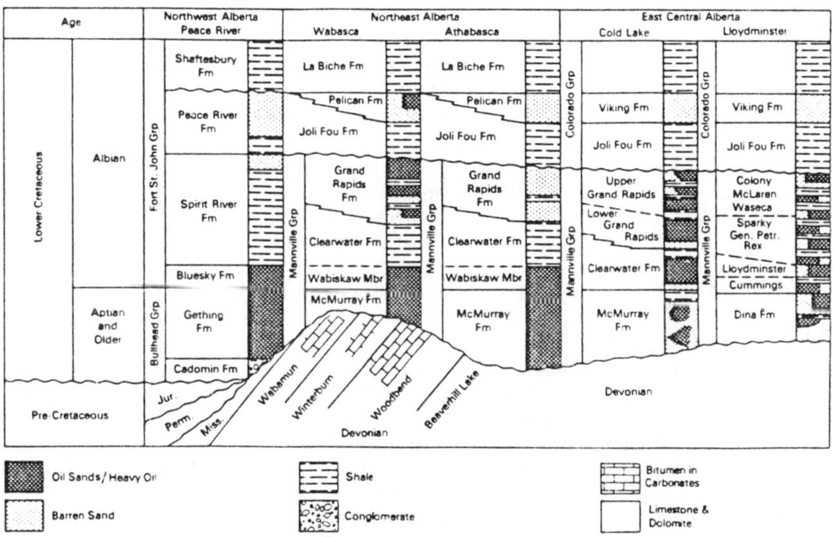

Figure 8. Table of formations indicating the main oil sand and heavy oil deposits in Alberta. From Wightman et al. (1989).

from the open ocean at various times throughout the Cretaceous, and attained $\delta^{18}O$ values as low as -5 to -4‰ during some cycles, whose timing coincided with Viking and Paddy deposition.

Another explanation is that marine sands deposited in shallow, inland seas are especially well situated for early incorporation of meteoric water. For example, early diagenetic minerals in a facies assemblage resulting from progradation into marine waters should show a systematic increase in $\delta^{18}O$ values as the freshwater phreatic lens dissipated seaward. Diagenetic siderite from the Cardium Formation sandstone, northwestern Alberta, shows such a pattern (Hart et al., 1992). Facies for which a greater freshwater influence is indicated from sedimentary features contain siderite with lower $\delta^{18}O$ values (and less Mg, Ca substitution for Fe; Mozley, 1989a,b) than more marine varieties. In addition, lowering of sea level, and development of emergent zones, should have facilitated post-depositional, early infiltration of meteoric water into shallow marine sands, given that large-scale migration of shorelines accompanied changes in relative sea level in the western Canada sedimentary basin. Two early phases of diagenetic siderite from marine shelf sandstones of the Upper Cretaceous Marshybank Formation, northwestern Alberta, may illustrate this effect (McKay, 1992). The earliest siderite has $\delta^{18}O$ values (+28‰) typical of precipitation from porewaters of marine origin, while a slightly later generation has much lower $\delta^{18}O$ values that can only be explained by mixing between meteoric water and pre-existing connate waters of dominantly marine origin. The cyclic nature of sea level fluctuation may have triggered many such invasions, the net result being a substantial flux of meteoric water during early diagenesis.

4. Natural and Artificial Diagenesis in Clearwater Formation Oil Sands, Alberta

In this section, stable isotope results for both naturally and anthropogenically-produced diagenetic minerals are described for the Lower Cretaceous Clearwater Formation (including the basal Wabiskaw Member) oil sands from the Cold Lake-Primrose areas of Alberta, Canada (Figs. 8, 9). These deposits have not been as deeply buried as units further west, and hence their early diagenetic character is more easily discerned. Present burial depths vary from ≈400 to 500 m, and post-Laramide erosion has removed no more than ≈1000 m of overlying strata (Hacquebard, 1977; Beaumont et al., 1985). Maximum temperatures of 60-70°C have been suggested from burial history reconstructions (Longstaffe et al., 1992b).

Despite shallow burial, the reactive framework mineralogy of these sands has induced substantial diagenesis that influenced their viability as commercial hydrocarbon reservoirs in at least two ways. First, the naturally occurring diagenetic assemblage has helped to determine the degree of initial hydrocarbon saturation. Second, the type and distribution of pore-lining and grain-coating minerals have largely controlled the mineral reactions, and hence reservoir behaviour, during enhanced recovery procedures used to produce the hydrocarbons. Both aspects are briefly discussed below: first, the natural diagenetic evolution of reservoir sands from the Clearwater Formation in the Cold Lake area, Alberta, and second, steam-induced mineral reactions that occurred in kaolinite-rich reservoir sands from the Wabiskaw Member in the Primrose area, Alberta.

4.1. THE MANNVILLE GROUP OIL SANDS, COLD LAKE-PRIMROSE AREA

The Alberta oil sands are of particular interest because they comprise one of the world's largest accumulations of hydrocarbons. Bitumen reserves in place are estimated to be about 266 x 10^9 m^3 (Alberta Energy Resources Conservation Board, 1985; Wightman et al., 1989). Oil sands stretch from the Athabasca area in the north through the Primrose area to the Cold Lake area in the south, and comprise one more or less continuous, 560 km long deposit (McIntosh and Beckie, 1989). Hydrocarbons are contained mostly within the Mannville Group; the Cold Lake area alone contains 35 x 10^9 m^3 of bitumen in place. While bitumen is present in all Mannville Group units in the Cold Lake area, Clearwater Formation reservoirs are the most prolific because of their thickness (mean pay zone ≈12 m) and lateral continuity (Wightman et al., 1989).

4.1.1. *Origin and Evolution the Hydrocarbons.* The origin of the enormous bitumen reserves in the Alberta oil sands has been debated for a long time (e.g., Ball, 1935; Hume, 1947; Link, 1951;

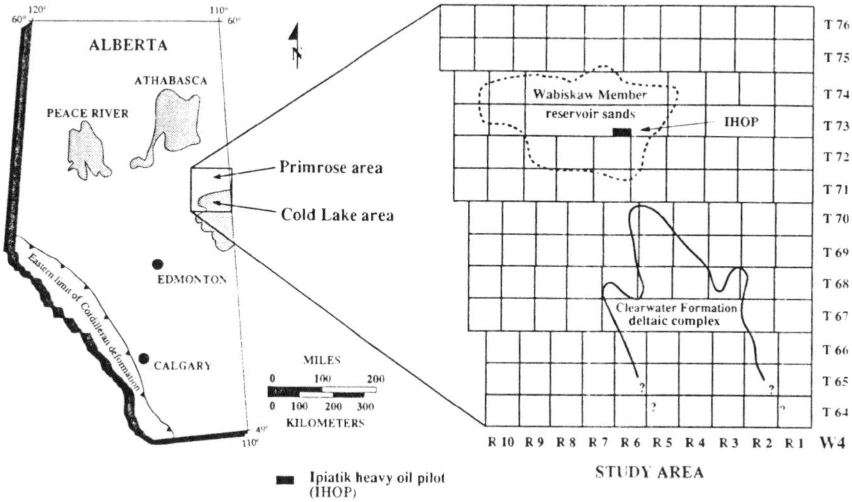

Figure 9. Alberta oil sand deposits, including the Clearwater Formation deltaic complex in the Cold Lake area, the Wabiskaw Member reservoir sands in the Primrose area, and the experimental heavy oil pilot operated at Ipiatik Lake (IHOP) by AEC Oil and Gas Company. The locations of the other major deposits located at Athabasca and Peace River are also shown. The enlarged area shows the locations of the Wabiskaw Member (Primrose area) and the Clearwater Formation sands (Cold Lake area) described here. The geographical grid is given by Township (north-south) and Range (east-west), west of the 4th meridian. Each subdivision is ≈9.6 x 9.6 km (6 x 6 miles).

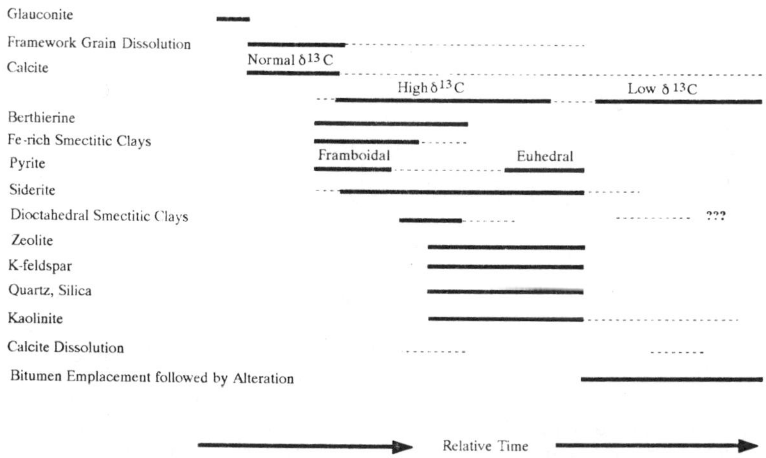

Figure 10. Generalized paragenetic sequence for the Clearwater Formation oil sands in the south-central portion of the deltaic complex (i.e., Townships 65-67, Ranges 2-5 W4). This locality is characterized by abundant berthierine and ^{13}C-rich calcite. Elsewhere in the deltaic sands, kaolinite is the main diagenetic clay mineral, and ^{13}C-rich calcite is much less common. After Longstaffe et al. (1989a,b) and Racki (1991).

Corbett, 1955). Vigrass (1968) suggested that the bitumen may be related to Lower Cretaceous conventional oils trapped elsewhere in Alberta. Chemical and microbiological studies have shown that subsurface pooled oils from western Canada can be altered by microbial action to produce high density crudes (Evans et al., 1971; Bailey et al., 1973). Several geochemical investigations have concluded that the Alberta oil sands formed by combined biodegradation and water washing of conventional crude oils that had migrated to near-surface traps (e.g., Deroo et al., 1974, 1977; Rubinstein et al., 1977; Hoffman and Strausz, 1986).

Mannville Group shales have been proposed as the main source rocks for the oil sands (e.g., Deroo et al., 1977; Jones, 1980). Moshier and Waples (1985), however, suggested that these units could produce only a tiny fraction of the hydrocarbon volume required. Hitchon and Filby (1984) used biomarker and trace element data to propose that Mississippian and Devonian source rocks were more appropriate sources. Garven (1989) suggested that the hydrocarbons originated not only from mature parts of the Mannville Group, but also from deeper organic-rich strata such as the Jurassic Nordeg shale, the Mississippian Banff shale, and the Upper Devonian Exshaw and Duvernay shales, travelling as part of the regional hydrodynamic system (Fig. 5). Creany and Allen (1990) and Allen and Creany (1991) used biomarker data to further indicate that most of the hydrocarbons in the oil sands were generated in organic-rich portions of the Nordegg and Exshaw shales, which lie below the Lower Cretaceous Joli Fou shales that cap the Mannville Group.

According to many investigators, hydrocarbon trapping in the Cold Lake and Primrose areas resulted from a combination of stratigraphic and structural features. Bitumen has been concentrated in the crest and southwest flank of a northwest-trending antiform created by dissolution of underlying Paleozoic salts (Vigrass, 1968). Salt removal caused a reversal in regional dip on the east side of the deposit (Wightman et al., 1989). Thick sands of reservoir quality continue into the structural lows east of the zone of major accumulation, but these areas are generally presumed to be water-bearing (a speculation that needs testing!). Stratigraphic confinement of the hydrocarbons has also been provided as the reservoir sands "shale out" to the north (Masters, 1984; Wightman et al., 1989).

4.1.2. *Hydrocarbon Production.* The Mannville Group deposits pose a special challenge for production because most of the hydrocarbons now consist of highly viscous, low API (8-15°) heavy oil and bitumen that are essentially immobile under normal reservoir conditions. Some reservoirs in the Athabasca area are sufficiently shallow that open-pit mining, followed by hot-water treatment of the bitumen-saturated sand, can be used to recover the hydrocarbons. Much of the resource, however, is too deeply buried for open-pit mining. In the Cold Lake-Primrose area *in situ* methods must be applied to recover the hydrocarbons. One commonly employed technology utilizes cyclic injection of steam and steam-condensate in the reservoir, heating and mobilizing the bitumen or heavy oil. However, the potential for damage to the formation (i.e., reduced ability to produce hydrocarbons) because of mineral reactions during steam stimulation is high.

Numerous published studies, not to mention countless confidential reports, have demonstrated that mineral/water interaction during steam/steam-condensate recovery of bitumen and heavy oil in the Alberta deposits can lead to substantial loss of permeability because of neoformation of smectitic clays, as well as physical migration of fines within the reservoir pore system (Capeling and Peggs, 1981; Shepherd, 1981; Sedimentology Research Group, 1981; Boon and Hitchon, 1983a,b; Hutcheon, 1984; Lefebvre and Hutcheon, 1986; Hebner et al., 1986; Kirk et al., 1987; Gunter and Bird, 1988; Gunter et al., 1988; Hutcheon et al., 1988, 1990; Abercrombie, 1989; Abercombie et al., 1989; Longstaffe et al., 1990, 1991). In addition, the stability during steaming of impermeable intervals (e.g., calcite-cemented sandstone or shale beds) that are used to channel and confine injected hot fluids to the appropriate parts of the reservoir must be considered. The challenge is to understand, predict and attenuate deleterious effects of artificial diagenesis on reservoir permeability in a situation where increased temperature replaces the roles that burial and time play in driving natural diagenesis.

Accurate assessment of the potential for formation damage in the bitumen/heavy oil reservoirs requires detailed data on their natural mineralogy, especially the nature and distribution of pore-lining and pore-filling minerals, and reactive framework grains (volcanic rock fragments, feldspars, shale clasts). Individual steam-injection pilot sites (pads) tap only a relatively small portion of any given reservoir. Technologies and production fluids designed to limit formation damage at one pad do not necessarily translate to an adjacent area because the mineralogy and diagenesis can be quite variable. In extensive deposits like the Alberta oil sands, the key to predicting reservoir mineralogy at particular localities lies with understanding the sedimentological, geochemical, hydrogeological and structural controls on diagenesis.

4.2. STRATIGRAPHY AND SEDIMENTOLOGY

In the Cold Lake and Primrose areas, the Mannville Group spans depths of ≈275 to 600 m (Wightman et al., 1989); regional dip is southwesterly at approximately 1.3 m/km. The sediments consist primarily of unconsolidated, siliciclastic materials of both continental and marine origin, and are mostly sands with interbedded siltstones and shales. They lie unconformably on the Paleozoic erosional surface and in turn are overlain conformably by marine shales of the Cretaceous Colorado Group. The Mannville Group in the Cold Lake area can be divided into three main units, the lowermost McMurray Formation, the overlying Clearwater Formation (including the basal Wabiskaw Member) and the uppermost Grand Rapids Formation, which is commonly split into Lower and Upper Members (Fig. 8).

The depositional setting in the Mannville Group is variable, including continental deposits of the McMurray Formation, the marine shelf environment of the Wabiskaw Member, marine shoreline deposits of the Clearwater Formation, and continental to marine shoreline sediments of the Grand Rapids Formations (Wightman et al., 1989). McMurray deposition was dominated by quartzose sediments derived from the Precambian Shield to the east; volcanic and feldspathic sediments originating mostly from the uplifted western Cordilleran region (Columbian orogeny) controlled deposition during Clearwater and Grand Rapids time (Minken, 1974).

4.2.1. *McMurray Formation.* The McMurray Formation is generally characterized by a mineralogically mature basal sand (quartz arenite and subarkosic sands) overlain by thinner interbedded sands and shales. Within the study area, it varies in thickness from 30 to 125 m. The sands contain mostly quartz with only minor feldspar, rock fragments and glauconite, although silt and clay content can be variable. The sedimentary environment varies from fluvial to lacustrine with deposition being controlled by the topography on the Paleozoic erosional surface (Vigrass, 1968; Wightman et al., 1989). In early McMurray time, quartzose sediments derived from the Canadian Shield were deposited as fluvial sheet sands, filling most of the topographic lows on the unconformity surface that separates the Paleozoic and Cretaceous sections. However, the thick channel deposits (up to 50 m) of the McMurray Formation that are prolific hydrocarbon reservoirs in the Athabasca area to the north are most commonly water saturated in the Cold Lake and Primrose areas to the south (Outtrim and Evans, 1977; Wightman et al., 1989).

4.2.2. *Wabiskaw Member, Primrose Area.* By late McMurray time, the boreal sea had transgressed sufficiently far south to periodically inundate the Cold Lake and Primrose areas, forming the upper zone of interbedded sands and shales dominated by the Wabiskaw Member in the more northerly portion of the study area. Following Wightman et al. (1991), the Wabiskaw Member is considered here as the lowermost portion of the Clearwater Formation (Fig. 8), and can be divided into two distinct facies.

The first facies, the reservoir sands, occurs in an east-west, elongated pod up to 40 m thick, located north of Township 70 and south of Township 75, between Ranges 4 and 10 W4 (Fig. 9). Dekker et al. (1987) interpreted these sands to be delta-front deposits, formed in a restricted, marginal marine environment. More recently, Wightman et al. (1991) has shown them to be marine estuarine sands that filled scours in the regional Wabiskaw surface. The very fine grained to fine grained reservoir sands are somewhat variable in composition, ranging from lithic arkoses to litharenites (Dekker et al., 1987; Beckie and McIntosh, 1988; Shier and Longstaffe, 1992). Elongate, concretionary sandstone intervals (up to several meters thick), cemented by calcite and/or siderite, are intercalated with the bitumen-saturated sands. These deposits are less mature mineralogically than the McMurray Formation quartz arenites and subarkosic sands that form the major reservoirs to the north, but are more mature (i.e., higher quartz and lower feldspar contents) than the Clearwater Formation reservoir sands in the Cold Lake area to the south.

The second facies, which is typical of the Wabiskaw Member regionally, contains muddy, bioturbated, bitumen-poor silts and silty sands (≈10 m thick), probably deposited in an open marine, shelf environment. Some stratigraphers consider this facies to be part of the regional lower Clearwater shale (5-18 m thick) that overlies the Wabiskaw reservoir sands (Beckie and McIntosh, 1988).

4.2.3. *Clearwater Formation (Exclusive of the Wabiskaw Member), Cold Lake Area.* Clearwater sands that overlie the basal Wabiskaw Member are part of a near-shore, marine, deltaic complex that prograded in a north-northeasterly direction into the boreal seaway as several overlapping lobes (Minken, 1974; Harrison et al., 1981; Wightman and Berezniuk, 1986) (Fig. 9). The deposits are composed of one or more very fine to fine-grained, coarsening upward sequences. The sands are mainly (feldspathic) litharenites, dominated by rock fragments (mostly volcanic, chert and shale). Thin calcite ± siderite-cemented intervals are ubiquitous. The Clearwater Formation sands in the Cold Lake area are mineralogically immature; they are poorer in quartz and richer in feldspar than the Wabiskaw Member sands in the Primrose area.

Within the area examined here, the Clearwater Formation has a relatively constant thickness of 35 to 45 m south of Township 64, but thickens to ≈65 m in Townships 64 to 66, Ranges 1 to 5W4 (Fig. 9). This latter zone contains the thickest intervals of bitumen-saturated sand (15 to 45 m). Wightman and Berezniuk (1986) suggested that channel switching caused deposition of discrete deltaic lobes. Abandoned lobes were submerged, but not to sufficient depths for substantial deposition of clay prior to growth of the next lobe, thus creating a thick sand at the depo-center. Further seaward towards the delta front, the sands thin, and shales become more prominent mineralogically. Immediately north of Township 66, the Clearwater Formation begins to thin.

Three sedimentary facies associations have been recognized in typical Clearwater reservoirs: delta-fringe silts and shales, lower delta-front sands and shales, and upper delta-front distributary channels, distributary-mouth bars and beach sands (Harrison et al., 1981; Wightman and Berezniuk, 1986; Wightman et al., 1989). The delta fringe consists mainly of silts and shales with minor, thin, discontinuous lenticular sands. Bitumen saturation is generally <6%. The lower delta-front sands are thicker, more continuous with sand > shale; average bitumen saturation is ≈8%. The upper delta front is composed mostly of relatively thick, uniform sands with bitumen saturation generally >10%; the best reservoir sands normally occur in the distributary channel, distributary-mouth bar and beach facies, which are best developed in the south-central portions of the complex.

Sedimentary facies, grain size, fines content and diagenetic clay minerals are the main controls on bitumen saturation and reservoir quality in these sands (Harrison et al., 1981). With decreasing grain size and increasing fines (silt plus clay) content, bitumen saturation generally decreases. However, for a given grain size or fines content, samples with abundant, diagenetic, grain-coating clays (most berthicrine and related trioctahedral, Fe-rich clays, Dean and Nahnybida, 1985; Longstaffe et al., 1992b) commonly have lower bitumen saturations than samples in which this clay is absent or poorly developed. Clay grain coats are abundant in some parts of the distributary channel, distributary-mouth bar and beach sand facies, which otherwise are the best reservoirs because of their coarser grain sizes and lower fines content relative to other facies. Accordingly, it has become quite important to understand how and why this somewhat peculiar diagenetic clay assemblage has developed.

Wightman and Berezniuk (1986) suggested that the stratigraphic position of a sand within the Clearwater deltaic complex has implications for *in situ* steam injection. The best reservoirs should occur in the upper portions of coarsening upwards sequences or associated channel facies, where lateral sand continuity would be best developed. Shale beds, which can lower vertical permeability to steam and isolate specific intervals, should be most continuous lower in the deltaic sequence. The carbonate-cemented sandstones that are ubiquitous throughout the Clearwater Formation (and the Wabiskaw Member) also play a role in controlling vertical permeability. However, predicting the distribution and lateral extent of these intervals, and whether they are facies dependent, requires understanding of the polygenic, diagenetic processes responsible for their formation.

Wightman and Berezniuk (1986) also noted that structural lows in the Clearwater sands (created by the solution of Paleozoic evaporites and by regional dip) contain a significant thickness of water-saturated sand (>5 m) below the bitumen-saturated sand. Conversely, in structural highs, bottom waters associated with the bitumen-saturated sands are very thin or absent. Water sands are preferential conduits of steam during *in situ* processes. Thick zones of such sands (the so-called 'thief zones') can divert much of the steam and steam-condensate from the intended target, and are to be avoided.

4.2.4. *Grand Rapids Formation.* Reservoir sands in the Grand Rapids Formation, which comprises the uppermost portion of the Mannville Group, were deposited in marine shoreline, deltaic and continental settings, variously described to include channel deposits, beaches and point bars (Kendall, 1977; Wightman et al., 1989). The lower Grand Rapids Member contains the largest amount of bitumen, mostly concentrated in areally restricted but laterally continuous shoreline sands (average pay zone $\approx$12 m; Wightman et al., 1989). The oil sands in the Upper Grand Rapids Member are more discontinuous, and thinner (5 to 10 m), because of the more dominant continental influence in this unit (Wightman et al., 1989). Marine shales are also more abundant in the Upper Grand Rapids Member. Finally, the Cold Lake area was inundated by a major transgression of the boreal sea, and the Cretaceous Joli Fou shales (Colorado Group) were deposited.

4.3. DIAGENESIS IN THE CLEARWATER FORMATION SANDS, SOUTH-CENTRAL COLD LAKE

Stable isotope data for diagenetic minerals from the Clearwater Formation in the Cold Lake area have provided some insight into the mineral-water-organic interactions that controlled the nature and distribution of these phases (Longstaffe et al., 1992b). The development of ^{13}C-rich calcite cements, first recognized by Dimitrakopoulos amd Muehlenbachs (1987), and their association with diagenetic berthierine (Fe-rich 1:1 layer clay mineral similar in chemical composition to chamosite), are of particular interest.

4.3.1. *Paragenesis.* The general paragenetic sequence for bitumen-saturated sands from the south-central portion of the Cold Lake area is illustrated in Figure 10 (Longstaffe et al., 1989a,b; Racki, 1991). Broadly similar results have been reported for localized regions of the Clearwater Formation (Abercrombie 1989; Hutcheon et al., 1989). Early diagenetic minerals in the bitumen-saturated sands include glauconite, calcite, siderite, pyrite, berthierine and Fe-rich smectitic clays. Alteration and dissolution of volcanic rock fragments and feldspars has also been substantial. Concretionary carbonate cementation (mostly calcite) also began very early in diagenesis; the earliest carbonates most commonly have δ^{13}C values of about 0‰. This calcite has completely filled pores and partially replaced framework grains; grain-coating clays have not been observed in these concretions. Formation of framboidal pyrite generally postdated early concretion development, and is relatively uncommon, except immediately below a major sequence boundary within the sands (Wickert et al., 1989).

Diagenetic berthierine and Fe-rich smectitic clays are confined almost entirely to the thick reservoir sands in the south-central portions of the Cold Lake area, between Townships 65 and 68 (section 4.2.3). Berthierine, in particular, is most abundant in sands from distributary-mouth bar and related facies. Within a given interval, it is not uncommon for the relative abundance of berthierine to decrease (and that of glauconite/illite or smectitic clay to increase) as concretionary intervals are approached. The appearance of high δ^{13}C, pore-lining or pore-filling calcite is preferentially concentrated in the berthierine-bearing intervals, although both phases rarely coexist. Where relationships can be discerned, it seems that this calcite crystallized after berthierine. The iron-poor nature of this calcite (relative to earlier and later varieties) may imply that Fe was preferentially sequestered by early Fe-clay formation (rather than pyrite).

Elsewhere in the bitumen-saturated sands, the main diagenetic clay mineral is pore-filling kaolinite. However, it is rare in the Fe-smectite and berthierine-rich portions of the deltaic complex, occurring only within calcite and/or siderite-rich zones of low bitumen-saturation, or in sands close to the bitumen-water contact.

Crystallization of K-feldspar and silica overgrowths, zeolites, siderite and euhedral pyrite occurred as burial proceeded, together with continued dissolution of feldspars and volcanic rock fragments. Kaolinite probably also continued to form in portions of the deltaic complex not dominated by the Fe-clays. The latest stages of diagenesis are most commonly marked by Fe-rich calcite, and in the fringes of the deltaic complex, perhaps also kaolinite. Emplacement of hydrocarbons (subsequently altered to bitumen or heavy oil) greatly attenuated further diagenetic

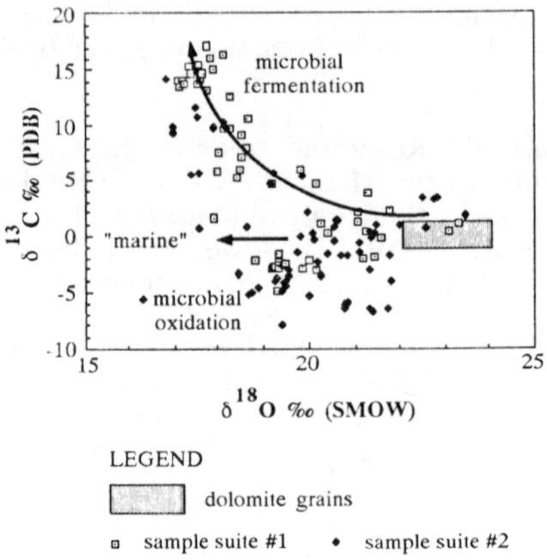

Figure 11. Carbon and oxygen isotope values for diagenetic calcite from Clearwater Formation sands and sandstones in the the Cold Lake area. The range for detrital dolomite grains that are abundant in some sand beds is also shown. Samples in Suite #1 were obtained from six *in situ* steam-injection pilots located in the south-central portion of the deltaic complex (Townships 65-67, Ranges 2-5 W4; Fig. 9). Suite #2 was collected throughout the deltaic complex (Townships 65-70, Ranges 1-6 W4), including the area sampled by suite #1. Diagenetic calcite with high $\delta^{13}C$ values is confined almost entirely to the thick reservoir sands in Townships 65-66 and Ranges 2-5, the same region in which berthierine is most abundant. After Longstaffe et al. (1992b).

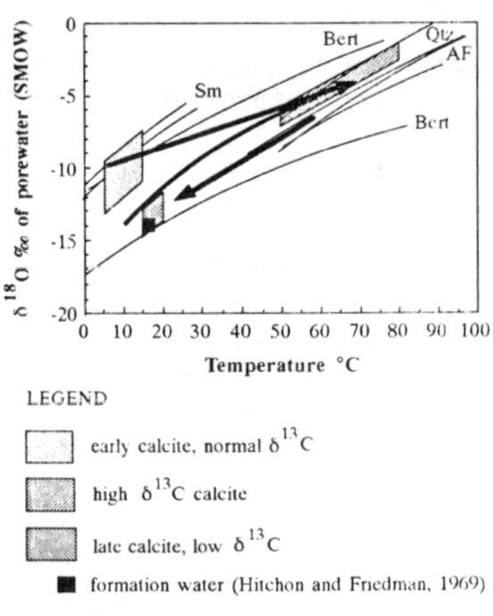

Figure 12. Possible porewater evolution path for berthierine-bearing sands from the Clearwater Formation in the south-central Cold Lake area. Conditions proposed for crystallization of calcite are indicated by the shaded areas. Sm = early smectitic clays; Bert = berthierine; Qtz = quartz; AF = K-feldspar. The following mineral-water fractionation curves were utilized: quartz-water, Clayton et al. (1972), as modified by Friedman and O'Neil (1969); calcite-water, O'Neil et al. (1969), as modified by Friedman and O'Neil (1977); alkali feldspar-water, O'Neil and Taylor (1967); smectite-water, Savin and Lee (1988) with appropriate modification for interstratification, and berthierine-water, Longstaffe et al. (1992b). After Longstaffe et al. (1992b).

mineralization.

Sands from the Clearwater Formation which remained water-saturated show the same general diagenetic pattern with a few important exceptions. First, grain-coating clays like berthierine are rare, and second, early diagenetic, grain-coating calcite with high $\delta^{13}C$ values is almost never encountered. Calcite crystallization apparently continued in the absence of bitumen.

4.3.2. *Isotopic Variations in Calcite.* Diagenetic calcite exhibits a large range in $\delta^{13}C$ values (-7.9 to +17.1‰), accompanied by a smaller variation in $\delta^{18}O$ values (+16.3 to +23.1‰) (Fig. 11). Samples with $\delta^{13}C$ values of 0±3‰, characteristic of "marine" carbon dominated by inorganic input, are most common in the earliest calcite concretions. The $\delta^{18}O$ values (+20 to +23‰) of this calcite, however, are significantly lower than expected (≈+30‰) for low temperature crystallization from open marine, Cretaceous seawater. Porewaters during very early diagenesis must have contained a sizeable fraction of meteoric water (brackish seawater?).

In the thicker sands that contain Fe-smectitic clays and berthierine, the $\delta^{13}C$ values of some bitumen-free concretions change from near 0‰ to +10‰ with continued growth of the concretion. In bitumen-stained sandstones and bitumen-saturated sands, this calcite has even higher $\delta^{13}C$ values. This calcite is also characterized by lower $\delta^{18}O$ values than the earlier, lower ^{13}C varieties (Fig. 11).

The lowering of calcite $\delta^{18}O$ with increasing $\delta^{13}C$ values probably resulted from increasing temperature as burial (and microbial procresses) proceeded. The specific reason for the ^{13}C enrichment is more equivocal. Strongly positive $\delta^{13}C$ values in diagenetic carbonates can be one by-product of microbial methanogenesis of organic matter (Irwin et al., 1977). However, Clayton (this volume) points out the general constraints on formation of ^{13}C-rich diagenetic carbonates. He notes that high $\delta^{13}C$ values for carbon dioxide associated with methane production are atypical, and that much lower values are the normal product of fermentation reactions. Only under conditions of extensive CO_2 reduction, during which much the ^{12}C-rich carbon dioxide is preferentially consumed, will ^{13}C-rich residual carbon dioxide become available for carbonate precipitation.

Dimitrakopoulos and Muehlenbachs (1987) suggested that ^{13}C-rich carbonate cements in the oil sands were derived from hydrocarbons through a two stage process that first required degradation of hydrocarbons by aerobic microbes, followed by microbial fermentation of the resulting biomass and its by-products. A difficulty with this model, however, is that the ^{13}C-rich calcite commonly predated local hydrocarbon emplacement. We can speculate that hydrocarbon emplacement in the Clearwater Formation occurred over a long period of time, and that biodegradation of hydrocarbons trapped elsewhere promoted crystallization of isotopically appropriate calcite cements in parts of the system that had not yet been bitumen saturated. However, it seems more probable that crystallization of the early diagenetic, ^{13}C-rich calcite is not specifically related to degradation of the hydrocarbons, and instead involves organic material originally trapped in the sediments during deposition.

A third, more poorly sampled generation of calcite cement (Fig. 10) demonstrates some tendency towards low $\delta^{13}C$ (to -8‰) and (in some cases) $\delta^{18}O$ values (Fig. 11). Its isotopic compositions suggest crystallization from formation waters that contained a sizeable fraction of meteoric water, and carried a mixture of carbonate ions, including some derived from low ^{13}C sources (see Clayton, this volume). The precise nature of the low ^{13}C source(s) cannot be deduced from the available data; contribution from oxidation of hydrocarbons during microbial action and water-washing cannot be ruled out.

4.3.3. *Porewater Evolution and the Origin of Berthierine.* Oxygen isotope results for diagenetic smectitic clays (+18.0 to +18.8‰), berthierine (+6.6 to +12.5‰), K-feldspar (+15.9 to +17.0‰) and quartz (+21.0 to +23.2‰), plus the results for calcite, have been interpreted in Figure 12 to produce an idealized curve for porewater evolution (heavy arrows). The porewater pathway was constrained by a maximum temperature of 70°C, the composition and temperature of local formation water, and the need to intersect the mineral-water curves (thin lines) in an order compatible with the paragenetic sequence.

The main observation arising from Figure 12 is that meteoric water was important throughout diagenesis. Crystallization of both calcite ($\delta^{13}C \approx 0$‰) and smectitic clays very early in diagenesis (10±5°C) required porewaters of about -10‰ (mixed meteoric-marine waters?). Even lower ^{18}O

porewaters may have been involved in crystallization of some berthierine. Continued berthierine crystallization during burial is consistent with the porewater pathway, as are the oxygen isotope compositions of associated high-^{13}C calcite, shown in Figure 12 for putative crystallization at the highest temperatures expected for these sediments. While some shift towards more ^{18}O-rich porewaters may be indicated by the compositions of quartz and alkali feldspar presumed to have crystallized near maximum burial, the porewater isotopic signature remained strongly meteoric.

Current formation waters show that a considerable lowering in $\delta^{18}O$ values and temperature occurred sometime following maximum burial. This change is most simply explained by increased infiltration of meteoric water as erosion of overlying units caused sands of the Clearwater Formation to re-enter the local hydrologic regime. The oxygen isotope results for the last generation of diagenetic calcite (from sands that were not saturated by bitumen) are consistent with crystallization from local meteoric water at current subsurface temperatures.

Depositional environment and the associated distribution of meteoric water has strongly influenced diagenesis in the Clearwater Formation sands. For example, the earliest (normal-^{13}C) calcite concretions, which have the highest $\delta^{18}O$ values (but still low relative to normal marine carbonate), are most abundant in distributary-mouth bar and related facies; such results are most simply explained by mixing between seawater and freshwater. The presence of syndepositional glauconite grains in these sands provides some further evidence for the transitional nature of this environment. The diagenetic evolution of berthierine and the ^{13}C-rich calcite can also be related to depositional environment, first through the delivery of Fe to the distributary channel and related facies (see below), then by the preservation of organic matter via rapid burial resulting from lobe switching (section 4.2.3), and finally by the sizeable fraction of meteoric input in the connate waters trapped by these sediments. The oxygen isotope results (Fig. 12) can be interpreted to indicate that berthierine formed preferentially to smectitic clays where the relative fraction of meteoric water in the connate porefluids was highest. This observation fits the pattern of increasing berthierine content away from normal-^{13}C calcite concretions. Alternatively, the lower porewater $\delta^{18}O$ values associated with berthierine may have resulted from abundant berthierine crystallization itself, given that the sediment-water system became closed and strongly reducing. We can speculate that dominance of meteoric water in the connate porefluids meant a restricted supply of sulphate, and hence a very limited zone of sulphate reduction. Fe was not consumed by sulphide precipitation; instead, it remained available for formation of berthierine. In addition, rather direct transition into the methanogenic zone of diagenesis during burial may have facilitated CO_2 reduction, an apparent requirement for precipitation of ^{13}C-rich diagenetic calcite (Clayton, this volume) .

Reports of berthierine are relatively uncommon in the sandstone diagenetic record for several reasons. First, because of its 7Å character, it is commonly mistaken for kaolinite in XRD. Second, special circumstances are needed for primary crystallization of grain-coating berthierine in sands: (i) an Fe supply, (ii) concentration of the Fe about sand grains, and (iii) a reducing environment, or at least one in which the oxygen supply has been greatly restricted (Curtis and Spears, 1968). Third, Curtis et al. (1985), among others, noted that Fe-rich chlorite replaces berthierine as temperatures rise. Jahren and Aagaard (1989), for example, described pore-filling berthierine from sandstones that were undergoing transformation to chamosite at temperatures as low as 70°C.

It appears that conditions were just right to both form and preserve berthierine, at least in some portions of the Clearwater Formation reservoirs. Dissolution of volcanic rock fragments provided abundant Fe for transport by river waters into distributary channels, perhaps ultimately to be accreted (as Fe-oxides and hydroxides) about sand grains by current movement in the distributary channel environment. Iron was remobilized when Fe-oxides and hydroxides became unstable as reducing conditions were established during burial. Because of the deficiency in sulphate (meteoric-dominated porewaters), rather direct transition into the methanogenic zone occurred, and microbial activity was probably instrumental in both iron and CO_2 reduction. Given the paucity of sulphate, Fe-fixation by diagenetic sulphides was limited, and Fe remained available for clays. It is even conceivable that Fe- and CO_2-reducing microbes preferentially concentrated ions required to form these clays on their cell membranes (Ferris et al., 1987). Finally, maximum burial temperatures in the study area probably never exceeded 70°C. As a result, temperatures required for transformation of berthierine to Fe-chlorite were never achieved.

4.4. ARTIFICIAL DIAGENESIS IN THE WABISKAW MEMBER SANDS, PRIMROSE AREA

Potential difficulties that can arise from steam injection of oil sand reservoirs have been described earlier. In this final section, the complicated nature of oil sand/steam interactions is illustrated using post-steam core from bitumen-saturated sands of the Wabiskaw Member (section 4.2.2) at Ipiatik Lake (Twp. 73 R6 W4), about 100 km north of Cold Lake, Alberta (Fig. 9). Because the post-steam core was obtained by sidetracking from an existing expendable well-bore (Fialka et al., 1993), mineralogical changes very near to the original site of steam injection and hydrocarbon recovery (0.6 m at the top of the reservoir to 1.8 m at the bottom) have been examined. Lowering of permeability in the "near well-bore" region can cause severe deterioration in the production of hydrocarbons. Direct observations on the stability of carbonate-cemented sandstones in the near well-bore region during steaming were also possible. The latter information is of particular value because the scheme to produce hydrocarbons at this site relied upon these impermeable units to channel and confine injected steam to specific reservoir intervals (section 4.1.2).

The experimental, cyclic steam injection pilot was operated from 1984 to 1990 by AEC Oil and Gas Company, at which time it was suspended because of deteriorating hydrocarbon productivity. The reservoir sands within the pilot area have average natural porosity of 30% and permeability ranging from 500 to 2000 md; the sands contain bitumen with an average API gravity of 11.0° (Fialka et al., 1993). AEC's Ipiatik Heavy Oil Pilot (IHOP) originally consisted of seven cyclic steam wells on a 1.46 ha spacing plus three observation wells, but a second phase of seven cyclic steam wells on a 2.42 ha spacing was developed in 1987 (Fig. 13). The post-steam core (#12; Fig. 13) was obtained during this second phase in 1989 as part of a programme to understand the causes of declining hydrocarbon productivity. The sands in the post-steam core had experienced four cycles of steaming over eighteen months. Cycles one and two were each of three months duration; cycles three and four each lasted six months. During each cycle maximum temperatures achieved near the base of the reservoir were ≈275-300°C declining to ≈100-125°C during each cycle. A thermal gradient probably existed between the lower and upper portions of the reservoir; maximum temperatures achieved near the top of the reservoir were ≤200°C declining to as low as 60-75°C at the end of a steam cycle.

4.4.1. *Pre-steam Mineralogy.* The depositional setting and general mineralogical nature of Wabiskaw Member lithic arkoses and litharenites were described in section 4.2.2. Within the IHOP area, framework grains are typically composed of quartz (35-65%, av. 50%), chert (0-15%, av. 7%), feldspar (plagioclase > K-feldspar; 10-22%, av. 16%), and sedimentary and volcanic rock fragments (0-21%, av. 10%) plus minor amounts of siderite, glauconite, micaceous minerals and carbonaceous material (1-5%, av. 1%). Clay matrix can be abundant, especially in some shalier sands (0-22%, av. 4%).

As for the overlying Clearwater Formation reservoirs, this reactive framework mineralogy is reflected by significant natural diagenetic alteration despite relatively low maximum burial temperatures. Three stages of diagenesis are apparent, all of which preceded bitumen saturation. The earliest stage is characterized by glauconite, mixed-layer clays (mostly C/S), fluorite, pore-filling calcite and siderite, and traces of siderite. Formation of diagenetic K-feldspar overgrowths and grain-coating calcite dominated the second stage, followed by kaolinite crystallization and partial dissolution of calcite during stage three. Framework grain dissolution and alteration occurred throughout diagenesis. Emplacement of hydrocarbons attenuated further diagenesis. Sands with the lowest content of matrix (detrital) clays and/or pore-filling diagenetic calcite generally have the highest bitumen contents.

Kaolinite is generally the most abundant diagenetic phase (0-10%, av. 6% at IHOP), occurring as a pore filling and as an replacement of partially leached feldspar grains. At IHOP, siderite (± calcite) cement is also common in the bitumen saturated sands (0-15%, av. 3%). Carbonate is most abundant, however, as discontinuous concretionary intervals completely cemented by calcite ± siderite (up to 40% of the sandstone at IHOP). The earliest generations of calcite and siderite have $\delta^{13}C$ (-4 to +4‰) and $\delta^{18}O$ values (+19 to +22‰) that indicate a dominantly inorganic source for the carbonate ions, and crystallization at low temperatures from porewaters containing a sizeable fraction of meteoric water (Shier and Longstaffe, 1992). Later calcite has much higher $\delta^{13}C$ (+10 to +13‰) and lower $\delta^{18}O$ values (+16 to +17‰), reflecting higher temperature, microbial processes

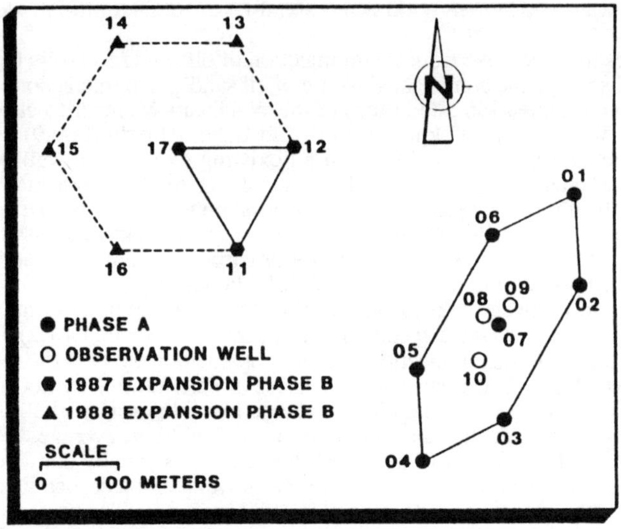

Figure 13. Well pattern for the Ipiatik Heavy Oil Pilot (IHOP) illustrating the locations of pre-steam core #16 and post-steam core #12. After Fialka et al. (1993).

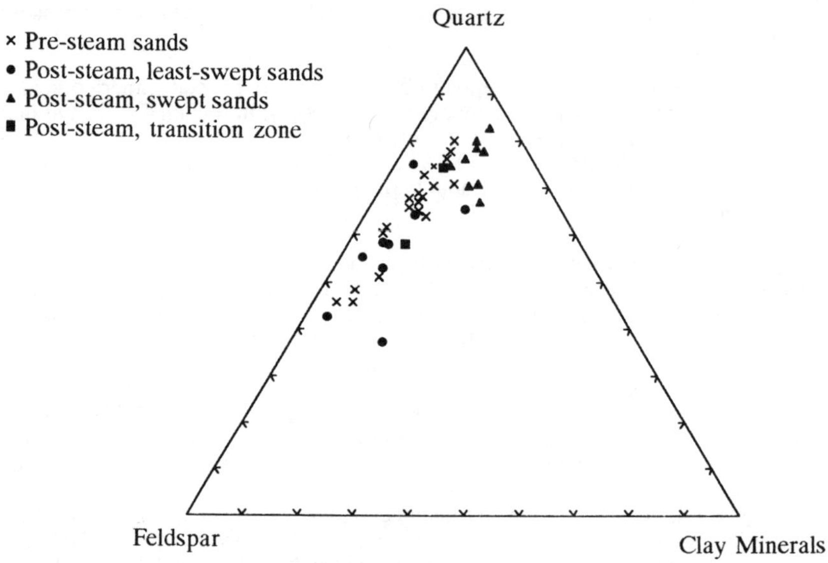

Figure 14. Comparison of quartz, feldspar and clay mineral contents (bulk XRD) in pre-steam versus post-steam reservoir sands (least-swept, swept and transition zones) from the Wabiskaw Member near Ipiatik Lake, Alberta (IHOP).

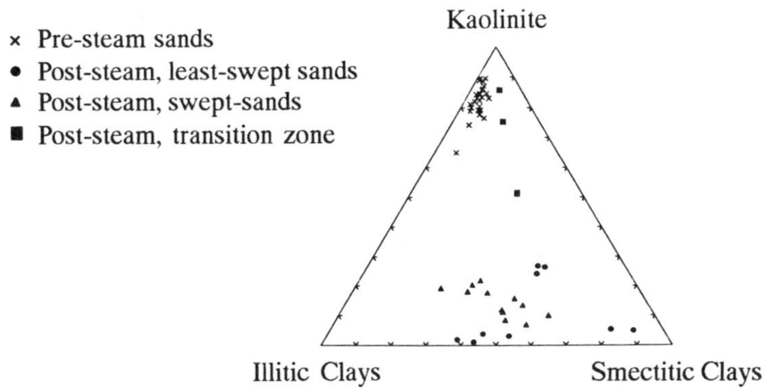

Figure 15. Comparison of kaolinite, illitic and smectitic clay contents (XRD, <2 μm size fraction) in pre-steam versus post-steam (least-swept, swept and transition zones) sands and sandstones from the Wabiskaw Member near Ipiatik Lake, Alberta (IHOP).

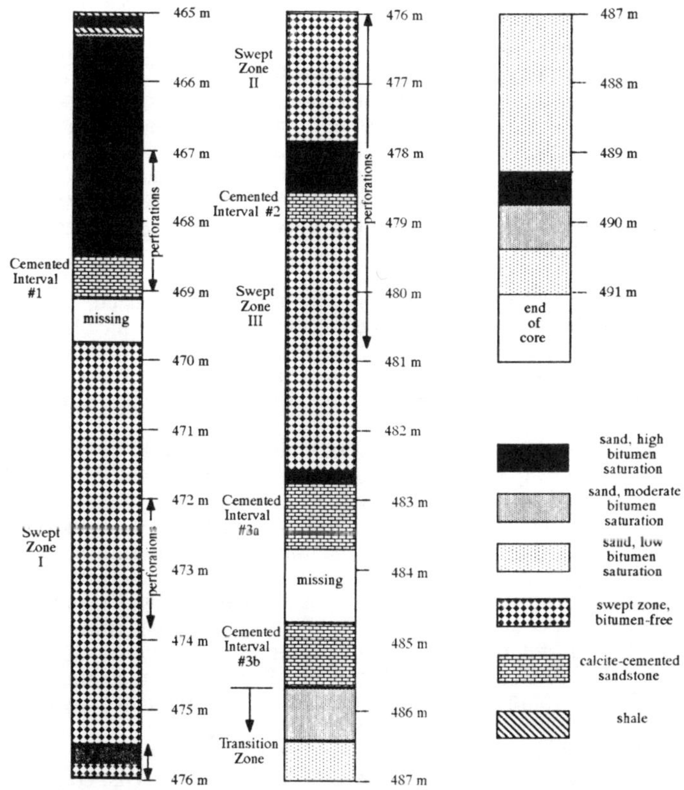

Figure 16. Schematic description of the post-steam core (#12) from the Wabiskaw Member near Ipiatik Lake, Alberta (IHOP). Portions of the well casing that were perforated to allow access of steam to the reservoir are indicated by the double-headed arrows.

within the methanogenic zone.

Ideally, detailed comparison of pre-and post-steam mineralogy requires "before" and "after" samples from virtually the same location within the pad. Regrettably, the sidetracking technique (Fialka et al., 1993) used to obtain the post-steam core required utilization of an expendable well-bore, and the site deemed expendable had not been cored prior to steaming. Fortunately, a suitable pre-steam core (#16) was available close to the site of the post-steam core (#12) (Fig. 13).

Pre-steam core #16 is typical of the IHOP site. The upper half consists of poorly consolidated, bitumen-rich sands that contain a few siderite-rich clasts, streaks and laminations, and occasional thin units of interlaminated silts, fine-grained siderite and organic matter. The upper half of the core is subdivided by the first major cemented sandstone, composed of calcite and siderite-cemented sandstone. A second major cemented unit, consisting of interbedded bitumen-bearing, calcite-cemented sandstone, bitumen-stained, siderite-cemented sandstone and bitumen-free, calcite-cemented sandstone, separates the upper and lower halves of the core. The lower half of the core consists of poorly consolidated, bitumen-bearing sands interbedded with cemented zones of varying induration. Poorly consolidated sands of lower bitumen saturation lie below the lowermost cemented interval, and represent the transition zone between the reservoir sands and underlying, water-saturated sands. The carbonate-cemented intervals are believed to be laterally continuous for several meters, but are not directly correlatable to other wells in the area, including the post-steam core.

The bulk mineralogy of bitumen-saturated sands and associated cemented sandstones from the pre-steam core was determined using x-ray powder diffraction methods (XRD). While abundances calculated from XRD are semiquantitative at best, such data are useful as a baseline for comparison with results obtained for the post-steam core using the same method. The XRD data are consistent with petrographic observations for this pad; the reservoir sands are dominated by quartz/chert and feldspars (plagioclase>K-feldspar), but calcite and siderite can be locally important (Fig. 14). Kaolinite and illite/glauconite are the most abundant clay minerals detected in the bulk samples.

Because reactions involving clay minerals can be so important during steam stimulation, the clay size-fraction (<2 μm) of the pre-steam core was also studied using XRD. As the petrographic and bulk rock XRD data suggested, the clay fraction is dominated by kaolinite (63-86%), with lesser amounts of illitic clay (8-28%), and minor amounts of chlorite (1-6%) and smectitic clay (1-6%) (Fig. 15). The relative abundances of these clay minerals do not vary significantly with depth from the top to the bottom of the reservoir interval.

4.4.2. *Post-steam Core.* An important reason for obtaining the post-steam core was to determine the physical modifications that had occurred immediately adjacent to the site of steam injection. The core can be subdivided into three main parts, an upper portion which remained heavily bitumen saturated, a middle portion from which bitumen was extracted, and a lower portion, below the base of the reservoir, which was not subjected to steam stimulation. Calcite-cemented sandstones separate each part (Fig. 16).

The upper portion of the core (≈3.5 m) was not extensively swept by steam ("least-swept"). This result was initially surprising given the proximity to the original well-bore (<1 m), two meters of which had been perforated to permit introduction of steam. Nevertheless, some bitumen was extracted; the pre-steam bitumen saturation of 65-70% declined to 48% (Fialka et al., 1993). The lack of significant hydrocarbon production from the upper portion of the reservoir was one of the important causes of poor hydrocarbon recovery from this pad. Ultimately, it was determined that hydraulic fracturing of the reservoir (conducted prior to steaming) was contained within the oil sands located between the upper and lower calcite-cemented sandstones (Fialka et al., 1993). As a result, the initial ability of steam to effectively penetrate the reservoir above the uppermost calcite-cemented zone (Fig. 16) was greatly restricted. The mineral reactions that ensued in this interval from albeit limited contact with steam (see below) undoubtedly caused the situation to deteriorate even further.

The middle portion of the core (≈14 m) represents reservoir sands that were swept of bitumen ("swept"). The core consists of a churned melange of sand, shaley clasts, clays and metal-casing fragments, consolidated in the core-sleeve by a matrix of fine sand and silt. Apparently, in steam-swept portions of the core, insufficient consolidation or lithification existed to withstand torque during drilling. Extensive calcite cementation, like that observed in other post-steam cores of

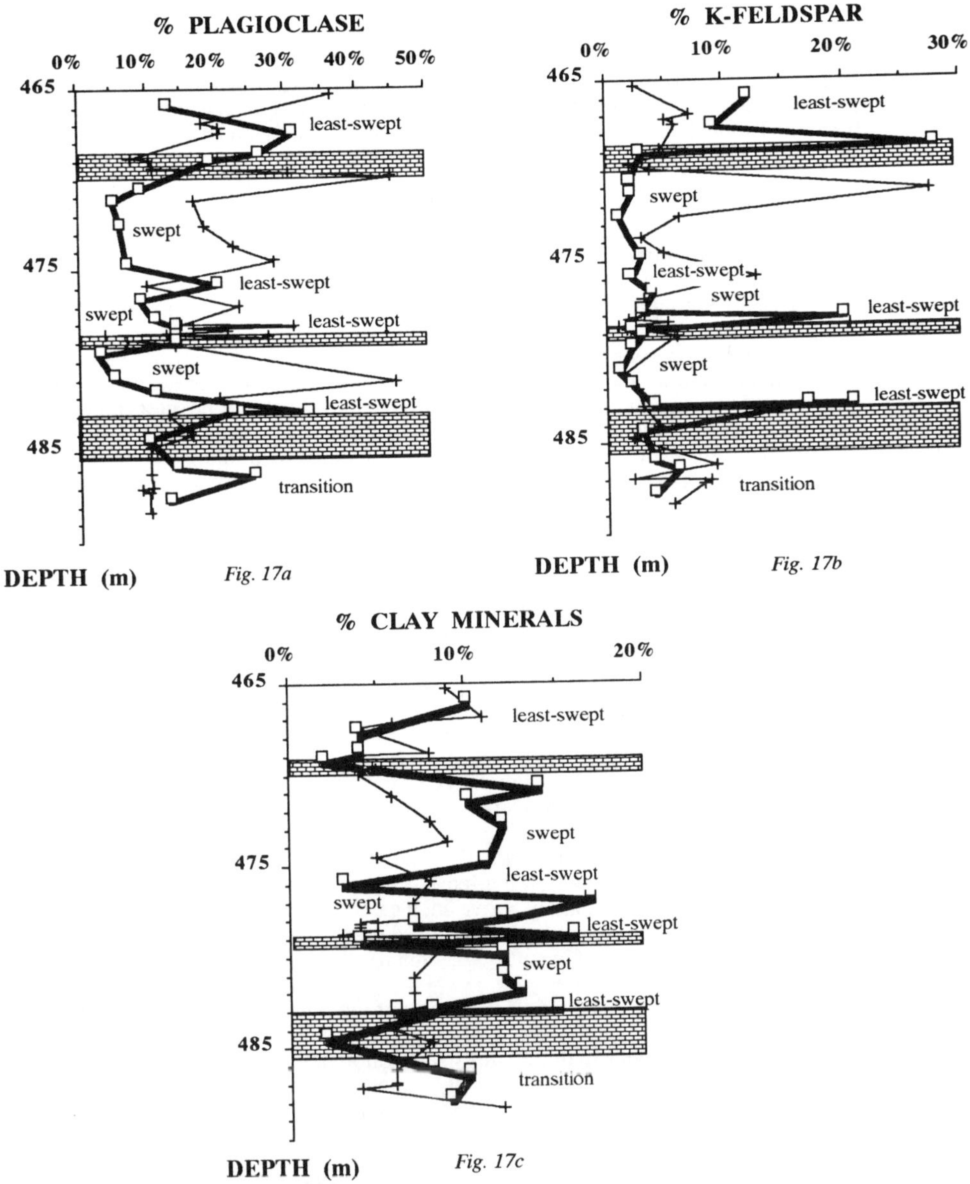

Figure 17. Comparison of bulk mineral content versus depth for pre- and post-steam cores from the Wabiskaw Member at IHOP. Mineral abundances were estimated using XRD. The crosses and light line represent variations in the pre-steam core (#16). The open squares and heavy line represent variations in the post-steam core (#12). Calcite-cemented intervals are indicated by the patterned area; least-swept, swept and transition zones are also shown. *Figure 17a:* plagioclase versus depth (m). *Figure 17b:* K-feldspar versus depth (m). *Figure 17c:* Total clay content versus depth (m).

Clearwater reservoirs (albeit not from the very near-well bore region; e.g., Hutcheon, 1984), had not occurred as a result of steaming.

The steam-swept portion of the core can be subdivided into three zones (Fig. 16). Zone I is immediately overlain by a calcite-cemented sandstone, which separates it from the upper, bitumen-saturated portion of the post-steam core. Zone II is separated from Zone I by a thin interval of sands which retained some bitumen saturation. Zone II is immediately underlain by another thin interval of bitumen-saturated sand which itself overlies a second calcite-cemented sandstone. Below this calcite-cemented interval lies Zone III, the lowermost steam-swept sands. Zone III is immediately underlain by bitumen-saturated sand (followed by a thick calcite-cemented interval). There is some pattern to this distribution of lithologies: calcite-cemented sandstones are immediately overlain by a thin interval of bitumen-saturated sand which, in turn, is overlain by a bitumen-swept zone. The bitumen-saturated sands that occur at the base of each steam-swept zone have bitumen saturations averaging 58%, not much lower than pre-steam values of 65-70% for these portions of the reservoir (Fialka et al., 1993).

The lower portion of the post-steam core (≈5.3 m), the transition zone, contains sands of moderate to poor bitumen saturation and higher water saturation, as is the case for the pre-steam core. Because these more permeable sands would tend to act as a "thief" zone, carrying injected steam away from the reservoir, the original well-bore was not perforated in this interval. The transition zone is further isolated from the overlying reservoir sands by a thick, calcite-cemented sandstone (Interval #3). The plan to limit contact with steam in this portion of the core succeeded; pre-steam and post-steam bitumen saturations remained the same at 45% (Fialka et al., 1993).

4.4.3. *Post-steam Bulk Mineralogy.* The bulk mineralogy of least-swept sands is very similar to pre-steam sands in relative abundances of quartz, feldspar and clay minerals (Fig. 14). The swept sands, however, are enriched in clays and depleted of feldspar relative to pre-steam material. Siderite, which is ubiquitous in the pre-steam sands, was not detected in either swept or least-swept, post-steam sands.

A depth-dependent comparison of pre- and post-steam bulk mineralogy for plagioclase, K-feldspar and total clay mineral (Figs.17a,b,c) confirms the general patterns in bulk mineralogy shown by Figure 14. Because pre- and post-steam cores are not from exactly the same location, direct comparisons for a given depth are not possible. Nevertheless, systematic depletion of plagioclase and K-feldspar, and enrichment of clay in the swept zones remains apparent, whereas in the least-swept and transitional zones, differences between pre- and post-steam samples are more or less random. This pattern holds for the thick, least-swept interval in the upper portion of the reservoir, as well as for the thinner intervals that most commonly appear just above a calcite-cemented sandstone. However, within the least-swept intervals, a significant increase in K-feldspar content seems to occur immediately above each calcite-cemented zone (Fig. 17b). This result suggests that K-feldspar was very stable at these locations; morphologies typical of neoformed K-feldspar have been observed in some SEM micrographs.

4.4.4. *Post-steam Clay Mineralogy.* The clay mineralogy of the post-steam core is very different from the pre-steam sands (Fig. 15). Kaolinite has been largely supplanted by smectitic and illitic (10Å) clay minerals. Kaolinite contents in the <2 µm size-fraction have declined from about 80% in the pre-steam sand to about 10-15% in both the swept and least-swept portions of the post-steam core (Fig. 18a). Only below the lowermost calcite-cemented interval, about 0.5 m into the transition zone do samples return to pre-steam clay contents and mineralogy (Figs. 18a,b,c). Smectitic clay is abundant throughout the swept and least-swept intervals; samples located immediately above calcite-cemented sandstones are especially rich in smectite (Fig. 18b). The increase in illitic clay content is most pronounced within the swept zones; the least-swept zones generally show less enrichment in illitic clay, except at the base of the lowermost swept zone (Fig. 18c).

A complex suite of minerals make up the smectitic clays in the the post-steam sands (Longstaffe et al., 1991). Least-swept portions commonly contain an interstratified swelling clay with a superstructure diffraction that varies from 27-28Å (Ca-saturated, 54% relative humidity) to 28-30Å (Ca-saturated, ethylene-glycol solvated) (Fig. 18c). This clay shows variable behaviour upon drying, rehydration and heating in the K-saturated form; the superstructure diffraction decreased to 25-26Å in one case, and disappeared entirely in others. Most abundant in the least-swept zones,

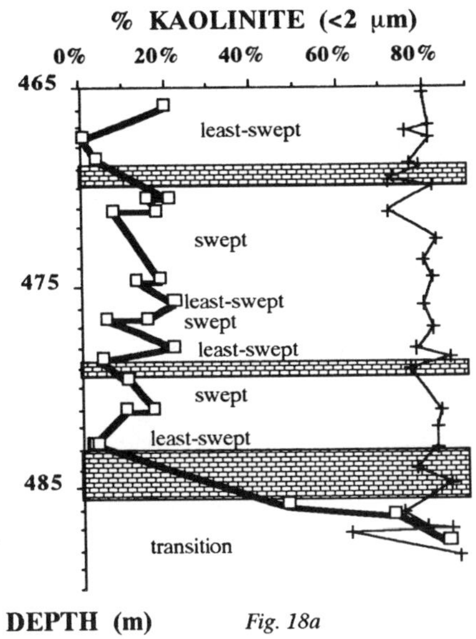

Fig. 18a

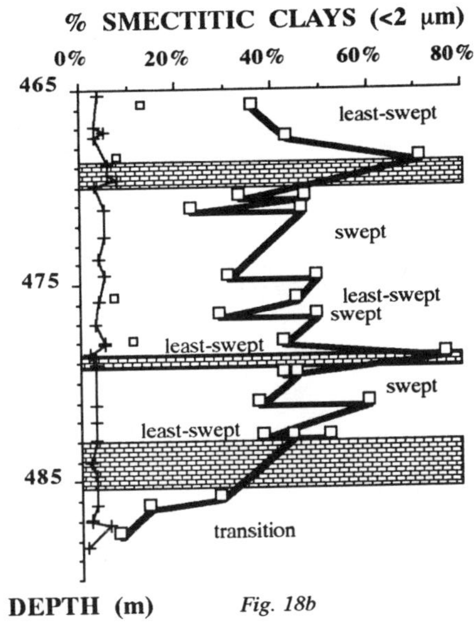

Fig. 18b

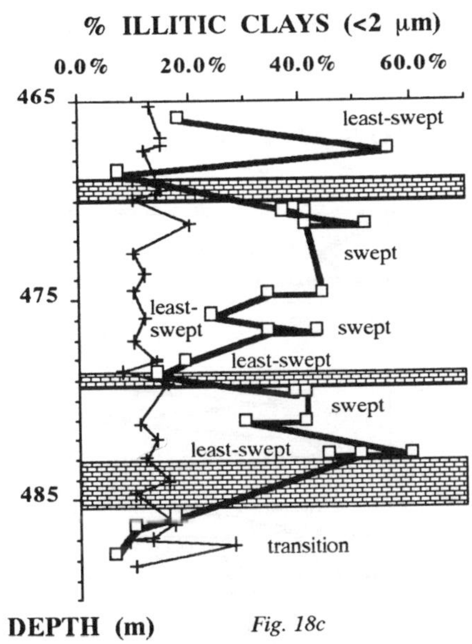

Fig. 18c

Figure 18. Comparison of clay mineralogy (<2 μm) versus depth for pre- and post-steam cores from the Wabiskaw Member at IHOP. Mineral abundances were estimated using XRD. The crosses and light line represent variations in the pre-steam core (#16). The big open squares and heavy line represent variations in the post-steam core (#12). Calcite-cemented intervals are indicated by the patterned area; least-swept, swept and transition zones are also shown. *Figure 18a:* kaolinite versus depth (m). *Figure 18b:* smectitic clay versus depth (m); small squares indicate a post-steam smectitic clay with a superstructure diffraction. *Figure 18c:* illitic clay versus depth (m).

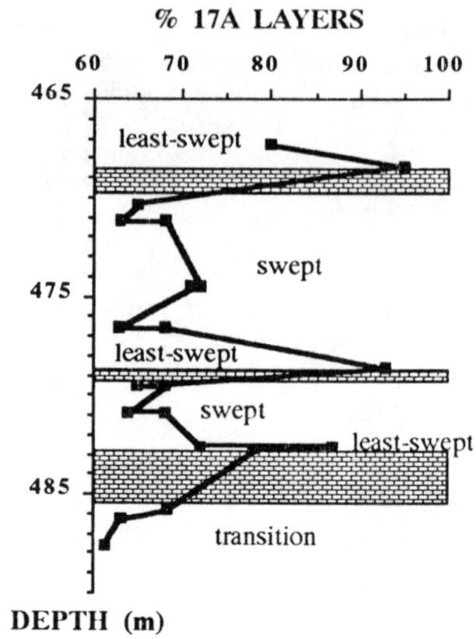

Figure 19. Comparison of percent 17Å layers in randomly interstratified smectitic clay (<2 μm) versus depth (m) for post-steam core from the Wabiskaw Member at IHOP. Calcite-cemented intervals are indicated by the patterned area; least-swept, swept and transition zones are also shown.

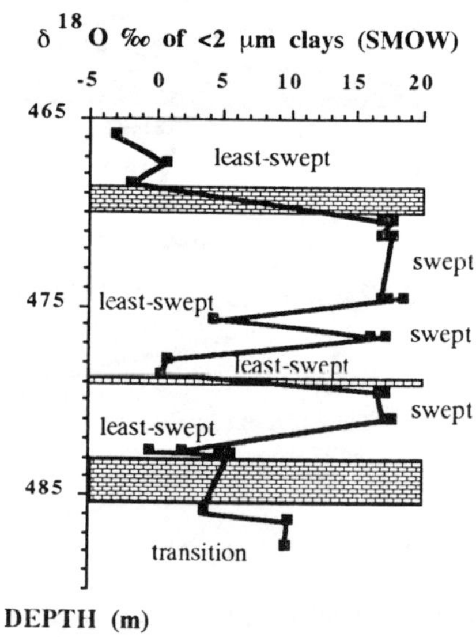

Figure 20. Comparison of $\delta^{18}O$ values for the <2 μm clay assemblage versus depth (m) for post-steam core from the Wabiskaw Member at IHOP. Calcite-cemented intervals are indicated by the patterned area; least-swept, swept and transition zones are also shown.

however, is a mixed-layer swelling clay containing 75 to 95% smectitic layers. Samples located immediately above calcite-cemented sandstones generally contain the highest percentage of smectite layers in the mixed-layer clay (Fig. 19). This mixed-layer clay expands to 16.3-16.7Å upon solvation with ethylene glycol (Ca-saturated), but generally fails to collapse completely to 10Å (K-saturated sample) until heated to ≥300°C. At lower temperatures, a broad peak is present between 10 and 11Å. This behaviour probably reflects variable development of poorly crystallized, Fe-hydroxy interlayer material in the smectitic clay. Dissolution of disseminated siderite within the bitumen-saturated sands during heating is a likely source for the iron.

In the swept zones, Ca-saturated, ethylene-glycol solvated samples of the smectitic clay are characterized by the typical smectite expansion to 16.8-17.0Å. However, upon K-saturation and drying to 107°C, complete collapse to 10Å does not always occur. Instead, low-angle asymmetry can be associated with the 10Å position, disappearing upon heating above 107°C. Like the smectitic clays from the least-swept sands, such behaviour may indicate at least minor incorporation of Fe hydroxy-interlayer material. However, unlike the smectitic clays from the least-swept zones, the content of smectitic layers (60-75%) is lower (Fig. 19), and the failure to collapse to 10Å upon heating is less pronounced. Much of the interstratification in the smectitic clay instead arises from the presence of 10Å layers.

It would be easy to assume that dissolution of K-feldspar, which was a consistent process only in the swept intervals (Fig. 18b), facilitated neoformation of illite and mixed-layered illite-smectite in the swept zones, given sufficiently high temperatures and an appropriate K^+/H^+ activity ratio. Addition of a proprietory additive to the injected steam, however, made this possibility more difficult to evaluate. The additive's intended purpose was to drive the injection-water composition in a direction that favoured neoformation of non-swelling clays (10Å phases), or at least limit the swelling of those that did form. Given the retained bitumen content of the least-swept intervals, it is clear that the additive-carrying steam did not significantly penetrate these intervals. In the swept intervals, though, a general increase in the abundance of illitic clay and the percentage of non-swelling, illitic layers in mixed-layer smectitic clays had to be evaluated as one possible outcome of the injection fluid chemistry, and the additive in particular (section 4.4.6). Another troubling observation was that experimental studies had demonstrated the propensity for neoformation of both smectite and zeolites (analcime, wairakite) during steaming of kaolinite-rich Cold Lake oil sands at 265°C (e.g., Hebner et al., 1986; Kirk et al., 1987). But zeolites were not detected in the post-steam core, not even in the swept zones where sufficiently high temperatures to encourage their crystallization had been reached. Why not?

4.4.5. *Stable Isotope Geochemistry of Post-steam, Calcite-cemented Sandstones.* In an attempt to further elucidate the nature of mineral reactions during steaming, the oxygen and carbon isotope compositions of calcite-cemented intervals, and the oxygen isotope values of post-steam clays (section 4.4.6), were obtained. The oxygen isotope results for the calcite strongly suggest that recrystallization under hydrothermal conditions has been insignificant. Direct comparison of pre- versus post-steam concretionary intervals is not possible. However, the range of $\delta^{18}O$ values for pre-steam (+16.0 to +21.6‰) versus post-steam calcite (+16.1 to +20.1‰) are virtually identical. If calcite had precipitated or recrystallized in a steam or steam-condensate dominated system, very much lower $\delta^{18}O$ values would have been expected. In experimental studies of a very similar system, Kirk et al. (1987) obtained post-steam values of -11 to -9‰ for calcite. In the Wabiskaw pilot, the oxygen isotope composition of the injection water (prior to contact with the reservoir) was -17.9‰. Calcite recrystallized in equilibrium with such water between 60-300°C should have $\delta^{18}O$ values of -12.5 to +4‰.

Some caution must be attached to this observation because the relative importance of isotopic fractionation during steam generation (and its subsequent condensation), and the magnitude of the fluid ^{18}O-shift resulting from interaction with the oil sands, are not known. However, both the $\delta^{18}O$ and $\delta^{13}C$ values of calcite from concretionary intervals in the post-steam core vary in exactly the same fashion as pre-steam concretions in the Wabiskaw Member and the Clearwater Formation (section 4.3.2). In other words, the naturally occurring isotopic fingerprint of the calcite-cemented sandstones has been preserved largely undisturbed.

This conclusion is consistent with the earlier observations concerning the formation of smectitic clays in response to steam injection, and in particular, the high smectite contents, the appearance of neoformed K-feldspar and the bitumen retention that occurred immediately above the calcite-cemented sandstones. In general, conversion of kaolinite to smectite releases much less acid than if zeolites were the main product. This diminishes the potential for calcite dissolution. In fact, the apparent neoformation of K-feldspar (rather than illitic clay) immediately above each calcite-cemented sandstones further suggests that the fluids accumulated at these sites were quite basic. One explanation is that steam (which tends to be acidic) rose within each reservoir compartment, whereas steam condensate (which tends to be basic) ponded at the calcite-cemented sandstone. A further consequence of enhanced smectite formation immediately above the calcite-cemented sandstones was a severe reduction in permeability very early in the steam-injection process; hence, bitumen was retained. In addition, the tops of the calcite-cemented sandstones became physically armoured from futher interaction with steam or steam condensate, limiting their degradation, and protecting their initial value as barriers or baffles to fluid movement within the reservoir.

4.4.6. *Oxygen Isotope Geochemistry of Post-steam Clays.* The oxygen isotope compositions obtained for post-steam clay minerals (<2 μm) are illustrated in Figure 20. Some of these results were unexpected, and have important implications for mineral reaction that occurred or did not occur as a result of steam stimulation.

In the least-swept zone at the top of the core, the smectitic clays have very low $\delta^{18}O$ values (-3.0 to +0.7‰). These results confirmed the mineralogical observations that suggested neoformation of smectitic clay via kaolinite and siderite dissolution. For example, the measured $\delta^{18}O$ value (-1.8‰) for one sample from just above the first cemented interval matches the predicted value (-2.0‰) calculated using the temperature at this site for the beginning of the first cycle of steam injection (81°C), and $\delta^{18}O$ value of the injected steam condensate (-17.9‰). This result further supports earlier arguments that the smectitic clay formed very early in the stimulation of these oil sands, and effectively limited later access of steam or steam condensate to this portion of the reservoir. Much higher temperatures (144-164°C) were reached at this site during later steam cycles; smectitic clays produced then would have had much lower $\delta^{18}O$ values (-9 to -8‰).

The mineralogy of post-steam Wabiskaw sand returned to pre-steam values within a very short interval (<0.5 m) below the lowermost calcite-cemented sandstone (Fig. 16). This behaviour is reflected in the oxygen isotope composition of clays from the transition zone. The unaltered clay assemblage (<2 μm) in the transition zone sands consists of about 85% diagenetic kaolinite ($\delta^{18}O$ = +8.4‰) and 15‰ detrital clays ($\delta^{18}O$ = +17.4‰). From these data, a $\delta^{18}O$ value of +9.75‰ can be predicted for the clay assemblage (<2 μm) not affected by steaming. This result compares very well with measured values (+9.67, +9.81‰) obtained for two samples located >0.5 m below the calcite-cemented sandstone (Fig. 20). In other words, the calcite-cemented interval has been an effective barrier to steam; at distances >0.5 m, reactions resulting from heating and/or penetration of steam or steam-condensate have been insignificant.

Just below the barrier, however, both the mineralogical and isotopic data indicate that some reaction has occurred (Figs. 18, 21). In the <2 μm fraction, the kaolinite content has declined to 48%, and been replaced by neoformed smectitic clay. The $\delta^{18}O$ value of this clay assemblage is +3.64‰. Assuming pre-steam $\delta^{18}O$ values for the diagenetic kaolinite and the inherited detrital clay assemblage, a $\delta^{18}O$ value of -8.1‰ can be calculated for the neoformed smectitic clay. The base of this calcite-cemented interval had an average temperature of ≈150°C during steaming. To form the -8.1‰ smectitic clay at 150°C requires a water of $\delta^{18}O$ = -17.9‰, exactly that measured for the injection fluid. This result may suggest that some steam (condensate?) accessed the transition zone just below the cemented barrier, but the penetration has been very limited. However, a second possibility is that groundwaters in the transition zone have a very similar oxygen isotope composition to the injection fluid (which was local meteoric waters), and that smectite formation occurred in equilibrium with naturally occurring groundwater, heated by conduction through the cemented sandstone. This scenario remains to be tested.

The most surprising results were obtained for clay minerals from the swept zones. The dramatic decrease in both feldspar and kaolinite contents, plus the disappearance of siderite, indicate that extensive dissolution of these phases has occurred. The concomitant increase in the total abundance of clay minerals, dominated by illitic and smectitic varieties, might therefore be considered to result from rock/water interaction during steam and steam-condensate injection, influenced perhaps by the proprietary additive (section 4.4.4).

The oxygen isotope data, however, demonstrate that such an origin for most of the clay mineral assemblage in the swept zones is unlikely. These clays have very high $\delta^{18}O$ values (+16.1 to +18.5‰) (Fig. 20), virtually identical to those of the detrital clay assemblage from Wabiskaw and Clearwater shales, mudstones and mudclasts (+16.0 to +18.2‰; av. 17.4‰). In retrospect, it has been noted that that the clay mineral assemblages in the swept zones and (pre-steam) shales and mudstones are also very similar (smectitic clay, 41 vs 36%; percent interstratification of smectitic clay, 68 vs 70%; illitic clay, 38 vs 40%; kaolinite, 20 vs 30%; chlorite, 6 vs 6%).

The conclusion that the clays in the swept zones are dominantly of detrital origin has important implications. First, neoformation of smectitic clays (or other phases like zeolites) was apparently very limited (to hydroxy-interlayers in some smectitic clay?). This suggests that these portions of the reservoir were contacted mostly by very acidic steam, which aggressively dissolved phases such as kaolinite and feldspar, and inhibited mineral precipitation. The lack of smectite formation during steaming undoubtedly facilitated extraction of bitumen. However, bitumen removal, dissolution of feldspar, kaolinite and siderite, plus the absence of significant hydrothermal mineral precipitation left the residual sand in a completely unconsolidated state. Hence, accumulation of detrital clays in this interval may reflect vertical condensation (matrix clays, clay laminae, clay clasts) as well as lateral migration of fines into the region during hydrocarbon production. Physical migration of fines, especially into the very near-well bore region, is not unexpected this close to the production well. Why the detrital clay aggregates were less reactive during steaming may be related more to their lower total surface area and contact time with steam (relative to phases such as pore-filling kaolinite) than to their composition.

The thin, least-swept intervals that are intercalated with the swept zones contain clays that have much lower $\delta^{18}O$ values (-0.5 to +5.6‰). These results are similar to those obtained for neoformed smectitic clay in the upper, least-swept portion of the core (Fig. 20). In these intervals, smectitic clay neoformation has occurred, and as a result, bitumen has been retained. Variable temperatures for smectitic clay neoformation, as well as changes in the injection fluid $\delta^{18}O$ value (to more ^{18}O-rich compositions) may be responsible for the range in oxygen isotope compositions. However, the possibility of contamination by detrital clay, churned into these intervals during drilling, cannot yet be excluded.

4.4.7. *Summary.* The mineralogical and isotopic data for the post-steam Wabiskaw sands illustrate that steam-sweep efficiency can be very low during *in situ* thermal recovery of hydrocarbons, even in the very near-well bore region. Such behaviour arose in the upper portions of the core because (i) the hydraulic fracture network intended to facilite penetration of steam was contained below this interval by calcite-cemented sandstone, and (ii) the natural mineral assemblage (siderite and diagenetic kaolinite, in particular) readily converted upon initial heating to smectitic clay. Smectitic clay formation further lowered permeability to steam and steam-condensate.

Neoformation of smectite and K-feldspar, and retention of bitumen, was particularly enhanced immediately above calcite-cemented sandstones within each perforated interval. This result probably reflects the more basic nature of the steam-condensate that accumulated at these sites. As a result, the top of each calcite-cemented sandstone was chemically and physically insulated from significant steam-induced dissolution. The very limited extent of mineralogical alteration below the lowermost calcite-cemented interval demonstrated the general effectiveness of the calcite-cemented sandstones as barriers to loss of steam/steam-condensate and heat.

Hydrothermal neoformation of the smectitic clays in the least-swept portions of the reservoir was confirmed by the oxygen isotope data. However, the clay mineral assemblage in the swept zone, rich both in smectitic and illitic clays, has oxygen isotope compositions characteristic of a detrital origin. Injected steam aggressively dissolved reactive phases in these intervals, smectitic clay neoformation was prevented, and as one result, bitumen was successfully extracted. In the swept zones, physical accumulation of clay fines in the near-bore region determined its clay mineralogy, rather than steam ± additive-induced clay neoformation and stabilization reactions.

5. Acknowledgements

Many of the observations summarized in this paper are derived from collaboration with Avner Ayalon, Cathy Connolly, Bob Fialka, Katsumi Marumo, Jennifer McKay, Mira Racki, Greg Robb and Barbara Tilley, to whom my appreciation is extended. Financial support for this research has been provided by the Natural Sciences and Engineering Research Council of Canada (Grant A7387), the Alberta Oil Sands Technology Research Authority (AOSTRA) and Alberta Energy Corporation, Ltd. I also wish to thank Drs. Andrew Parker and Bruce Sellwood for their efforts in organizing the NATO-ASI meeting, and for editing this volume.

6. References

Abercrombie, H. J. (1989) Water-rock interaction during diagenesis and thermal recovery, Cold Lake, Alberta. Ph.D. thesis, University of Calgary, Calgary, Alberta, Canada.

Abercrombie, H. J., Shevalier, M. and Hutcheon, I. E. (1989) Natural diagenesis during steam-assisted recovery of heavy oil, Cold Lake, Alberta, Canada. In: Proceedings of the Sixth International Symposium on Water-Rock Interaction, D. L. Miles, ed., 1-4.

Aharon, P. (1983) Analysis of the anomalous $^{18}O/^{16}O$ and D/H isotope ratios in the tropical rainfall over the western Pacific Ocean. Transactions of the American Geophysical Union 64, 196.

Allan, J. and Creaney, S. (1991) Oil families of the western Canada basin. Bulletin of Canadian Petroleum Geology 39, 107-122.

Alberta Energy Resources Conservation Board (1985) Atlas of Alberta's crude bitumen reserves.

Ayalon, A. and Longstaffe, F.J. (1988) Oxygen-isotope studies of diagenesis and porewater evolution in the western Canada sedimentary basin: evidence from the Upper Cretaceous basal Belly River sandstone, Alberta. Journal of Sedimentary Petrology 58, 489-505.

Baertschi, P. (1976) Absolute ^{18}O content of standard mean ocean water. Earth and Planetary Science Letters 31, 341-344.

Bailey, N.J.L., Krouse, H.R., Evans, C.R. and Rogers, M.A. (1973) Alteration of crude oil by waters and bacteria - evidence from geochemical and isotope studies. American Association of Petroleum Geologists Bulletin 57, 1276-1290.

Ball, M. W. (1935) Athabasca oil sands: Apparent examples of local origin of oil. American Association of Petroleum Geologists Bulletin 19, 153-171.

Beaumont, C. (1981) Foreland basins. Geophysical Journal of the Royal Astronomical Society 65, 291-329.

Beaumont, C., Boutilier, R., Mackenzie, A.S. and Rullkotter, J. (1985) Isomerization and aromatization of hydrocarbons and the paleothermometry and burial history of Alberta foreland basin. American Association of Petroleum Geologists Bulletin 69, 546-566.

Becker, R.H. and Clayton, R.N. (1976) Oxygen isotope study of a Precambrian banded iron-formation, Hamersley Range, Western Australia. Geochimica et Cosmochimica Acta 40, 1153-1165.

Beckie, K. N. and McIntosh, R. A. (1988) Geology and resources of the Primrose crude bitumen deposits, northeastern Alberta. In: Fourth UNITAR/UNDP Conference on Heavy Crude and Tar Sands, R. F. Meyer and E. J. Wiggins, eds., Paper No. 71.

Benzel, W.M. and Graf, D.L. (1984) Studies of smectite membrane behaviour: importance of layer thickness and fabric in experiments at 20°C. Geochimica et Cosmochimica Acta 48, 1769-1778.

Bigeleisen, J. and Mayer, M.G. (1947) Calculation of equilibrium constants for isotopic exchange reactions. Journal of Chemical Physics 15, 261-267.

Bird M.I. (1988) Isotopically depleted rainfall and El Niño. Nature 331, 489-490.

Bird, M.I. and Chivas, A.R. (1988) Stable-isotope evidence for low-temperature kaolinitic weathering and post-formational hydrogen-isotope exchange in Permian kaolinites. Chemical Geology (Isotope Geosciences Section) 72, 249-265.

Bird, M.I., Longstaffe, F.J. and Fyfe, W.S. (1993) Oxygen-isotope fractionation in titanium-oxide minerals at low temperature. Geochimica et Cosmochimica Acta 57, 3083-3091.

Bird, M.I., Longstaffe, F.J., Fyfe, W.S. and Bildgen, P. (1992a) Oxygen-isotope systematics in a multiphase weathering system in Haïti. Geochimica et Cosmochimica Acta 56, 2831-2838.

Bird, M.I., Longstaffe, F.J., Fyfe, W.S., Kronberg, B.I. and Kishida, A. (1992b) An oxygen-isotope study of weathering in the eastern Amazon Basin, Brazil. In: Continental Isotopic Indicators of Climate, Chapman Conference Volume (in press).

Bloch, J. (1990) Stable isotopic composition of authigenic carbonates from the Albian Harmon Member (Peace River Formation): evidence of early diagenetic processes. Bulletin of Canadian Petroleum Geology 38, 39-52.

Boles, J.R. (1978) Active ankerite cementation in the subsurface Eocene of southwest Texas. Contributions to Mineralogy and Petrology 68, 13-22.

Boles, J.R. (1981) Clay diagenesis and effects on sandstone cementation (case histories from the Gulf Coast Tertiary). In: Clays and the Resource Geologist. F.J. Longstaffe, ed., Mineralogical Association of Canada Short Course 7, 148-168.

Boles, J.R. and Franks, S.G. (1979) Clay diagenesis in Wilcox sandstones of southwest Texas: implications of smectite diagenesis on sandstone cementation. Journal of Sedimentary Petrology 49, 55-70.

Boon, J.A. and Hitchon, B. (1983a) Application of fluid-rock studies to *in situ* recovery from oil sand deposits, Alberta, Canada-I. Aqueous phase results for an experimental-statistical study of water-bitumen-shale reactions. Geochimica et Cosmochimica Acta 47, 235-248.

Boon, J.A. and Hitchon, B. (1983b) Application of fluid-rock studies to *in situ* recovery from oil sand deposits, Alberta, Canada-II. Mineral transformations during an experimental-statistical study of water-bitumen-shale reactions. Geochimica et Cosmochimica Acta 47, 249-257.

Capeling, R. R. and Peggs, J. K. (1981) Experimental steamflood, Cold Lake oil sands. In: The Future of Heavy Crude and Tar Sands, R.F. Meyer and C.T. Steele, eds., 361-368.

Capuano, R.M. (1992) The temperature dependence of hydrogen isotope fractionation between clay minerals and water: Evidence from a geopressured system. Geochimica et Cosmochimica Acta 56, 2547-2554.

Carothers, W.W., Adami, L.H. and Rosenbauer, R.J. (1988) Experimental oxygen isotope fractionation between siderite-water and phosphoric acid liberated CO_2-siderite. Geochimica et Cosmochimica Acta 52, 2445-2450.

Carpenter, A.B. (1978) Origin and chemical evolution of brines in sedimentary basins. Oklahoma Geological Survey Circular 79, 60-77.

Chase, C.G. and Perry, E.C. (1972) The oceans: growth and oxygen isotope evolution. Science 177, 992-994.

Clayton, R.N. (1959) Oxygen isotope fractionation in the system calcium carbonate-water. Journal of Chemical Physics 30, 1246-1250.

Clayton, R.N., Friedman, I., Graf, D.L., Mayeda, T.K., Meents, W.F. and Shimp, N.F. (1966) The origin of saline formation waters, 1. Isotopic composition. Journal of Geophysical Research. 71, 3869-3882.

Clayton, R.N., Muffler, L.J.P. and White, D.E. (1968) Oxygen isotope study of calcite and silicates of the River Ranch No. 1 well, Salton Sea geothermal field, California. American Journal of Science 266, 968-979.

Clayton, R.N., O'Neil, J.R. and Mayeda, T.K. (1972) Oxygen isotope exchange between quartz and water. Journal of Geophysical Research 77, 3057-3067.

Connolly, C.A., Walter, L.M., Baadsgaard, H. and Longstaffe, F.J. (1990a) Origin and evolution of formation waters, Alberta basin, western Canada sedimentary basin. I. Chemistry. Applied Geochemistry 5, 375-395.

Connolly, C.A., Walter, L.M., Baadsgaard, H. and Longstaffe, F.J. (1990b) Origin and evolution of formation waters, Alberta basin, western Canada sedimentary basin. II. Isotope systematics and water mixing. Applied Geochemistry 5, 397-413.

Coplen, T.B. and Hanshaw, B.B. (1973) Ultrafiltration by a compacted clay membrane - I. Oxygen and hydrogen isotope fractionation. Geochimica et Cosmochimica Acta 37, 2295-2310.

Corbett, C.S. (1955) In-situ origin of McMurray oil of northeastern Alberta and its relevance to general problem of origin of oil. American Association of Petroleum Geologists 39, 1601-1649.

Craig, H. (1957) Isotopic standards for carbon and oxygen and correction factors for mass-spectrometric analysis of carbon dioxide. Geochimica et Cosmochimica Acta 12, 133-149.

Craig, H. (1961a) Standards for reporting concentrations of deuterium and oxygen-18 in natural waters. Science 133, 1833-1834.

Craig, H. (1961b) Isotopic variations in meteoric waters. Science 133, 1702-1703.

Craig, H. (1965) The measurement of oxygen isotope paleo-temperatures. In: Stable Isotopes in Oceanographic Studies and Paleotemperatures. E. Tongiorgi, ed., Consiglio Nazionale delle Richerche, Laboratorio de Geologia Nucleare, Pisa, 161-182.

Craig, H. and Gordon, L.I. (1965) Deuterium and oxygen-18 variations in the oceans and the marine atmosphere. In: Proceedings of the Spoleto Conference on Stable Isotopes in Oceanographic Studies and Paleotemperatures. Laboratorio di Geologia Nucleare, Pisa, Italy, 9-13.

Creaney, S. and Allan, J. (1990) Hydrocarbon generation and migration in the western Canada sedimentary basin. In: Classic Petroleum Provinces, J. Brooks, ed., Special Publication of the Geological Society, Blackwell Scientific.

Crerar, D.A., Namson, J., Chyi, M.S., Willians, L. and Feigenson, M.D. (1982) Manganiferous chert of the Franciscan assemblage: I. General geology, ancient and modern analogues, and implications for hydrothermal conversion at oceanic spreading centers. Economic Geology 77, 519-540.

Curtis, C.D. and Spears, D.A. (1968) The formation of sedimentary iron minerals. Economic Geology 24, 257-270.

Curtis, C.D., Hughes, C.R., Whiteman, J.A. and Whittle, C.K. (1985) Compositional variations within some sedimentary chlorites and some comments on their origin. Mineralogical Magazine 49, 375-386.

Dansgaard, W. and Tauber, H. (1969) Glacier oxygen-18 content and Pleistocene ocean temperatures. Science 166, 499-502.

Dean, R.S. and Nahnybida, C. (1985) Authigenic trioctahedral clay minerals coating Clearwater Formation sand grains in Cold Lake, Alberta - Extended abstract. Applied Clay Science 1, 237-238.

Degens, E.T. and Epstein, S. (1962) Relationship between $^{18}O/^{16}O$ ratios in coexisting carbonates, cherts and diatomites. American Association of Petroleum Geologists Bulletin 46, 534-542.

Deines, P. (1980) The isotopic composition of reduced organic carbon. In: Handbook of Environmental Isotope Geochemistry, Volume 1: The Terrestrial Environment, A. P. Fritz and J.Ch. Fontes, eds., Elsevier, 329-406.

Demir, I. (1988) Studies of smectite membrane behaviour: electrokinetic, osmotic, and isotopic fractionation processes at elevated pressures. Geochimica et Cosmochimica Acta 52, 727-737.

Dekker, F., Visser, C. and Dankers, P. (1987) The Primrose-Kirby area in southern Athabasca, Alberta, Canada: a detailed geological investigation. American Association of Petroleum Geologists, Special Studies in Geology #25, 507-520.

Deroo, G., Tissot, B., McCrossan, R. G. and Der, F. (1974) Geochemistry of the heavy oils of Alberta. In: Oil Sands, Fuels of the Future, L. Hills, ed., Canadian Society of Petroleum Geologists, Memoir 3, 148-167.

Deroo, G., Powell, T. G., Tissot, B. and McCrossan, R. G. (1977) The origin and maturation of petroleum in the western Canadian Sedimentary Basin, Alberta. Geological Survey of Canada Bulletin 262, 136p.

Dickinson, W.W. (1987) An oxygen isotope model for interpreting carbonate diagenesis in nonmarine rocks (Green River Basin, Wyoming, U.S.A.). Chemical Geology (Isotope Geoscience Section) 65, 103-116.

Dimitrakopoulos, R.G. and Muehlenbachs, K. (1987) Biodegradation of petroleum as a source of ^{13}C-enriched carbon dioxide in the formation of carbonate cement. Chemical Geology (Isotope Geoscience Section) 65, 283-291.

Domenico, P.A. and Robbins, G.A. (1985) The displacement of connate water from aquifers. Bulletin of the Geological Society of America 96, 328-335.

Eslinger, E.V. and Yeh, H-W. (1981) Mineralogy, O^{18}/O^{16} and D/H ratios of clay-rich sediments from Deep Sea Drilling Project site 180, Aleutian Trench. Clays and Clay Minerals 29, 309-315.

Evans, C.R., Rogers, M.A. and Bailey, N.J.L. (1971) Evolution and alteration of petroleum in

western Canada. Chemical Geology 8, 147-170.
Faure, G. (1986) Principles of Isotope Geology, second edition. John Wiley and Sons, New York, 589 pp.
Ferris, F.G., Fyfe, W.S. and Beveridge, T.J. (1987) Bacteria as nucleation sites for authigenic minerals in a metal-contaminated lake sediment. Chemical Geology 63, 225-232.
Fialka, B.N., McClanahan, R.K., Robb, G.A and Longstaffe, F.J. (1993) The evaluation of cyclic steam stimulation in an oil sand reservoir using post-steam core analysis. Canadian Journal of Petroleum Technology 32, 56-62.
Friedman, I. and Hardcastle, K. (1973) Interstitial water studies, Leg 15 - Isotopic composition of water. In: Initial Reports of the Deep Sea Drilling Project 20. B.C. Heezen, I.D. MacGregor et al., eds., U.S. Government Printing Office, Washington, D.C., 901-903.
Friedman, I. and Hardcastle, K. (1974) Deuterium in interstitial water from Red Sea cores. In: Initial Reports of the Deep Sea Drilling Project 23. R.B. Whitmarsh, O.E. Weser et al., eds., U.S. Government Printing Office, Washington, D.C., 969-970.
Friedman, I. and O'Neil, J.R. (1977) Compilation of stable isotope fractionation factors of geochemical interest. In: Data of Geochemistry, sixth edition. M. Fleischer, ed., United States Geological Survey Professional Paper 440-KK, 12 pp. + figures.
Garven, G. (1989) A hydrogeologic model for the formation of the giant oil sands deposits of the western Canada sedimentary basin. American Journal of Science 289, 105-166.
Gat, J.R. (1980) The isotopes of hydrogen and oxygen in precipitation. In: Handbook of Environmental Isotope Geochemistry, Volume 1: The Terrestrial Environment, A. P. Fritz and J.Ch. Fontes, eds., Elsevier, 21-47.
Gieskes, J.M. and Lawrence, J.R. (1981) Alteration of volcanic matter in deep sea sediments: evidence from the chemical composition of interstitial waters from deep sea drilling cores. Geochimica et Cosmochimica Acta 45, 1687-1703.
Girard, J-P., Savin, S.M. and Feng, X. (1991) Intracrystalline fractionation of oxygen isotopes between hydroxyl and non-hydroxyl sites in kaolinite: thermal dehydroxylation and partial fluorination approaches. Annual Meeting, Clay Mineral Society 28, 58.
Gonfiantini, R., Panichi, C. and Tongiorgi, E. (1968) Isotopic disequilibrium in travertine deposition. Earth and Planetary Science Letters 5, 55-58.
Graf, D.L. (1982) Chemical osmosis, reverse chemical osmosis, and the origin of subsurface brines. Geochimica et Cosmochimica Acta 46, 1431-1448.
Graf, D.L., Friedman, I. and Meents, W.F. (1965) The origin of saline formation water, II. Isotopic fractionation by shale micropore systems. Illinois Geological Survey Circular 393, 32 pp.
Graf, D.L., Meents, W., Friedman, I. and Shimp, N. (1966) The origin of saline formation waters, III: calcium chloride waters. Illinois Geological Survey Circular 397.
Graham, C.M., Viglino, J.A. and Harmon, R.S. (1987) An experimental study of hydrogen isotope exchange between aluminous chlorite and water. American Mineralogist 72, 566-579.
Gregory, R.T. and Taylor, H.P., Jr. (1981) An oxygen isotope profile in a section of Cretaceous oceanic crust, Samail Ophiolite, Oman: evidence for $\delta^{18}O$ buffering of the oceans by deep (>5 km) seawater-hydrothermal circulation at mid-ocean ridges. Journal of Geophysical Research 86, B4, 2737-2755.
Gregory, R.T. and Criss, R.E. (1986) Isotopic exchange in open and closed systems. In: Stable Isotopes in High Temperature Processes. J.W. Valley, H.P. Taylor Jr. and J.R. O'Neil, eds., Reviews in Mineralogy Volume 16, Mineralogical Society of America, 91-127.
Gunter, W.D. and Bird, G.W. (1988) CO_2 production in tar sand reservoirs under *in situ* steam temperatures: reactive calcite dissolution. Chemical Geology 70, 301-311.
Gunter, W.D., Bird, G.W., Aggarwall, P.K. and Leone, J.A. (1988) Modeling of smectite synthesis in reservoir sands: Comparison of path predictions to autoclave experiments. In: The Fourth UNITAR/UNDP Conference on Heavy Crude and Tar Sands, R.F. Meyer and E.J. Wiggins, eds., Paper 29.
Hacquebard, P.A. (1977) Rank of coal as an index of organic metamorphism for oil and gas in Alberta. In: The Origin and Migration of Petroleum in the Western Canadian Sedimentary Basin, Alberta. G. Deroo, T.G. Powell, B. Tissot and G. McCrossan, eds., Geological Survey of Canada Bulletin 262, 11-22.

Hagemann, R., Nief, G., Roth, E. (1970) Absolute isotopic scale for deuterium analysis of natural waters. Absolute D/H ratio for SMOW. Tellus 22, 712-715.
Hamza, M.S. and Epstein, S. (1980) Oxygen isotopic fractionation between oxygen of different sites in hydroxyl-bearing silicate minerals. Geochimica et Cosmochimica Acta 44, 173-182.
Harrison, D.B., Glaister, R.P. and Nelson, H.W. (1981) Reservoir description of the Clearwater oil sand, Cold Lake, Alberta Canada. In: The Future of Heavy Crude Oils and Tar Sands, R.F. Meyer and C. T. Steele, eds., McGraw Hill, New York, 264-279.
Hart, B.S., Longstaffe, F.J. and Plint, A.G. (1992) Evidence for relative sea level change from isotopic and elemental composition of siderite in the Cardium Formation, Rocky Mountain Foothills. Bulletin of Canadian Petroleum Geology 40, 52-59.
Haydon, P.R. and Graf, D.L. (1986) Studies of smectite membrane behaviour: temperature dependence, 20-180°C. Geochimica et Cosmochimica Acta 50, 115-121.
Hebner, B.A., Bird, G.W. and Longstaffe, F.J. (1986) Fluid/pore mineral transformation during simulated steam injection: Implications for reduced permeability damage. Journal of Canadian Petroleum Technology 25, 1-6.
Herzig, P.M., Becker, K.P., Stoffers, P., Backer, H. and Blum, N. (1988) Hydrothermal silica chimney fields in the Galapagos Spring Center at 86°W. Earth and Planetary Science Letters 89, 261-272.
Hitchon, B. (1969a) Fluid flow in the western Canada sedimentary basin: 1. Effect of topography. Water Resources Research 5, 186-195.
Hitchon, B. (1969b) Fluid flow in the western Canada sedimentary basin; 2. Effect of geology. Water Resources Research 5, 460-469.
Hitchon, B. (1984) Geothermal gradients, hydrodynamics, and hydrocarbon occurrences, Alberta, Canada. American Association of Petroleum Geologists Bulletin 68, 713-743.
Hitchon, B. and Filby, R. H. (1984) Use of trace elements for classification of crude oils into families - Example from Alberta, Canada. American Association of Petroleum Geologists Bulletin 68, 838-849.
Hitchon, B. and Friedman, I. (1969) Geochemistry and origin of formation waters in the western Canada sedimentary basin - I. Stable isotopes of hydrogen and oxygen. Geochimica et Cosmochimica Acta 33, 1321-1349.
Hoefs, J. (1987) Stable isotope geochemistry, third edition. Springer-Verlag, New York, Heidelberg, Berlin, 241 pp.
Hoffman, C.F. and Strausz, O.P. (1986) Bitumen accumulation in Grosmont Platform Complex, Upper Devonian, Alberta, Canada. American Association of Petroleum Geologists Bulletin 70, 1113-1128.
Holser, W.T. (1979) Trace elements and isotopes in evaporites. In: Marine Minerals, R.G. Burns, ed., Reviews in Mineralogy, Mineralogical Society of America, 295-346.
Hower, J., Eslinger, E.V., Hower, M.E. and Perry, E.A. (1976) Mechanism of burial metamorphism of argillaceous sediment: 1. Mineralogical and chemical evidence. Geological Society of America Bulletin 87, 725-737.
Hume, G.S. (1947) Results and significance of drilling operations in the Athabasca bituminous sands. Canadian Institute of Mining and Metallurgy, Engineering Transactions 50, 298-324.
Hutcheon, I. (1984) A review of artificial diagenesis during thermally enhanced recovery. In: Clastic Diagenesis, D.A. McDonald and R.C. Surdam, eds., American Association of Petroleum Geologists, Memoir 37, 413-429.
Hutcheon, I., Abercrombie, H.S. and Krouse, H.R. (1990) Inorganic origin of carbon dioxide during low temperature thermal recovery of bitumen: Chemical and isotopic evidence. Geochimica et Cosmochimica Acta 54, 165-171.
Hutcheon, I., Abercrombie, H., Putnam, P., Gardner, R. and Krouse, R. (1989) Sedimentology and diagenesis of the Clearwater Formation at Tucker Lake. Bulletin of Canadian Petroleum Geology 37, 83-97.
Hutcheon, I., Abercrombie, H., Shevalier, M. and Nahnybida, C. (1988) A comparison of formation reactivity in quartz-rich and quartz-poor reservoirs during steam assisted recovery. In: Fourth UNITAR/UNDP Conference on Heavy Crude and Tar Sands., R.F. Meyer and E.J. Wiggins, eds., Paper 235.
Irwin, H., Curtis, C. and Coleman, M. (1977) Isotopic evidence for source of diagenetic

carbonates formed during burial of organic-rich sediments. Nature 269, 209-213.

Jahren, J.S. and Aagaard, P. (1989) Compositional variations in diagenetic chlorites and illites, and relationships with formation-water chemistry. Clay Minerals 24, 157-170.

James, A.T. and Baker, D.R. (1976) Oxygen isotope exchange between illite and water at 22°C. Geochimica et Cosmochimica Acta 40, 235-239.

Jones, R.W. (1980) Some mass balance and geological constraints on migration mechanisms. In: Problems of Petroleum Migration, W.H. Roberts III and R.J. Cordell, eds., American Association of Petroleum Geologists, Studies in Geology 10, 45-68.

Karhu, J. and Epstein, S. (1986) The implication of the oxygen isotope records in coexisting cherts and phosphates. Geochimica et Cosmochimica Acta 50, 1745-1756.

Keith, M.L. and Weber, J.N. (1964) Carbon and oxygen isotopic composition of selected limestones and fossils. Geochimica et Cosmochimica Acta 28, 1787-1816.

Keith, M.L. and Weber, J.N. (1965) Systematic relationships between carbon and oxygen isotopes in carbonates deposited by modern corals and algae. Science 150, 498-501.

Kendall, G.H. (1977) Importance of reservoir description in evaluating in-situ recovery methods for Cold Lake heavy oil. Part 1-- Reservoir description. Journal of Canadian Petroleum Technology, 41-47.

Kharaka, Y.K. and Carothers, W.W. (1986) Oxygen and hydrogen isotope geochemistry of deep basin brines. In: Handbook of Environmental Isotope Geochemistry, Volume 2. P. Fritz and J.Ch. Fontes, eds., Elsevier, Amsterdam, Oxford, New York, Tokyo, 305-360.

Kharaka, Y.K., Berry, A.F. and Friedman, I. (1973) Isotopic composition of oil-field brines from Kettleman North Dome, California and their geological implications. Geochimica et Cosmochimica Acta 37, 1899-1908.

Kharaka, Y.K., Lico, M.S., Wright, V.A. and Carothers, W.W. (1979) Geochemistry of formation waters from Pleasant Bayou No. 2 well and adjacent areas in coastal Texas. Proceedings of the 4th Geopressured-Geothermal Energy Conference, Austin, Texas, 168-193.

Kirk, J.S., Bird, G.W. and Longstaffe, F.J. (1987) Laboratory study of the effects of steam-condensate flooding in the Clearwater Formation: High temperature flow experiments. Bulletin of Canadian Petroleum Geology 35, 34-47.

Knauth, L.P. (1988) Origin and mixing history of brines, Palo Duro Basin, Texas, U.S.A. Applied Geochemistry 3, 455-474.

Knauth, L.P. and Epstein, S. (1976) Hydrogen and oxygen isotope ratios in nodular and bedded cherts. Geochimica et Cosmochimica Acta 40, 1095-1108.

Knauth, L.P. and Lowe, D.R. (1978) Oxygen isotope geochemistry of cherts from the Onverwacht Group (3.4 billion years), Transvaal, South Africa, with implications for secular variations in the isotopic composition of cherts. Earth and Planetary Science Letters 41, 209-222.

Knauth, L.P. and Beeunas, M.A (1986) Isotope geochemistry of fluid inclusions in Permian halite with implications for the isotopic history of ocean water and the origin of saline formation waters. Geochimica et Cosmochimica Acta 50, 419-433.

Kyser, T.K. (1987) Equilibrium fractionation factors for stable isotopes. In: Stable Isotope Geochemistry of Low Temperature Fluids. T.K. Kyser, ed., Mineralogical Association of Canada Short Course, Volume 13, 1-84.

Kyser, T.K., Caldwell, W.G.E., Whittaker, S.G. and Cadrin, A.J. (in press) Paleoenvironment and geochemistry of the northern portion of the Western Interior Seaway during Late Cretaceous time. In: Evolution of the Western Interior Foreland Basin, W.G.E. Caldwell and E.G. Kaufmann, eds., Geological Association of Canada Special Paper.

Lambert, S.J. and Epstein, S. (1980) Stable isotope investigations of an active geothermal system in Valles Caldera, Jemez Mountains, New Mexico. Journal of Volcanological and Geothermal Research 8, 111-129.

Land, L.S. (1984) Frio sandstone diagenesis, Texas Gulf Coast: a regional isotopic study. In: Clastic Diagenesis. R.C. Surdam and D.A MacDonald, eds., American Association of Petroleum Geologists, Memoir 37, 47-62.

Land, L.S. and Dutton, S.P. (1978) Cementation of a Pennsylvanian deltaic sandstone: isotopic data. Journal of Sedimentary Petrology 48, 1167-1176.

Land, L.S. and Prezbindowski, D. (1981) The origin and evolution of saline formation waters: Lower Cretaceous carbonates, south central Texas, U.S.A. Journal of Hydrology 54, 51-74.

Lawrence, J.R. (1973) Interstitial water studies, Leg 15 - oxygen and carbon isotope variations in water, carbonates and silicates from the Venezuela Basin (site 149) and the Aves Rise (site 148). In: Initial Reports of the Deep Sea Drilling Project 20, B.C. Heezen, I.D. MacGregor et al., eds., U.S. Government Printing Office, 891-899.

Lawrence, J.R. (1974) Stable oxygen and carbon isotope variations in the pore waters, carbonates and silicates, Sites 225 and 228, Red Sea. In: Initial Reports of the Deep Sea Drilling Project 23, R.B. Whitmarsh, O.D. Weser et al., eds., U.S. Government Printing Office, 939-942.

Lawrence, J.R. and Taylor, H.P., Jr. (1971) Deuterium and oxygen-18 correlation: clay minerals and hydroxides in Quaternary soils compared to meteoric waters. Geochimica et Cosmochimica Acta 35, 993-1003.

Lawrence, J.R. and Taylor, H.P., Jr. (1972) Hydrogen and oxygen isotope systematics in weathering profiles. Geochimica et Cosmochimica Acta 36, 1377-1393.

Lawrence, J.R. and Taviani, M. (1988) Extreme hydrogen, oxygen and carbon isotope anomalies in the pore waters and carbonates of the sediments and basalts from the Norwegian Sea: Methane and hydrogen from the mantle? Geochimica et Cosmochimica Acta 52, 2077-2083.

Lawrence, J.R., Gieskes, J.M. and Broecker, W.S. (1975) Oxygen isotope and cation composition of DSDP pore waters and the alteration of layer II basalts. Earth and Planetary Science Letters 27, 1-10.

Lawrence, J.R., Gieskes, J.M. and Anderson, T.F. (1976) Oxygen isotope material balance calculations, Leg 35. In: Initial Reports of the Deep Sea Drilling Project, Volume 35. P. Worstell, ed., U.S. Government Printing Office, Washington, D.C., 507-512.

Lawrence, J.R., Drever, J.I., Anderson, T.F. and Brueckner, H.K. (1979) Importance of alteration of volcanic material in the sediments of Deep Sea Drilling Site 323: chemistry, $^{18}O/^{16}O$ and $^{87}Sr/^{86}Sr$. Geochimica et Cosmochimica Acta 43, 573-588.

Lefebvre, R. and Hutcheon, I. (1986) Mineral reactions in quartzose rocks during thermal recovery of heavy oil, Lloydminster, Saskatchewan, Canada. Applied Geochemistry 1, 395-405.

Link, T. A. (1951) Source of oil in "Tar Sands" of Athabasca River, Alberta Canada. American Association of Petroleum Geologists Bulletin 35, 854-864.

Liu, K-K. and Epstein, S. (1984) The hydrogen isotope fractionation between kaolinite and water. Isotope Geoscience 2, 335-350.

Lloyd, R.M. (1966) Oxygen isotope enrichment of sea water by evaporation. Geochimica et Cosmochimica Acta 30, 801-814.

Longstaffe, F.J. (1983) Diagenesis, IV. Stable isotope studies of diagenesis in clastic rocks. Geoscience Canada 10, 44-58.

Longstaffe, F.J. (1984) The role of meteoric water in diagenesis of shallow sandstones: stable isotope studies of the Milk River aquifer and gas pool. In: Clastic Diagenesis. R.C. Surdam and D.A. MacDonald, eds., American Association of Petroleum Geologists, Memoir 37, 81-98.

Longstaffe, F.J. (1986) Oxygen isotope studies of diagenesis in the basal Belly River sandstone, Pembina I-Pool, Alberta. Journal of Sedimentary Petrology 56, 78-88.

Longstaffe, F.J. (1987) Stable isotope studies of diagenetic processes. In: Stable Isotope Geochemistry of Low Temperature Fluids. T.K. Kyser, ed., Mineralogical Association of Canada Short Course, Volume 13, 187-257.

Longstaffe, F.J. (1988) Mineralogical and oxygen-isotope studies of clastic diagenesis: implications for fluid flow in sedimentary basins. In: Hydrogeology of Sedimentary Basins. B. Hitchon, ed., Proceedings of the Third Annual Canadian/American Conference on Hydrogeology, National Water Well Association, 204-220.

Longstaffe, F.J. (1989) Stable isotopes as tracers in clastic diagenesis. In: Burial Diagenesis, I.E. Hutcheon, ed., Mineralogical Association of Canada Short Course, Volume 15, 201-277.

Longstaffe, F.J. (in press) Meteoric water and sandstone diagenesis in the western Canada sedimentary basin. In: Diagenesis and Basin Development. A.G. Robinson, ed., American Association of Petroleum Geologists Memoir.

Longstaffe, F.J. and Ayalon, A. (1987) Oxygen-isotope studies of clastic diagenesis in the Lower Cretaceous Viking Formation, Alberta: implications for the role of meteoric water. In: The Diagenesis of Sedimentary Sequences. J.D. Marshall, ed., Geological Society Special Publication 36, 277-296.

Longstaffe, F.J. and Ayalon, A. (1990) Hydrogen-isotope geochemistry of diagenetic clay minerals from Cretaceous sandstones, Alberta, Canada: evidence for exchange. Applied Geochemistry 5, 657-668.

Longstaffe, F.J. and Ayalon, A. (1991) Mineralogical and O-isotope studies of diagenesis and porewater evolution in continental sandstones, Cretaceous Belly River Group, Alberta, Canada. Applied Geochemistry 6, 291-303.

Longstaffe, F.J., Ayalon, A. and Racki, M.A. (1989a). Natural diagenesis of Clearwater Formation reservoirs in the Cold Lake area, Alberta, Part I: Mineralogical studies. Canadian Society of Petroleum Geologists, Exploration Update '89, Calgary, Programs and Abstracts, 130.

Longstaffe, F.J., Ayalon, A. and Racki, M.A. (1989b). Oxygen- and carbon-isotope studies of diagenesis in heavy oil deposits of the Clearwater Formation, northeastern Alberta. Geological Association of Canada, Program with Abstracts, v. 14, p. A85.

Longstaffe, F.J., Ayalon, A. and Racki, M.A. (1992b) Stable isotope studies of diagenesis in berthierine-bearing oil sands, Clearwater Formation, Alberta. In: Proceedings of the 7th International Symposium on Water-Rock Interaction, Y.K. Kharaka and A.S. Maest, eds., A.A. Balkema Publishers, 955-958.

Longstaffe, F.J., Fialka, B.N. and Robb, G.A. (1991) Mineralogical modifications of oil sands during steam stimulation: post-steam core from the Wabiskaw Member, northeastern Alberta. Annual Meeting, Clay Mineral Society 28, 102.

Longstaffe, F.J., Robb, G.A. and Fialka, B.N. (1990). Clay mineralogical studies of post-steam core from the near-well bore region in an oil sands reservoir, Alberta. Canadian Society of Petroleum Geologists Annual Meeting, Calgary, 91.

Longstaffe, F.J., Tilley, B.J., Ayalon, A. and Connolly, C.A. (1992a) Controls on porewater evolution during sandstone diagenesis, western Canada sedimentary basin: an oxygen isotope perspective. In: Origin, Diagenesis, and Petrophysics of Clay Minerals in Sandstones, D.W. Houseknecht and E. Pittman, eds., SEPM Special Publication No. 47, 13-34.

Machemer, S.D. and Hutcheon, I. (1988) Geochemistry of early carbonate cements in the Cardium Formation, central Alberta. Journal of Sedimentary Petrology 58, 136-147.

Marumo, K., Longstaffe, F.J. and Matsubaya, O. (in review) Mineralogical and stable isotopic studies of interlayer and structural water in clay minerals from fossil and active hydrothermal systems, southwestern Hokkaido, Japan.

Marumo, K., Nagasawa, K. and Kuroda, Y. (1980) Mineralogy and hydrogen isotope geochemistry of clay minerals in the Ohnuma geothermal area, northeast Japan. Earth and Planetary Science Letters 47, 255-262.

McIntosh, R.A. and Beckie, K.N. (1989) Western Canada heavy oil and tar sand resources. Oil and Gas Journal 87(29), 73-75.

Masters, J.A. (1984) Lower Cretaceous oil and gas in western Canada. In: Elmworth - Case Study of a Deep Basin Gas Field, J.A. Masters, ed., American Association of Petroleum Geologists, Memoir 38, 1-33.

McKay, J.L. (1992) Diagenesis of the Upper Cretaceous Marshybank Formation, northwestern Alberta - northeastern Britsh Columbia. M.Sc. thesis, University of Western Ontario, London, Ontario, Canada.

McKay, J.L., Longstaffe, F.J. and Plint, A.G. (1989) Geochemical variations across marine-nonmarine transitions within the Bad Heart Formation, northwestern Alberta and northeastern British Columbia. Geological Association of Canada, Programme with Abstracts 14, A14.

Minken, D. F. (1974) The Cold Lake oil sand: Geology and a reserves estimate. In: Oil Sands, Fuel of the Future, L. Hill, ed., Canadian Society of Petroleum Geologists, Memoir 3, 84-99.

Moshier, S.O. and Waples, D. (1985) Quantitative evaluation of Lower Cretaceous Mannville Group as source rock for Alberta's oil sands. American Association of Petroleum Geologists Bulletin 69, 161-172.

Mozley, P.S. (1989a) Relation between depositional environment and the elemental composition of early diagenetic siderite. Geology 17, 704-706.

Mozley, P.S. (1989b) Complex compositional zoning in concretionary siderite: implications for geochemical studies. Journal of Sedimentary Petrology 59, 815-818.

Muehlenbachs, K. and Bowring, S.A. (1992) Is the imprint of primordial ocean evident in the 3.96

Ga Acasta gneiss? Transactions of the American Geophysical Union 73 (supplement), 324.
Muehlenbachs, K. and Clayton, R.N. (1976) Oxygen isotope composition of the oceanic crust and its bearing on seawater. Journal of Geophysical Research 81, 4365-4369.
Nabelek, P.I. (1987) General equations for modeling fluid/rock interaction using trace elements and isotopes. Geochimica et Cosmochimica Acta 51, 1765-1769.
O'Neil, J.R. (1968) Hydrogen and oxygen isotope fractionation between ice and water. Journal of Physical Chemistry 72, 3683-3684.
O'Neil, J.R. (1979) Stable isotope geochemistry of rocks and minerals. In: Lectures in Isotope Geology. E. Jäger and J.C. Hunziker, eds., Springer-Verlag, Berlin, 264-273.
O'Neil, J.R. (1987) Preservation of H, C, and O isotopic ratios in the low temperature environment. In: Stable Isotope Geochemistry of Low Temperature Fluids. T.K. Kyser, ed., Mineralogical Association of Canada Short Course, Volume 13, 85-128.
O'Neil, J.R. and Kharaka, Y.F. (1976) Hydrogen and oxygen isotope exchange reactions between clay minerals and water. Geochimica et Cosmochimica Acta 40, 241-246.
O'Neil, J.R. and Taylor, Jr., H.P. (1967) The oxygen isotope and cation-exchange chemistry of feldspars. American Mineralogist 52, 1414-1437.
O'Neil, J.R., Clayton, R.N. and Mayeda, T.K. (1969) Oxygen isotope fractionation in divalent metal carbonates. Journal of Chemical Physics 51, 5547-5558.
O'Neil, J.R., Johnson, C.M., White, L.D. and Roedder, E. (1986) The origin of fluids in the salt beds of the Delaware Basin, New Mexico and Texas. Applied Geochemistry 1, 265-271.
Outtrim, C.P. and Evans, R.G. (1977) Alberta's oil sands and their evaluation. In: The Oil Sands of Canada-Venezuela. D.A. Redford and A.G. Winestock, eds., Canadian Institute of Mining and Metallurgy, Special Volume 17, 36-66.
Park, R., and Epstein, S. (1960) Carbon isotope fractionation during photosynthesis. Geochimica et Cosmochimica Acta 21, 110-126.
Perry, E.C., Jr. (1967) The oxygen isotope chemistry of ancient cherts. Earth and Planetary Science Letters 3, 62-66.
Perry, E.C., Jr. (1990) Comment on "The implication of the oxygen isotope records in coexisting cherts and phosphates" by J. Karhu and S. Epstein. Geochimica et Cosmochimica Acta 54, 1175-1179.
Perry, E.C., Jr. and Tan, F.C. (1972) Significance of oxygen and carbon isotope variations in early Precambrian cherts and carbonate rocks of southern Africa. Geological Society of America Bulletin 83, 647-664.
Perry, E.C., Jr., Ahmad, S.N. and Swulius, T.M. (1978) The oxygen isotope composition of 3,800 m.y. old metamorphosed chert and iron formation from Isukasia, West Greenland. Journal of Geology 86, 223-239.
Pierre, C., Ortlieb, L. and Person, A. (1984) Supratidal evaporitic dolomite at Ojo de Liebre lagoon: mineralogical and isotopic arguments for primary crystallization. Journal of Sedimentary Petrology 54, 1049-1061.
Racki, M. (1991) Diagenesis of the Clearwater Formation, Cold Lake, Alberta. M.Sc. thesis, University of Western Ontario, London, Ontario, Canada.
Rubinstein, I., Strausz, O.P., Spyckerelle, C., Crawford, R.J. and Westlake, D.W.S. (1977) The origin of the oil sand bitumens of Alberta: a chemical and a microbiological simulation study. Geochimica et Cosmochimica Acta 41, 1341-1353.
Savin, S.M. (1967) Oxygen and hydrogen isotope ratios in sedimentary rocks and minerals. Ph.D. thesis, California Institute of Technology, Pasadena, 220 pp.
Savin, S.M. and Epstein, S. (1970a) The oxygen and hydrogen isotope geochemistry of clay minerals. Geochimica et Cosmochimica Acta. 34, 25-42.
Savin, S.M. and Epstein, S. (1970b) The oxygen and hydrogen isotope geochemistry of ocean sediments and shales. Geochimica et Cosmochimica Acta 34, 43-63.
Savin, S.M. and Lee, M. (1988) Isotopic studies of phyllosilicates. In: Hydrous Phyllosilicates (exclusive of micas). S.W. Bailey, ed., Reviews in Mineralogy 19, 189-223.
Savin, S.M. and and Yeh, H-W. (1981) Stable isotopes in ocean sediments. In: The Sea, Volume 7, The Oceanic Lithosphere. C. Emiliani, ed., John Wiley and Sons, New York, 1521-1554.
Schwartz, F.W. and Longstaffe, F.J. (1988) Ground water and clastic diagenesis. In: Hydrogeology, W. Back, J.S. Rosenshein and P.R. Seaber, eds., Geological Society of

America, The Geology of North America, Volume O-2, 413-434.
Sedimentology Research Group (1981) The effects of in-situ steam injection on Cold Lake oil sands. Bulletin of Canadian Petroleum Geology 29, 447-478.
Shackleton, N.J. (1967) Oxygen isotope analyses and Pleistocene temperatures reassessed. Nature 215, 15-17.
Shepherd, D.W. (1981) Steam stimulation recovery of Cold Lake bitumen. In: The Future of Heavy Crude and Tar Sands, R.F. Meyer, J.C. Wynn and J.C. Olson, eds., McGraw Hill, N.Y., 349-360.
Shier, W. and Longstaffe, F.J. (1992) Diagenesis of Wabiskaw Member (Lower Cretaceous Formation) oil sands, northeastern Alberta. American Association of Petroleum Geologists, Program with Abstracts, Calgary, 118-119.
Sofer, Z. and Gat, J.R. (1975) The isotopic composition of evaporating brines: effect of the isotopic activity ratio in saline solutions. Earth and Planetary Science Letters 26, 179-186.
Suchecki, R.K. and Land, L.S. (1983) Isotopic geochemistry of burial-metamorphosed volcanogenic sediments, Great Valley sequence, northern California. Geochimica et Cosmochimica Acta 47, 1487-1499.
Suzuoki, T. and Epstein, S. (1976) Hydrogen isotope fractionation between OH-bearing minerals and water. Geochimica et Cosmochimica Acta 40, 1229-1240.
Syers, J.K., Chapman, S.L., Jackson, M.L., Rex, R.W. and Clayton, R.N. (1968) Quartz isolation from rocks, sediments and soils for determination of oxygen isotopes composition. Geochimica et Cosmochimica Acta 32, 1022-1025.
Taylor, H.P., Jr. (1974) The application of oxygen and hydrogen isotope studies to problems of hydrothermal alteration and ore deposition. Economic Geology 69, 843-883.
Taylor, H.P., Jr. and Epstein, S. (1961) Relationships between O^{18}/O^{16} ratios in coexisting minerals of igneous and metamorphic rocks, Part 2. Application to petrologic problems. Geological Society of America Bulletin 73, 675-694.
Taylor, R.S., Mathews, W.H. and Kupsch, W.O. (1964) Tertiary. In: Geological History of Western Canada, R.G. McCrossan and R.P. Glaister, eds., Alberta Society of Petroleum Geologists, 190-194.
Taylor, T.R. (1990) The influence of calcite dissolution on reservoir porosity in Miocene sandstones, Picaroon Field, offshore Texas Gulf Coast. Journal of Sedimentary Petrology 60, 322-334.
Thode, H.G., Monster, J. and Dunford, H.B. (1961) Sulphur isotope geochemistry. Geochimica et Cosmochimica Acta 25, 150-174.
Tilley, B.J. (1988) Diagenesis and porewater evolution in Cretaceous sedimentary rocks of the Alberta Deep Basin. Ph.D. thesis, University of Alberta, Edmonton, Alberta, Canada.
Tilley, B.J. and Longstaffe, F.J. (1989) Diagenesis and isotopic evolution of porewaters in the Alberta Deep Basin: the Falher Member and Cadomin Formation. Geochimica et Cosmochimica Acta 53, 2529-2546.
Tilley, B.J., Nesbitt, B.E. and Longstaffe, F.J. (1989) Thermal history of Alberta Deep Basin: comparative study of fluid inclusion and vitrinite reflectance data. American Association of Petroleum Geologists Bulletin 73, 1206-1222.
Tóth, J. (1978) Gravity-induced cross-formational flow of formation fluids, Red Earth region, Alberta, Canada: analysis, patterns, evolution. Water Resources Research 14, 805-843.
Tóth, J. (1980) Cross-formational gravity-flow of groundwater: a mechanism of the transport and accumulation of petroleum (the generalized hydraulic theory of petroleum migration). In: Problems of Petroleum Migration, W.H. Roberts III and R.J. Cordell, eds., American Association of Petroleum Geologists Studies in Geology No.10, 121-167.
Urey, H.C. (1947) The thermodynamic properties of isotopic substance. Journal of the Chemical Society (London), 562-581.
Veizer, J. and Hoefs, J. (1976) The nature of O^{18}/O^{16} and C^{13}/C^{12} secular trends in sedimentary carbonate rocks. Geochimica et Cosmochimica Acta 40, 1387-1395.
Veizer, J., Compston, W., Hoefs, J. and Nielsen, H. (1982) Mantle buffering of the early oceans. Naturwissensch 69, 173-180.
Vigrass, L.W. (1968) Geology of Canadian heavy oil sands. American Association of Petroleum Geologists Bulletin 52, 1984-1999.

Wadleigh, M.A. and Veizer, J. (1992) $^{18}O/^{16}O$ and $^{13}C/^{12}C$ in lower Paleozoic articulate brachiopods: Implications for the isotopic composition of seawater. Geochimica et Cosmochimica Acta 56, 431-443.

Wenner, D.B. and Taylor, H.P. Jr. (1971) Temperature of serpentinization of ultramafic rocks based upon O^{18}/O^{16} fractionation between coexisting serpentine and magnetite. Contributions to Mineralogy and Petrology 32, 165-185.

White, D.E. (1965) Saline waters of sedimentary rocks. In: Fluids in Subsurface Environments. A. Young and J.E. Galley, eds., American Association of Petroleum Geologists, Memoir 4, 342-366.

Whitney, G. and Northrop, H.R. (1988) Experimental investigation of the smectite to illite reaction: dual reaction mechanisms and oxygen-isotope systematics. American Mineralogist 73, 77-90.

Wickert, L.M., Longstaffe, F.J. and Pemberton, S.G. (1989) A diagenetic investigation of sequence stratigraphy in the lower Cretaceous Clearwater Formation, Cold Lake oil sands, east-central Alberta. Geological Association of Canada, Program with Abstracts 14, p. A86.

Wightman, D.M. and Berezniuk, T. (1986) Resource characterization and depositional modelling of the Clearwater Formation, Cold Lake oil sands deposit, east-central Alberta. In: Proceedings of the 1986 Tar Sands Symposium, J.D. Westhoff and L.C. Marchant, eds., Jackson, Wyoming.

Wightman, D.M., MacGillivray, J.R., McPhee, D., Berhane, H. and Berezniuk, T. (1991) McMurray Formation and Wabiskaw Member (Clearwater Formation): Regional perspective derived from the north Primrose area, Alberta, Canada. In: Fifth UNITAR/UNDP International Conference on Heavy Crude and Tar Sands (in press).

Wightman, D., Rottenfusser, B., Kramers, J. and Harrison, R. (1989) Geology of the Alberta oil sands deposits. In: AOSTRA Technical Handbook on Oil Sands, Bitumens and Heavy Oils, L. G. Hepler and C. Hsi, eds., AOSTRA Technical Publication Series, Volume 6, 1-9.

Wilson, M.R., Kyser, T.K., Mehnert, H.H. and Hoeve, J. (1987) Changes in the H-O-Ar isotope composition of clays during retrograde alteration. Geochimica et Cosmochimica Acta 51, 869-878.

Yeh, H-W. (1980) D/H ratios and late-stage dehydration of shales during burial. Geochimica et Cosmochimica Acta 44, 341-352.

Yeh, H-W. and Epstein, S. (1978) Hydrogen isotope exchange between clay minerals and sea water. Geochimica et Cosmochimica Acta 42, 140-143.

Yeh, H-W. and Eslinger, E.V. (1986) Oxygen isotopes and the extent of diagenesis of clay minerals during sedimentation and burial in the sea. Clays and Clay Minerals 34, 403-406.

Yeh, H-W. and Savin, S.M. (1976) The extent of oxygen isotope exchange between clay minerals and seawater. Geochimica et Cosmochimica Acta 40, 743-748.

Yeh, H-W. and Savin, S.M. (1977) Mechanism of burial metamorphism of argillaceous sediments: 3. O-isotope evidence. Geological Society of America Bulletin 88, 1321-1330.

Yurtsever Y. and Gat J.R. (1981) Atmospheric waters. In: Stable Isotope Hydrology: Deuterium and Oxygen-18 in the Water Cycle, J.R. Gat and R. Gonfiantini, eds., IAEA Technical Report Series 210, 103-142.

Zheng, Y-F. (1991) Calculation of oxygen isotope fractionation in metal oxides. Geochimica et Cosmochimica Acta 55, 2299-2307.

Index